United Kingdom Balance of Payments
The Pink Book 2006

Editor: John Bundey

Office for National Statistics

© Crown copyright 2006

Published with the permission of the Controller of Her Majesty's Stationery Office (HMSO).

You may re-use this publication (excluding logos) free of charge in any format for research, private study or internal circulation within an organisation. You must re-use it accurately and not use it in a misleading context. The material must be acknowledged as Crown copyright and you must give the title of the source publication. Where we have identified any third party copyright material you will need to obtain permission from the copyright holders concerned.

This publication is also available at the National Statistics website: www.statistics.gov.uk

For any other use of this material please apply for a Click-Use Licence for core material at www.opsi.gov.uk/click-use/system/online/pLogin.asp or by writing to:

Office of Public Sector Information
Information Policy Team
St Clements House
2–16 Colegate
Norwich NR3 1BQ
Fax: 01603 723000
E-mail: hmsolicensing@cabinet-office.x.gsi.gov.uk

First published 2006 by
PALGRAVE MACMILLAN
Houndmills, Basingstoke, Hampshire RG21 6XS and
175 Fifth Avenue, New York, NY 10010, USA
Companies and representatives throughout the world.

PALGRAVE MACMILLAN is the global academic imprint of the Palgrave Macmillan division of St. Martin's Press, LLC and of Palgrave Macmillan Ltd. Macmillan® is a registered trademark in the United States, United Kingdom and other countries. Palgrave is a registered trademark in the European Union and other countries.

ISBN 1-4039-9387-4
ISSN 0950–7558

This book is printed on paper suitable for recycling and made from fully managed and sustained forest sources.

A catalogue record for this book is available from the British Library.

10 9 8 7 6 5 4 3 2 1
15 14 13 12 11 10 09 08 07 06

Printed and bound in Great Britain by Hobbs the Printer Ltd, Totton, Hampshire.

A National Statistics publication

National Statistics are produced to high professional standards set out in the National Statistics Code of Practice. They are produced free from political influence.

About the Office for National Statistics

The Office for National Statistics (ONS) is the government agency responsible for compiling, analysing and disseminating economic, social and demographic statistics about the United Kingdom. It also administers the statutory registration of births, marriages and deaths in England and Wales.

The Director of ONS is also the National Statistician and the Registrar General for England and Wales.

Contact points

For enquiries about this publication, contact the Editor, John Bundey

Tel: 020 7533 6078

E-mail: john.bundey@ons.gsi.gov.uk

For general enquiries, contact the National Statistics Customer Contact Centre.

Tel: 0845 601 3034
            (minicom: 01633 812399)

E-mail: info@statistics.gsi.gov.uk

Fax: 01633 652747

Letters: Room 1015, Government Buildings,
            Cardiff Road, Newport NP10 8XG

You can also find National Statistics on the Internet at www.statistics.gov.uk

Contents

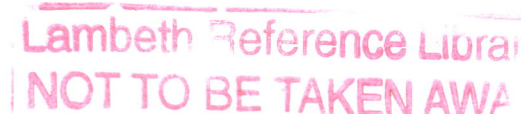

		Page
	Introduction	1

Part 1: Current account

1	Summary of balance of payments	20
2	Trade in goods	30
3	Trade in services	40
4	Income	54
5	Current transfers	70

Part 2: Capital account, financial account and international investment position

6	Capital account	76
7	Financial account	78
8	International investment position	96

Part 3: Geographical breakdown

9	Geographical breakdown of current account	116
10	Geographical breakdown of International investment position	154

Part 4: Supplementary information

Balance of payments and the relationship to national accounts	162
Methodological notes	165
Further information on UK balance of payments	185
Glossary of terms	187
Index	193

List of contributors

Authors:
Geraldine Davies
Angie Francis
Perry Francis
Caroline Lakin
John Lowes
Tom Orford
Ellie Turner

Production team:
Jeremy Brocklehurst
Geraldine Davies
Alistair Dent
Michelle Franco
Charles Jumbo
Deborah Kennion
Andy Leach
Phil Lewin
Carole Rennie

Chart typesetting by the Desktop Publishing Unit, ONS Titchfield

Preface

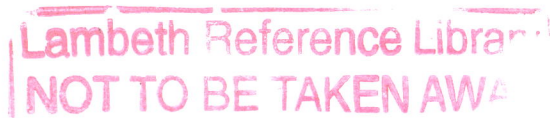

The annual National Statistics *Pink Book* contains estimates of the balance of payments of the United Kingdom. The presentation of the accounts is based on the *IMF Balance of Payments Manual 5th edition (BPM5)*.

Pink Book data in computer-readable form

Free access to National Statistics data is available online at www.statistics.gov.uk

Access around 40,000 time series, of primarily macro-economic data, drawn from the main tables in a range of our major economic and labour market publications. Download complete releases, or view and download your own customised selection of individual time series.

Also access cross sectional data and metadata from across the Government Statistical Service (GSS), organised by theme and subject. Download many datasets, in whole or in part, or consult catalogue information for all GSS statistical resources, including censuses, surveys, periodicals and enquiry services. Information is posted as PDF electronic documents, or in XLS and CSV formats, compatible with most spreadsheet packages

Complete copies of this publication are available to download free of charge on the following web page: www.statistics.gov.uk/products/p1140.asp

Quarterly estimates

Quarterly estimates of the main components of the balance of payments for the last two years are published in a quarterly National Statistics First Release and in more detail in *UK Economic Accounts*.

Long run quarterly and annual estimates consistent with the *Pink Book* are published in the *Economic Trends Annual Supplement*. The latest estimates are also given in summary form in the *Monthly Digest of Statistics and in Financial Statistics*

Comments and enquiries

The Office for National Statistics (ONS) is keen to receive comments on this publication and suggestions for improvements, which can be considered for future editions of the *Pink Book*. Comments can be sent in writing to:

> **John Bundey**
> Pink Book Editor
> Balance of Payments & Financial Sector Division
> Office for National Statistics
> Room D3/20
> 1 Drummond Gate
> London SW1V 2QQ
> Tel: 020 7533 6078, Fax: 020 7533 5189
> E-mail: bop@ons.gsi.gov.uk

Enquiries regarding balance of payments estimates should be directed to the following:

Trade in goods:
Caroline Lakin 020 7533 6070
(caroline.lakin@ons.gsi.gov.uk)

Trade in services, current transfers and capital account:
Tom Orford 020 7533 6095
(tom.orford@ons.gsi.gov.uk)

Income, financial account and international investment position:
Perry Francis 020 7533 6075
(perry.francis@ons.gsi.gov.uk)

An introduction to the United Kingdom balance of payments

Introduction

The balance of payments is one of the UK's key economic statistical series. It measures the economic transactions between United Kingdom residents and the rest of the world. It also draws a series of balances between inward and outward transactions, provides a net flow of transactions between UK residents and the rest of the world and reports how that flow is funded. Economic transactions include:

(i) exports and imports of goods, such as oil, agricultural products, other raw materials, machinery and transport equipment, computers, white goods and clothing;

(ii) exports and imports of services such as international transport, travel, financial and business services;

(iii) income flows, such as dividends and interest earned by foreigners on investments in the UK and by the UK investing abroad;

(iv) financial flows, such as direct investment, investment in shares, debt securities, loans and deposits; and

(v) transfers, which are offsetting entries to any one-sided transactions listed above, such as foreign aid and funds brought by migrants to the UK.

The international investment position measures the levels of financial investment with the rest of the world, inward and outward.

International statistical standards

The Office for National Statistics (ONS) follows the international standards relating to balance of payments and international investment position statistics. There are several reasons for this. First, domestic and foreign analysts will be assured that the UK's official balance of payments and international investment position statistics comply with objective, coherent international standards that reflect current, global analytic needs. Second, the UK is a member of the international community and international users need comparable data for comparison between countries. Third, the UK, as a member of the European Union, as well as organisations such as the IMF and OECD, needs to compile its various economic statistics in conformity with standards set by those organisations. Fourth, the UK can compare and reconcile its data with those of other countries. Statistics need to be as comparable as possible in order to carry out this validation.

To facilitate such consistency and to provide guidelines for its members, the IMF issued the *Balance of Payments Manual*. The first edition appeared in 1948 and the most recent (fifth) edition in 1993. The conceptual framework of the UK balance of payments corresponds to that underlying the fifth edition of the IMF Manual, referred to as *BPM5*. *BPM5* was implemented in the UK's balance of payments accounts and international investment position statistics in September 1998.

A process of reviewing the existing international standards started in the mid 1980s with the specific objective of harmonising, to the maximum extent possible, the statistical concepts, definitions, statistical units, classifications and terminology. Release of the revised standards started in 1993 with *BPM5* and the third edition of the *System of National Accounts (SNA93)*. *BPM5* was prepared by the IMF in close co-operation with national compilers and with the Statistical Office of the European Communities, the OECD, the United Nations and the World Bank. Those five organisations jointly published *SNA93*. In 1995, the EU produced its own version of *SNA93*, the *European System of Accounts (ESA95)*, upon which the UK's national accounts are based and which is consistent with *BPM5*. Both *SNA93* and *BPM5* were amended in 2000 to give more consistent guidance on the treatment of financial derivatives.

The United Nations Statistics Commission and the IMF Board of Directors have subsequently approved the comprehensive and parallel updating of the National Accounts and Balance of Payments manuals, in order to ensure their consistency and achieve greater harmonisation. ONS has developed the following webpage to inform users of progress and to invite their input:

www.statistics.gov.uk/about/Consultations/NA/default.asp

Conceptual framework definitions

Balance of payments

Broadly speaking, the UK balance of payments is a statistical statement designed to provide a systematic record of the UK's economic transactions with the rest of the world. It may be described as a system of consolidated accounts in which the accounting entity is the UK economy and the entries refer to economic transactions between residents of the UK and residents of the rest of the world (non-residents).

The balance of payments accounts are concerned not only with payments made but also any economic transactions during a period that give rise to a payment in an earlier or later period, e.g. goods may change ownership in one period, though payment may be made in an earlier period (pre-payment) or in a later period (trade credit). They also include transactions for which there may never be a payment, e.g. goods shipped under foreign aid or goods shipped between related enterprises. There is also more than one 'balance': the balance of payments is a system of accounts in which many balances can be derived, such as the balance of goods and services, the balance on current account, and the balance on capital and financial account.

Balance of payments statements cover a wide range of economic transactions which include:

(i) goods, services, income and current transfers; and

(ii) capital transactions, such as capital transfers; and

(iii) financial transactions involving the UK claims on, and liabilities to, non-residents.

Category (i) is shown in the current account, category (ii) in the capital account and category (iii) in the financial account.

International investment position

The UK's international investment position is a closely related set of statistics. It can be viewed as the balance sheet recording the UK's stock (or level) of foreign financial assets and liabilities at a particular date. The net international investment position is the difference between the stock of foreign financial assets and foreign liabilities at a particular date.

Viewed more broadly, the international investment position can be shown as a reconciliation statement of the stock of investment at two different points in time by showing financial transactions and other changes (non-transaction changes) such as price changes, exchange rate variations and other adjustments that occurred during the period. Financial transactions which are included in the reconciliation statement are equivalent to the transactions measured in the financial account of the balance of payments. ONS does not currently publish a full reconciliation of the international investment position showing price, exchange rate and other changes.

Classifications such as assets and liabilities, type of investment (direct, portfolio and other investment and reserve assets), and instrument of investment, are used consistently in both the balance of payments and the international investment position.

Concepts of territory and residence

In compiling the UK balance of payments and international investment position, the UK economy is conceived as comprising the economic entities that have a closer association with the territory of the UK than with any other territory. Each such economic entity is described as a resident of the UK. Any economic entity which is not regarded as a resident of the UK is described as a non-resident. The concept of residency is not based on nationality.

The UK's economic territory is defined to include the territories lying within its political frontiers and territorial seas, and in the international waters over which it has exclusive jurisdiction. It also includes its territorial enclaves abroad holding embassies, consulates, military bases, scientific stations, information or immigration offices, aid agencies etc., whether owned or rented by the UK governments with the formal agreement of the countries where they are located.

The UK offshore islands – Jersey, Guernsey and Isle of Man – are classified as non-resident to the UK. Thus transactions between UK residents and the islands are in the balance of payments, but transactions between islanders are not counted in the UK balance of payments. The islands are not part of the EU, so statistics relating to them are not required under *ESA95* and they have to be excluded from the UK's economic territory to ensure full UK consistency with *ESA95*. This treatment is also technically consistent with *BPM5* recommendations which states that 'In a maritime country, economic territory includes islands that belong to the country and are subject to the same fiscal and monetary authorities as the mainland; goods and persons move freely to and from the mainland and islands...'. The offshore islands are subject to their own fiscal authorities and have their own tax systems. Furthermore, there are impediments to taking up residency on the Channel Islands.

For balance of payments purposes, residents of an economy are generally deemed to have a centre of economic interest in the economy and to be resident for at least one year. The residents of the UK comprise:

(i) Resident general government institutions including the Scottish Parliament, Welsh Assembly, Northern Ireland Assembly and local government authorities and statutory bodies. The UK embassies, consulates, military establishments, etc. physically located abroad are included in the UK's economic territory and are therefore residents; similar entities of other countries physically located within the UK are outside the UK's economic territory and are therefore non-residents.

(ii) Resident financial and trading enterprises which include all enterprises engaged in the production of goods and services on a commercial or equivalent basis within the territory of the UK. Enterprises may be incorporated or unincorporated; privately or government owned and/or controlled; and locally or foreign owned and/or controlled. The definition of an enterprise in terms of the territory in which it is located often makes it necessary to divide a single legal entity into a head office operating in one economy and a branch operating in another economy. Resident enterprises include UK branches of foreign companies and exclude foreign branches of UK companies.

(iii) Resident non-profit bodies, those in which individuals and/or enterprises combine, as owners, to produce goods and services within the territory of the UK for purposes other than to provide a financial return for themselves. Examples are churches, charitable organisations and representative business organisations such as Chambers of Commerce.

(iv) Resident households and individuals which broadly encompass all persons residing in the territory of the UK for one year or more, whose general centre of economic interest is considered to be the UK. The UK's official diplomatic and consular representatives, the UK's armed forces, other UK government personnel stationed abroad and their dependants, and UK students studying abroad are also included even though they may all be abroad for one year or more. They are treated as UK residents since their centre of interest is considered to be the UK. Generally, the centre of economic interest of persons visiting the UK for less than one year is considered to be outside the UK and they are therefore regarded as non-residents, but if they stay for one year or more they are considered to be residents for balance of payments purposes. Irrespective of their length of stay, non-residents also include foreign diplomatic, consular, military and other government personnel, their dependants, and foreign students studying in the UK.

Double-entry system

Rules for the UK double entry system

Credit entries, changes in all economic resources provided by the UK to non-residents, including:	**Debit entries,** changes in all economic resources received by the UK from non-residents, including:
Exports of goods and services	Imports of goods and services
Income accruing on the resources to UK from residents	Income accruing on the resources to non-residents from UK
Financial liabilities of the UK to non-residents	Financial claims of UK on non-residents
Transfers which are offsets to debit entries	Transfers which are offsets to credit entries

Examples of UK double entry recording	Credits	Debits
1. Sales of goods (value 100) to non-residents for foreign exchange (i.e. goods provided and bank payment (a bank deposit) received in an account held abroad)		
Goods	100	
Bank deposits, foreign currency assets		100
2. Purchase of goods (value 120) from a non-resident using trade Credit (i.e. goods received and a claim on a resident (trade credit liability) provided)		
Goods		120
Trade credit liabilities	120	
3. Food aid (value 5) provided to non-residents (i.e. goods provided and transfer imputed)		
Goods	5	
Current transfers		5
4. payment of a loan (value 25) by a resident company to a non-resident lender (i.e. liability to a non-resident reduced and a reduction in bank deposits held abroad)		
Loan repayment	−25	
Bank deposits, foreign currency, assets		−25

Conceptually, an economic transaction has two sides: something of economic value is provided and something of equal value is received. The balance of payments reflects this in a double-entry recording system of credits and debits. When an economic value is provided (e.g. UK exports a car) a credit entry is made, and when the corresponding economic value is received (e.g. a payment for the car) a debit entry is made. For example, when an exporter sells (provides) goods to a non-resident, the exporter may receive cash (a financial asset) or another type of financial asset (e.g. a trade credit claim) in return. The export is represented by a credit entry and the financial asset acquired is represented by an offset debit entry. Similar entries are made when an importer buys a car (debit) and pays for it (credit). So a credit entry represents a change in rest of world ownership of any sort of UK asset (real or financial); a debit entry represents a change in UK ownership of rest of world assets.

An understanding of the double-entry recording system is necessary for a complete understanding of balance of payments statistics.

Under the double-entry system, by definition credit entries must equal debit entries. Credit entries are required for exports of goods and services, income receivable, and changes in financial liabilities. Likewise, debit entries are required for imports of goods and services, income payable, and changes in financial assets. Where something of economic value is provided without something of economic value in exchange (i.e. without a *quid pro quo*) the double-entry system requires an offset to be imputed (a transfer entry) of equivalent value. For example, food exported as aid requires a credit entry for the goods provided and a debit transfer as the aid offset.

Sign convention in the UK balance of payments statistics

The sign convention used in presenting the UK balance of payments statistics is to give a positive sign to an increase in either credit or debit entries and a negative sign to a decrease in credit or debit entries. Balances (calculated as credits less debits) or items which are net credits have no sign, while balances which are net debits have a negative sign.

When considering making international comparisons it should be borne in mind that there is no unique or correct sign convention and other countries/institutions use variations. In particular the convention used by the IMF in their publications gives no sign to credit entries and a minus sign to all debit entries (e.g. imports and acquisitions of assets).

Errors and omissions

It follows that, in principle, under a double-entry accounting system, the difference between the sum of credit and debit entries must be zero. In practice, some transactions are not measured accurately (i.e. errors) and some are not measured at all (i.e. omissions). Data sources used to compile the accounts often measure the credit and debit sides from different data sources and may not always do so consistently. There could be many reasons why these sources may not measure the acquisition side of the transaction and the corresponding payments, either in the same accounting period or at the same value. To restore the equality of credit and debit entries, a net errors and omissions item is included in the balance of payments accounts. The item indicates whether credit or debit transactions would be needed to balance the accounts, but does not show where the discrepancy lies. Usually the financial account is considered to be the most likely source.

Valuation

It is important that the balance of payments and international investment position statistics carry values that have economic meaning to enable useful analysis, and to provide meaningful indicators of cross-border economic activity. It is also important for the double-entry accounting system that a uniform valuation is adopted. This means that the credit and debit entries of each transaction – which in practice may be derived from independent sources – should be valued at the same price. In addition, a uniform valuation is essential to sum different types of transactions on a consistent and comparable basis. The use of a uniform valuation principle aids understanding by users. Moreover, statistics for different countries will not be comparable unless both parties to a transaction adopt the same valuation principle. It is also important to use a principle which is consistent with national accounting principles. For all these reasons, market price is used in UK economic statistics for valuing transactions.

Market price is the amount of money that a willing buyer pays to acquire something from a willing seller, when such an exchange is between independent parties and involves only commercial considerations. In practice, one or more of the conditions needed to establish a market price may be absent and other valuations may be used.

For the most part, the price at which a transaction is recorded in the accounts of the transactors or in the administrative records used as data sources will be the market price or a very close approximation of it. This valuation is known as the transactions price and is the practical valuation basis used in the balance of payments, both because it aids consistent recording of credits and debits and because of its usual proximity to the ideal market valuation. The following paragraph discusses a special case of transactions where market prices may not apply, namely transfer pricing between affiliated enterprises in different countries.

Transfer pricing

Where transactions are between affiliated enterprises in different countries, the prices adopted in their books for recording transactions in goods and services and any associated indebtedness and interest – referred to as transfer prices – may not correspond to prices that would be charged to independent parties. There will be some departure from the market price principle if transfer prices are different from those charged to enterprises outside the group. However there are practical difficulties in identifying and suitably adjusting individual cases. Transfer pricing to avoid tax is illegal in the UK so the distortions in the international accounts caused by transfer pricing are not considered widespread. For both reasons, adjustments to account for transfer pricing are rarely made in practice.

Assets and liabilities

As with all international investment position statistics, foreign financial assets and liabilities should, in principle, be valued at their current market price at the reference date. In practice this is not always possible and valuation guidelines are adopted in order to approximate market valuation, particularly for those financial assets and liabilities that are only rarely transacted. For example, in measuring the value of direct investment in equity capital, much of which is never traded or is traded infrequently, market value is approximated by one of the following methods: a recent transaction price; directors' value; or net asset value. Over time, this is likely to underestimate the true market value of Foreign Direct Investment.

Unit of account and conversion

Transactions and stock positions originally denominated in foreign currencies need to be converted to pounds sterling using market rates of exchange prevailing at the time of the transaction (balance of payments) or at the reference date (international investment position). Transactions should be converted at the mid-point of the buying and selling exchange rates applying at the time of transaction. Stocks should be converted at the mid-point of the buying and selling exchange rates applying at the beginning or end of the period. In practice, the actual rate used varies according to the source of the transaction or stock data.

Time of recording

Transactions

The time of recording of transactions in balance of payments and international investment position statistics is, in principle, the time of change of ownership (either actual or imputed). Under the double-entry system, both sides of a transaction should be recorded in the same period. This is consistent with the principle of accrual accounting, which requires that transactions be recorded when economic value is created, transformed, exchanged, transferred or extinguished.

Change of ownership is considered to occur when legal ownership of goods changes, when services are rendered and when income accrues. In the case of transfers, those which are imposed by one party on another, such as taxes and fines, should ideally be recorded at the moment at which the underlying transactions or other flows occur which give rise to the liability to pay; other transfers should be recorded when the goods, services etc. change ownership.

For financial transactions, the time of change of ownership is taken to be the time when transactions are entered in the books of the transactors. That is taken to be the time when a foreign financial asset or liability is acquired, relinquished by agreement, sold or repaid. The commitment or pledging of an asset does not constitute an economic transaction, and no entry should be shown unless a change of ownership actually occurs in the period covered. Likewise, the entries for loan drawings should be based on actual disbursements and not on commitments or authorisations. Entries for loan repayments should be recorded at the time they are due rather than on the actual payment date.

Both sides of a transaction should be recorded in the same period. In practice the time of recording of transactions in the balance of payments and international investment position statistics will reflect the practices in data sources, and may diverge from the principle of time of change of ownership. For the UK, transactions in goods credits (exported goods) are mainly recorded at the time when goods are shipped as this is assessed to be a generally good practical approximation of the time when ownership changes. Goods debits (imported goods) are recorded when customs records relating to the movement of the goods across the frontier are processed, again in the expectation that this is the best practical approximation to change of ownership that can be generally achieved. For the remainder of the current account, the time of the recording of transactions generally complies with the time of change of ownership. Exceptions occur mainly because the record-keeping practices of some data providers may not be on this basis. Financial account transactions usually are recorded appropriately, that is, when the parties record transactions in their books. However, some transactions may be derived from information supplied by intermediaries that are not party to the transactions and may not be aware of the time of change of ownership. Also, some enterprises may adopt accounting practices that lead to inconsistent time of recording; a simple example is that different enterprises may close off their accounts at different times of day.

Stock

The time of recognising the stock of a foreign financial asset or liability follows naturally from the time of recording of a transaction in that asset or liability. For example, if a transaction is undertaken to acquire a foreign financial asset, there will also be a consequential increase in the stock of foreign financial assets at the end of that period. Of course, if the asset is disposed of before the end of the period, it will not contribute to the stocks statistics to be recorded for the period, but the disposal will have given rise to another transaction to be recorded for the period.

Types of transactions in the balance of payments

An economic transaction occurs when something of economic value is provided by one party to another. Transactions that are considered to have economic value comprise those in goods, services, income and financial assets and liabilities. The transactions recorded in a balance of payments statement stem from dealings between two parties, one being a resident and the other a non-resident. The types of transactions included in the balance of payments are exchanges, one-sided transactions and imputed transactions.

Exchanges

Exchanges are the most important and numerous type of transaction. They include transactions in which one transactor provides something of economic value to another transactor and receives in return something of equal value.

Special cases of imputation/estimation

Migrants' transfers

A special statistical treatment is required when a person migrates, that is when the person's status changes from non-resident to resident (or vice versa). When this change occurs, the property owned by the migrant becomes the property of a resident instead of that of a non-resident (or vice versa). This change of ownership of net worth between economies is included in the balance of payments. For example, any financial assets held abroad by the migrant become claims by the UK on the rest of the world.

Offset entries are made corresponding to the transfer of net worth and, by their nature, these are included as transfers in the capital account. This treatment amounts to envisaging a transfer of property from the person in their capacity as a non-resident to the person in their capacity as a resident (or vice versa). In principle, this transaction embraces all the migrant's property, whether or not it accompanies the migrant.

Reinvested earnings

A number of special cases of imputed transactions feature in balance of payments compilation. One case involves the reinvestment of earnings in resident enterprises by their non-resident direct investors. These *reinvested earnings* are regarded as being paid out as investment income and then reinvested in the enterprises from which they originated. They are therefore recorded both as a component of investment income in the current account and as a component of direct investment in the financial account. It is considered analytically useful to identify these transactions separately in economic statistics because of the substantial contribution they make to the stock of direct investment finance in a country.

Financial services

A further case relates to estimation for the implicit fees (financial services) associated with foreign exchange trading. Estimates of the implicit service fees being earned on foreign exchange trading with non-resident counterparties are made by splitting the total service fees reported by exchange traders into resident/non-resident shares using a number of assumptions and other published information.

Exceptions to change of ownership

In economic statistics, transactions are considered to occur when the goods and financial assets change ownership between transactors, when services are provided by one transactor to another, or when income is earned by one transactor from another. However, there are certain situations in which no change of ownership legally occurs, but where transactions are nonetheless considered to have occurred for balance of payments purposes. The situations include financial leases, goods imported into or exported from the UK for processing and return, and transactions between a head office in one country and a branch in another.

Financial leases

A financial lease is regarded as a method of obtaining all the rights, risks and rewards of ownership of real resources without holding legal ownership. Although legal ownership remains with the lessor during the term of the lease, all the risks and responsibilities apply to the lessee. In these cases, the basic nature of the transaction is given precedence over its legal form, by imputing a change of ownership of the resource to the lessee. As a result of this imputation, a financial liability is recognised and lease payments are classified as partly loan repayments in the financial account and partly interest in the current account, rather than as services in the current account.

Goods for processing

In economic statistics, the value of goods entering or leaving the UK for processing and returning to the country of origin after processing should be recorded on a gross basis, i.e. recording the goods both when they enter (as imports) and when they leave (as exports), even though there is no legal change of ownership of those goods. Thus a good entering the UK to be processed and returned to the country of origin is recorded as an import at the appropriate value and subsequently as an export – recorded by the customs system at the original value plus the added value of the processing. A symmetrical treatment should be applied to UK goods exported for processing and return. The basis for this treatment is that such goods lose their identity during processing by being transformed or incorporated into different goods. On the other hand, for goods undergoing repairs only the value of the repair, not the gross value of the goods, is included in the goods credits or debits.

Branches

In economic statistics, it is usually necessary to split the activities of a legal entity and recognise two units, a head office in one country and a branch in another. Flows of goods, services, income and finance between the branch and its head office are therefore treated as transactions, even though they are legally part of the same unit. For example, goods and services sent from the head office to its branch are to be treated as exports of goods and services by the head office.

There are two cases where such splitting becomes necessary. The first occurs when production of goods and services is undertaken by the personnel, plant and equipment of the legal entity in an economic territory outside the economic territory of the head office, provided certain conditions apply. These conditions include: the intention to operate in the separate economy indefinitely or over a long period (12 months is used as a rule of thumb); keeping a set of accounts of the branch's activity (i.e. income statement, balance sheet, transactions with the parent entity); eligibility to pay income tax in the host country; having a substantial physical presence; and receiving funds for the branch's work which are paid into its own bank account.

The second case occurs when a person or legal entity resident in one economy owns land and buildings located in another economy. Ownership of immovable assets is always attributed in balance of payments and international investment position statistics to residents of the economy in which the assets are located. Thus land in the domestic territory, which is in fact owned by a non-resident, is treated as being owned by a notional resident entity, which in turn has a foreign direct investment liability to the real owner. It should also be recalled that the territorial enclaves associated with embassies, military bases etc. are regarded as part of the economic territory of the economy they represent. When these institutions buy and sell the land in these enclaves they are effectively adding to and subtracting from the economic territory of their government. Such transactions in land owned by foreign embassies are recorded in the capital account as the acquisition/disposal of non-produced, non-financial assets.

Other changes in the international investment position

In addition to the financial transactions included in the balance of payments, the international investment position reconciliation statement includes the other changes which contribute to differences between opening and closing positions for a period.

Other changes in position may occur through price changes, exchange rate changes and other adjustments. Price changes are valuation changes that occur because of changes in the market price of a financial instrument, such as a change in the price of a share or debt security, or through revaluing a company's net worth.

Exchange rate changes are due to fluctuations in the value of the pound, in which the accounts are compiled, relative to the currencies in which foreign assets and liabilities are denominated.

Other adjustments can arise from a number of causes such as write-off of bad debts, classification changes, monetisation/demonetisation of gold, and the allocation/cancellation of Special Drawing Rights. A reclassification would occur where a foreign investor's equity investment in an enterprise increased during the reporting period and the increase was sufficient to change the classification of the investor's total equity holding at the end of the period from portfolio investment to direct investment. Monetisation of gold occurs when the Bank of England monetises commodity stocks of gold and adds these to its monetary gold holdings as part of the UK's official reserve assets. Special Drawing Rights in the IMF are also included in the UK's official reserve assets. Allocations and cancellations of these instruments are included as other adjustments.

Gross and net recording

Entries for current and capital account items are generally treated so that credits for each component are recorded separately from debits. Current and capital account transactions, in this context, are described as being recorded gross.

Gross recording contrasts to the recording of transactions in the financial account, which is mainly on a net basis, although for long-term trade credits and loans, gross drawings and repayments are included in the financial account. The net recording of other financial account items means that, for each item, credit transactions are combined with debit transactions to arrive at a single result – either a net credit or net debit – reflecting the net effect of all increases and decreases in holdings of that type of asset or liability during the recording period. There are several types of netting in the financial account, e.g. the netting of purchases and sales within an instrument in an asset position, and netting of assets and liabilities as in the case of direct investment.

Standard balance of payments classification

Balance of payments and international investment position statistics need to be arranged in a coherent structure to facilitate their use and adaptation for purposes such as policy formulation, analytical studies, projections, bilateral comparisons, and regional and global aggregations. *BPM5* contains a *standard classification* and list of *standard components* of the balance of payments and international investment position. These standards were developed taking into account the views of national compilers and analysts, and the requirement to harmonise concepts and definitions with related international statistical standards and classifications. The classification also reflects the separation of categories that may exhibit different economic behaviour, may be important in a number of countries, are readily collectable, and are needed for harmonising with other bodies of statistics.

The standard balance of payments classification comprises two main groups of accounts – the *current account* and the *capital and financial account*. Transactions classified to the *current account* include goods and services, income and current transfers. Within the *capital and financial account,* the *capital account* includes capital transfers and the net acquisition or disposal of non-produced, non-financial assets. The *financial account* includes transactions in financial assets and liabilities.

Transactions in *current account* and *capital account* items are generally shown on a gross basis (gross debits and credits separately). Transactions in *financial account* items are mainly recorded on a net basis.

Current account

Table A (opposite) shows the standard classification of the *current account*. Each of the broad categories is described briefly below, while individual component items are described in detail in subsequent chapters.

Goods and services are divided into separate accounts for *goods and services. Goods* comprise most movable goods that change ownership between UK residents and non-residents.

Services comprise services provided between UK residents and non-residents, together with some transactions in goods where, by international agreement, it is not practical to separate the goods and services components (e.g. goods purchased by travellers are classified to services).

Income refers to income earned by UK residents from non-residents and vice versa. Income covers compensation of employees and investment income. *Compensation of employees* comprises wages, salaries and other benefits earned by individuals from economies other than those in which they are residents, as well as earnings from extraterritorial bodies such as foreign embassies, which often employ staff from the economy in which they are located. *Investment income* comprises income earned from the provision of financial capital and is classified by direct, portfolio and other investment income and income earned on the UK's reserve assets.

Transfers represent offsets to the provision of resources between residents and non-residents with no quid pro quo in economic value (for example, the provision of food aid). *Current transfers* are distinguished from *capital transfers,* which are included in the *capital account*. *Current transfers* represent the offset to the provision of resources that are normally consumed within a short period (less than twelve months) after the transfer is made. In the example of food aid, the food is presumed to be consumed within twelve months of it being received. The classification of current transfers is by general government and other sectors.

A Summary of balance of payments in 2005

£ million

	Credits	Debits
1. Current account		
A. Goods and services	322 298	366 540
1. Goods	211 175	278 473
2. Services	111 123	88 067
2.1. Transportation	17 974	20 101
2.2. Travel	16 868	32 806
2.3. Communications	3 036	2 664
2.4. Construction	522	455
2.5. Insurance	1 578	880
2.6. Financial	23 260	4 866
2.7. Computer and information	5 832	2 110
2.8. Royalties and licence fees	7 313	4 986
2.9. Other business	30 738	15 973
2.10. Personal, cultural and recreational	1 966	788
2.11. Government	2 036	2 438
B. Income	187 037	157 166
1. Compensation of employees	1 211	1 137
2. Investment income	185 826	156 029
2.1 Direct investment	79 146	34 574
2.2 Portfolio investment	45 275	44 164
2.3 Other investment (including earnings on reserve assets)	61 405	77 291
C. Current transfers	16 313	28 492
1. Central government	4 071	13 499
2. Other sectors	12 242	14 993
Total current account	**525 648**	**552 198**
2. Capital and financial accounts		
A. Capital account	4 178	1 776
1. Capital transfers	3 964	1 094
2. Acquisition/disposal of non-produced, non-financial assets	214	682
B. Financial account	733 093	715 617
1. Direct investment	87 725	56 539
Abroad		56 539
1.1. Equity capital		16 672
1.2. Reinvested earnings		40 597
1.3. Other capital[1]		−730
In United Kingdom	87 725	
1.1. Equity capital	76 531	
1.2. Reinvested earnings	11 095	
1.3. Other capital[2]	99	
2. Portfolio investment	127 021	160 710
Assets		160 710
2.1. Equity securities		64 766
2.2. Debt securities		95 944
Liabilities	127 021	
2.1. Equity securities	2 670	
2.2. Debt securities	124 351	
3. Financial derivatives (net)		2 451
4. Other investment	518 347	495 261
Assets		495 261
4.1 Trade credits		439
4.2 Loans		134 814
4.3 Currency and deposits		359 797
4.4 Other assets		211
Liabilities	518 347	
4.1. Trade credits	–	
4.2. Loans	238 122	
4.3. Currency and deposits	279 551	
4.4. Other liabilities	674	
5. Reserve assets		656
5.1. Monetary gold		–
5.2. Special drawing rights		−8
5.3. Reserve position in the IMF		−1 911
5.4. Foreign exchange		2 230
Total capital and financial accounts	**737 271**	**717 393**
Total current, capital and financial accounts	**1 262 919**	**1 269 591**
Net errors and omissions	6 672	

1 Other capital transaction on direct investment abroad represents claims on affiliated enterprises less liabilities to affiliated enterprises

2 Other capital transactions on direct investment in the United Kingdom represents liabilities to direct investors less claims on direct investors

Capital account

The *capital account* comprises both capital transfers and the acquisition and disposal of non-produced, non-financial assets (such as copyrights). The latter includes land purchases and sales associated with embassies and other extraterritorial bodies. Capital transfers entries are required where there is no quid pro quo to offset the transfer of ownership of fixed assets, or the transfer of funds linked to fixed assets (e.g. aid to finance capital works), or the forgiveness of debt. It also includes the counterpart to the transfer of net wealth by migrants, referred to as migrants' transfers.

Financial account

The *financial account* comprises transactions associated with changes of ownership of the UK's foreign financial assets and liabilities. The main classifications used in the financial account are discussed in conjunction with the international investment position classification below.

The *international investment position* measures the UK's stock of external financial assets and liabilities, whereas the *balance of payments financial account* measures transactions in these assets and liabilities. Hence the classifications used in the *financial account* and *international investment position* need to be essentially the same.

Major classifications of the financial account and international investment position

Items in the financial account and international investment position statement are classified on a number of bases. The main ones are *type of investment, assets and liabilities, instrument of investment, sector,* and *original contractual maturity of financial instruments.*

A comparison of the international investment position statement and the balance of payments financial account shows one minor difference. In the category of direct investment in the financial account, reinvested earnings are shown separately whereas, in the international investment position statement, where no separate market price valuation of reinvested earnings can exist, the reinvested earnings are grouped into a composite category for equity and reinvested earnings.

Type of investment

The type of investment used in the UK's balance of payments and international investment position consists of five broad categories:

(i) *Direct investment capital* refers to capital provided to or received from an enterprise, by an investor in another country (i.e. an individual, enterprise or group of related individuals or enterprises) who is in a direct investment relationship with that enterprise. A *direct investment* relationship exists if the investor has an equity interest in an enterprise, resident in another country, of 10 per cent or more of the ordinary shares or voting stock. The *direct investment* relationship extends to branches, subsidiaries and to other businesses where the enterprise has significant shareholding.

(ii) *Portfolio investment* refers to transactions in equity and debt securities (apart from those included in direct investment and reserve assets). Debt securities comprise bonds and notes and money market instruments. In comparison with direct investment, it indicates investment where the investor is not assumed to have any appreciable say in the operation of the enterprise (e.g. less than 10 per cent of the ordinary share or voting stock).

(iii) *Financial derivatives* cover any financial instrument the price of which is based upon the value of an underlying asset (typically another financial asset). Financial derivatives include options (on currencies, interest rates, commodities, indices, etc.), traded financial futures, warrants and currency and interest swaps. Under *BPM5,* transactions in derivatives are treated as separate transactions, rather than being included as integral parts of underlying transactions to which they may be linked as hedges. Only estimates for the settlement receipts/payments on UK banks' interest rate swaps and forward rate agreements are included in financial derivatives.

(iv) *Other investment* is a residual category that captures transactions not classified to direct investment, portfolio investment, financial derivatives or reserve assets of the compiling economy. *Other investment* covers trade credits, loans (including financial leases), currency and deposits, and a residual category for any other assets and liabilities.

(v) *Reserve assets* refer to those foreign financial assets that are available to, and controlled by, the monetary authorities such as the Bank of England for financing or regulating payments imbalances. Reserve assets comprise: monetary gold, Special Drawing Rights, reserve position in the IMF, and foreign exchange held by the Bank.

Assets and liabilities

A financial *asset* is generally in the form of a financial claim on the rest of the world that is either represented by a contractual obligation (such as a loan) or is evidenced by a security (such as a share certificate). Two financial assets – monetary gold and Special Drawing Rights in the IMF – are not claims on the rest of the world. They are, however, included in international investment assets because they are readily available for payment of international obligations. A financial *liability* represents a financial claim of the rest of the world on the UK. Assets and liabilities in the international investment position statement are components of the balance sheet of an economy with the rest of the world. In the financial account the asset and liability classifications in essence reflect, respectively, transactions in claims on non-residents (assets) and in claims by non-residents (liabilities).

In the international investment position, the difference between assets and liabilities is the *net international investment position,* also referred to as the *net liability position/net asset position,* depending on the balance.

For *direct investment,* in both the financial account and international investment position, the main classification is by direction of investment, i.e. *direct investment abroad* and *direct investment in the UK. Direct investment abroad* is derived by netting liabilities of the UK *direct investors* to their *direct investment enterprises* against claims on their direct investment enterprises abroad. Similarly, *direct investment in the UK* is derived after netting claims of the UK direct investment enterprises against their liabilities to those direct investors abroad.

Instrument of investment

Several instruments of investment are also identified. Some of these are only applicable to one type of capital, i.e. the instrument *reinvested earnings* is only applicable to direct investment, while *monetary gold* and *Special Drawing Rights* are only used for reserve assets.

The major instruments and grouping of instruments identified in balance of payments and international investment statistics include:

(i) monetary gold;

(ii) Special Drawing Rights;

(iii) foreign exchange;

(iv) reserve position in IMF;

(v) equity;

(vi) reinvested earnings;

(vii) debt securities;

(viii) financial derivatives;

(ix) trade credit;

(x) loans;

(xi) currency and deposits; and

(xii) other assets/liabilities.

Financial derivatives data are presented as an annex to the international investment chapter.

Similar instruments may be combined into groups or combined with certain types of investment to make statistical presentations less cluttered.

For example:

(i) trade credit, loans, deposits, and other forms of finance including all debt securities, but excluding equity capital and reinvested earnings, between non-financial enterprises in a direct investment relationship, are combined and shown only as *other direct capital*. Similar aggregation applies to finance between a financial enterprise and a non-financial enterprise and between financial enterprises only in case of permanent debt;

(ii) bonds, bills, notes and money market instruments within portfolio investment are shown separately but under a heading of *debt securities;* and

(iii) a number of financial assets, held as part of the UK's reserves assets (currency and deposits, bills, bonds, notes and money market instruments), are grouped under the category *foreign exchange* within the reserve assets category.

Foreign equity and debt

At a broader level, instruments may be combined to show foreign equity and foreign debt. Foreign equity includes equity capital, reinvested earnings and equity securities. Foreign debt is a residual item containing all other instruments. They may be compiled on a gross basis (e.g. foreign debt/assets and liabilities) or on a net basis (e.g. net foreign debt).

Sectorisation

Transactor units within an economy may be grouped together into *institutional sectors*. Units within the same *institutional sector* may be expected to behave similarly in their financial and other dealings and in response to differing economic and political stimuli. The principle of classification by sector, or sectorisation, in the financial account and international investment position is to identify the sector of the domestic creditor for assets and the sector of the domestic debtor for liabilities.

Four sectors are generally distinguished in the standard components of the ONS balance of payments and international investment statistics: *monetary financial institutions; central government; public corporations;* and *other*.

Within the current and capital accounts, sectorisation is also applied to current and capital transfers, where a split between *general government* and *other* is used.

Original contractual maturity

The fifth edition of the balance of payments manual looks to distinguish between long-term or short-term investment. Investment longer than one year is deemed to be long-term and investment less than one year is deemed to be short-term.

Other financial classifications

Other classifications in the financial account and international investment position include the domicile of liabilities issued by residents, drawings and repayments for long-term liabilities in the form of both trade credits and loans and the currency of assets and liabilities.

Country classification

The general principles applying to the compilation of a global balance of payments statement for the UK can be applied to the preparation of a statement for the UK's transactions with an individual country or a group of countries.

Reliability of estimates

All the value estimates are calculated as accurately as possible; however they cannot always be regarded as being absolutely precise to the last digit shown. Similarly, the index numbers are not necessarily absolutely precise to the last digit shown. Some figures are provisional and may be revised later; this applies particularly to many of the detailed figures for the latest years.

Revisions since ONS *Pink Book* 2005

The current account balance is revised from 1992 onwards.

Goods – the data are revised from 2001 to reflect later data from HM Revenue & Customs and other data suppliers. Estimates of aviation fuel procured in foreign airports (oil imports) have been revised upwards on the basis of an improved split into goods and services based on more detailed information provided by the Civil Aviation Authority. These revisions are therefore offset in the goods and services estimates by balancing downwards revisions to estimates of services imports.

The estimates for trade associated with Missing Trader Intra-Community (MTIC) VAT fraud had implicitly been deflated by different deflators i.e. the import and export deflators for the appropriate products. This meant that the net effect on the balance of trade in volume terms was not neutral. This has been changed so that the part of exports thought to be related to VAT MTIC fraud is now deflated by the relevant import deflators so the overall effect is neutral. This has affected estimates of the volumes and prices of exports with volumes being revised upwards and prices downwards in recent periods. The methodology for the estimates of the volume and price for imports of fuels other than oil has also been changed. A smaller proportion of the import cost is now thought to be due to freight charges than had been previously assumed because of the switch from coal to other energy products such as gas and electricity. As a result, the estimates of the volume of goods imports have been revised upwards and that for services has been revised downwards.

Services – are revised back to 1992. The earliest revisions result from a general reassessment of data during the annual supply and use balancing process. Revisions from 1995 to imports and exports of financial services reflect new estimates of banks' spread earnings on trading activity in foreign exchange, derivatives and securities. Revisions to exports of insurance services from 1999 onwards reflect the use of data from the International Trade in Services Survey replacing projections. Other revisions from 2003 onwards are mainly caused by the use of the final results from ONS's annual International Trade in Services Survey and financial surveys, the Chamber of Shipping's annual balance of payments survey for 2004 and the revisions mentioned in the section on Trade in Goods above.

Income, Financial Account and IIP – figures are revised from 1995 following implementation of improved methodology for estimating UK investment in property abroad, making use of information collected through the Survey of English Housing. More information on the new methodology is contained in an article published in the August 2005 edition of *Economic Trends* (Aspden, 2005). In addition, UK banks' investment income receipts and payments have been revised to reflect the inclusion of new data. Further revisions from 2003 onwards are largely the result of later and corrected data from annual direct investment and financial surveys.

Current transfers – the data is revised from 1999. These revisions are the result of an improved methodology for estimating UK receipts of agricultural subsidies from the EU. Revisions for 2003 onwards also reflect the latest results from the Expenditure and Food survey which are used to derive net non-life insurance claims and the latest data from annual direct investment inquiries which provide estimates of taxes on investment income.

Capital transfers – the data is revised from 1997. These revisions reflect the use of improved data sources for public corporations' debt forgiveness. Revisions from 2004 onwards also reflect the final results from the annual International Trade in Services survey for 2004.

Symbols and conventions used in the tables

Rounding

As figures have been rounded to the nearest final digit, there may be slight discrepancies between the sums of the constituent items and the totals as shown.

Symbols

The following symbols are used throughout:

    ..   = not available
    -    = nil or less than a million

References

The internationally agreed framework for the presentation of the Balance of Payments and the National Accounts are described in the following publications:

Balance of Payments Manual (5th edition 1993), International Monetary Fund (ISBN 1-55775-339-3). www.imf.org/external/np/sta/bop/BOPman.pdf

Balance of Payments Textbook (1996), International Monetary Fund (ISBN 1-55775-570-1). www.imf.org/external/np/sta/bop/BOPtex.pdf

Balance of Payments and International Investment Position, Australia: Concepts, Sources and Methods (1998) Australian Bureau of Statistics (ISBN 0-642-25670-5). www.abs.gov.au/Ausstats/abs@.nsf/0/09998F91F5A8A7BFCA25697E0018FB0A?Open

European System of Accounts (ESA 1995), Office for Official Publications of the European Communities (ISBN 92-827-7954-8)

System of National Accounts (1993), (ISBN 92-1-161352-3). http://unstats.un.org/unsd/sna1993/introduction.asp

The United Nations Statistics Commission and the IMF Board of Directors have approved the comprehensive and parallel updating of the National Accounts and Balance of Payments manuals, in order to ensure their consistency and achieve greater harmonisation. The ONS has developed the following webpage to inform users of progress and to invite their input: www.statistics.gov.uk/about/Consultations/NA/default.asp

Current account

Part 1

Chapter 1
Summary of balance of payments

Current account

The UK has recorded a current account deficit in every year since 1984. Prior to 1984, the current account recorded a surplus in 1980 to 1983. Since the last surplus was recorded in 1983, there have been three main phases in the development of the current account. In the first phase, from 1984 to 1989, the current account deficit increased steadily to reach a high of £26.3 billion in 1989 (equivalent to -5.1 per cent of GDP); during the second phase, from 1990 until 1997, the current account deficit declined to a low of £0.8 billion in 1997; in the third phase, since 1998, the current account deficit has widened sharply. The deficit in 2005, at £26.6 billion, is the highest recorded in cash terms but only equates to -2.2 per cent of GDP.

The profile for the current account has historically followed that of trade in goods, its biggest and most cyclical component. For a while, at the end of the 1990s, that pattern changed, but in recent years the pattern has remerged and the increasing deficit on trade in goods is mirrored by an increase in the current account deficit. The last trade in goods surplus, recorded in 1982, was the main driver of a current account surplus. Following 1982, the goods balance went into deficit and this increased to a peak of £24.7 billion in 1989, while the current balance deteriorated to a deficit of £26.3 billion. From 1989 until the late-1990s, both the trade in goods and current account deficits fell and then subsequently rose. From 1999 to 2003 the goods deficit continued to grow but the current account deficit stabilised, due to a widening income surplus. From 2004, the deficit on trade in goods has increased steadily, matched by a rise in the current account deficit.

Figure 1.1
Current account balance

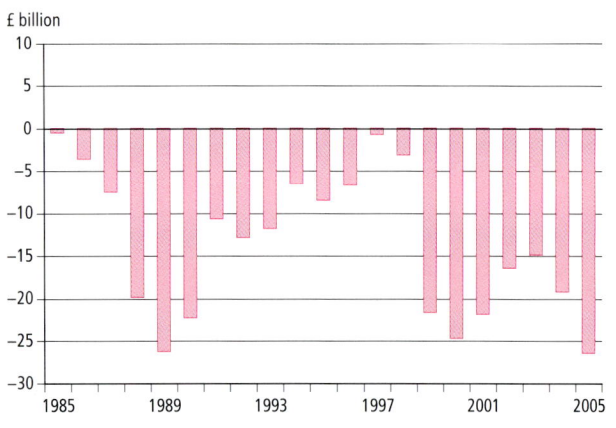

Figure 1.2
Current balance as a percentage of GDP

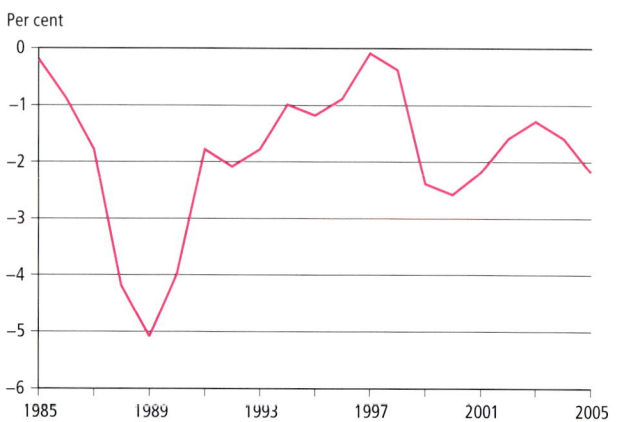

Trade in goods and services

The trade in goods account recorded net surpluses in the years 1980, 1981 and 1982, largely as a result of growth in exports of North Sea oil. Since then, however, the trade in goods account has remained in deficit. The deficit grew significantly in the late 1980s to reach a peak of £24.7 billion in 1989, before narrowing in the 1990s to levels of around £10 billion to £14 billion. In 1998 the deficit jumped by about £10 billion, and it has continued to rise since, reaching a cash record of £67.3 billion in 2005.

Figure 1.3
Trade in goods and services
Credits less debits

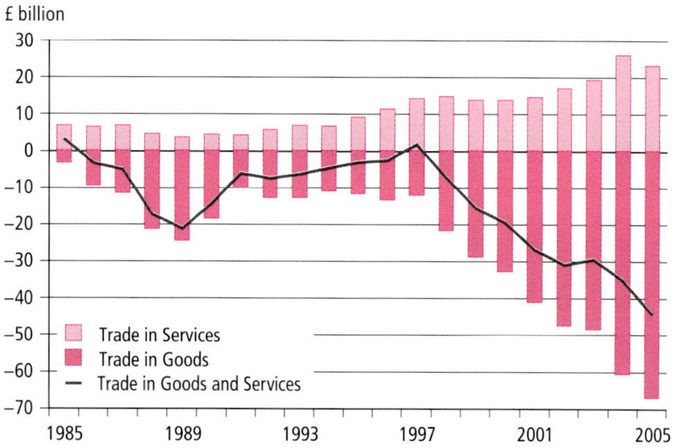

The trade in services account has shown a surplus for every year since 1966. The surplus on services increased fairly steadily until 1987 during which time it broadly offset the deficit on trade in goods. From 1988 to 1994 the surplus was reasonably steady at around £5 billion annually. From 1995 to 1997 the services surplus increased significantly, to around £14 billion. It remained at this level until 2001, after which time the surplus rose again to reach a peak of £25.9 billion in 2004 before falling back slightly in the latest year to £23.1 billion.

Income

The income account consists of compensation of employees and investment income, the latter dominating the account. Historically the balance on compensation of employees has generally been in deficit, but it moved into surplus in the late 1990s where it has generally remained.

The investment income balance has generally shown a surplus (since records began in 1946 there have only been nine years which have shown a deficit) although it was not until 1994 that the surplus exceeded £3 billion annually. Surpluses on direct investment income have been partly offset by deficits on other investment – principally banks' net interest payments on loans and deposits. Since 1994 there has been a substantial improvement in the investment income balance, largely due to an increasing surplus on direct investment. By sector, the improvement in the investment income balance has been driven by monetary financial institutions, moving from a deficit of £6.1 billion in 1993 to a surplus of around £10–£12 billion for the latest four years.

Figure 1.4
Investment income
Credits less debits

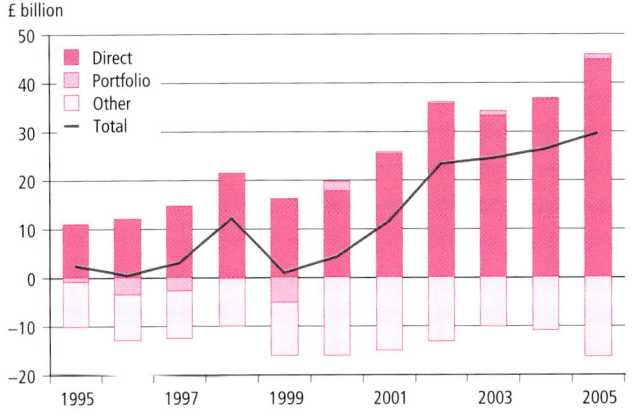

Current transfers

The transfers account has shown a deficit in every year since 1960. The deficit increased steadily to reach £4.9 billion in 1990. In 1991, the deficit reduced to £1.2 billion, reflecting £2.1 billion receipts from other countries towards the UK's cost of the first Gulf conflict. The deficit has since increased, to reach a record £12.2 billion in 2005. Separate data for central government and other sectors are available from 1986 and show that both have been consistently in deficit since 1992. The majority of payments to and receipts from EU institutions are recorded as other sector transactions as they relate to the original payee or ultimate recipient of the payment/receipt. The volatility in this account is driven by fluctuating net contributions to EU Institutions.

Figure 1.5
Current transfers
Credits less debits

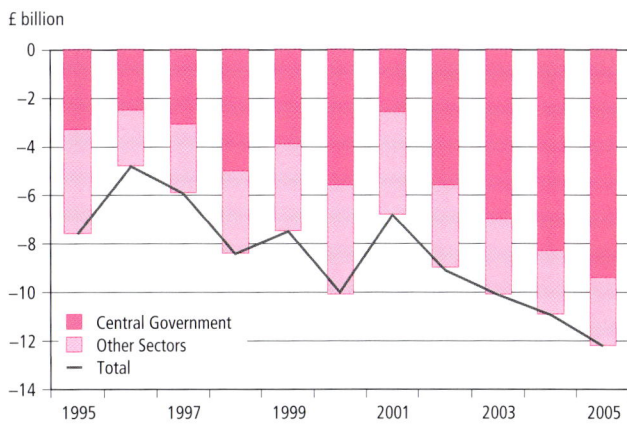

Revisions since *Pink Book 2005*

The current account balance has been revised back to 1992 in this publication as compared with data published in the 2005 *Pink Book*, reflecting the incorporation of GDP balancing adjustments for Trade in Services, annual inquiry results for 2003, 2004 and 2005 and the inclusion of methodological improvements and new data sources. Details of the sources of these changes are given on pages 17–18 of the Introduction; the impact of the changes can be seen in figure 1.6 and in table 1.1R.

Figure 1.6
Revisions since *Pink Book 2005*
Credits less debits

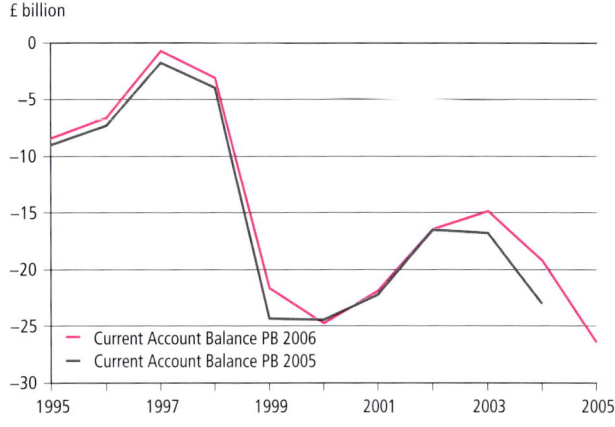

Investment flows, levels and income

One important set of relationships within the balance of payments is the link between the financial account (investment flows), the international investment position (levels or balance sheets), and the income deriving from the balance sheets. This is explained in more detail in the Introduction. Although a reconciliation statement between opening and closing levels and flows is not officially compiled in the UK, table 1.3 shows the rudiments of this relationship over the years for which consistent detailed data are available. Within the three main categories of investment (direct, portfolio and other), as well as reserve assets, it can be seen that the difference in the values of the balance sheet at the end of one year and the previous year is approximately equal to the value of financial transactions in that year. The difference between the two amounts is explained by valuation, exchange rate and other effects, for example, company write-offs, etc.

Figure 1.7

International investment position and income

Credits less debits

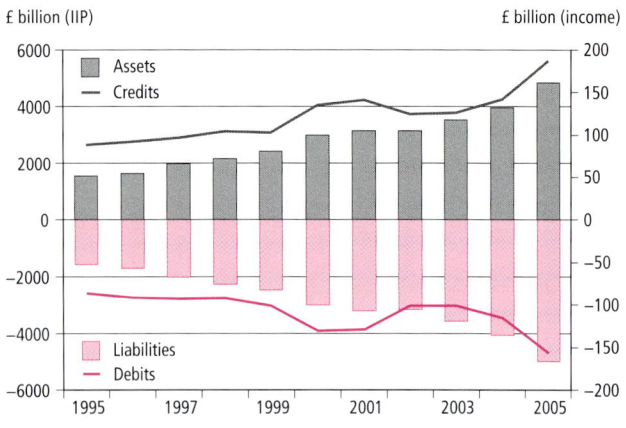

The value of both external assets and liabilities in the international investment position has been rising steadily since 1980, reflecting both the increased global investment and the increasing prices of external assets and liabilities. The UK's external assets exceeded external liabilities in every year until 1990 however since then external liabilities have tended to exceed external assets. Since 1995, there has been a more than threefold increase in both the levels of external assets and liabilities with the latter exceeding £5.0 trillion at the end of 2005. Between 1995 and 2002 the international investment position was fairly stable, with the exception of 1998, but from the 2003 onwards the net liability position has increased steadily to end 2005 at £168.9 billion.

Implied 'rates of return'

Another important relationship is that which exists between investment income and the international investment position. This can be considered most easily by looking at the implied 'rates of return' for both assets and liabilities. In total the implied rate of return on liabilities was higher than assets from the late 1970s until around 1993 to 1994. Since this time, although the return on assets has been higher, both have been at relatively low levels. Because other investment constitutes around half of the value of the balance sheets it is not surprising that the rates of return have reflected the movements in interest rates on loans and deposits such as LIBOR.

Figure 1.8

Implied rates of return on assets

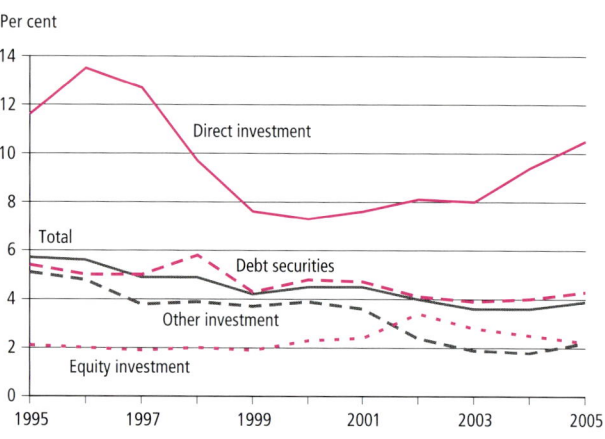

Figure 1.9

Implied rates of return on liabilities

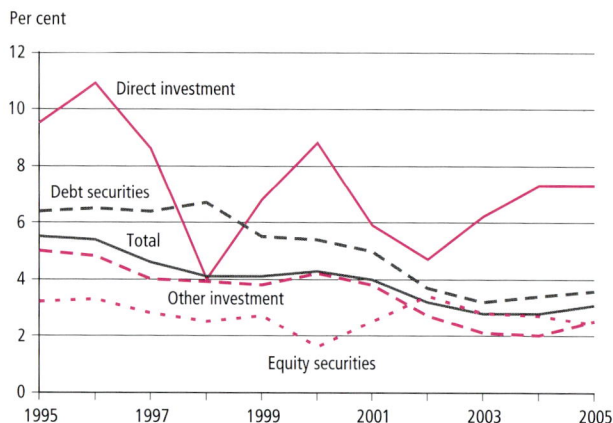

The rates of return for direct investment are significantly higher than for other forms of investment. This is probably a consequence in part of comparatively lower valuations since direct investment levels are at book value rather than market value used elsewhere, but may also reflect the higher return required to make the longer term investment worthwhile.

Within portfolio investment, rates of return on debt securities have been falling, in line with the downward trend in official interest rates. Rates of return on equity have fluctuated, largely reflecting valuation changes in balance sheets caused by the rise and fall in stock market prices. The difference between rates of return on debt and equity has narrowed considerably in recent years.

Rates of return on other investment were similar to returns on debt in the early 1990s. They have, in general, been in decline since then, and in recent years are equal to or lower than returns on equity.

1.1 Summary of balance of payments
Balances (credits less debits)

£ million

				Current account								
	Trade in goods	Trade in services	Total goods and services	Compensation of employees	Investment income	Total income	Current transfers	Current balance	Current balance as % of GDP[1]	Capital account	Financial account	Net errors & omissions
	LQCT	KTMS	KTMY	KTMP	HMBM	HMBP	KTNF	HBOG	AA6H	FKMJ	HBNT	HHDH
1946	−101	−274	−375	−20	76	56	166	−153	..	−21	181	−7
1947	−358	−197	−555	−19	140	121	123	−311	..	−21	552	−220
1948	−152	−64	−216	−20	223	203	96	83	0.7	−17	−58	−8
1949	−137	−43	−180	−20	206	186	29	35	0.3	−12	−103	80
1950	−54	−4	−58	−21	378	357	39	338	2.6	−10	−447	119
1951	−692	32	−660	−21	322	301	29	−330	−2.3	−15	426	−81
1952	−272	123	−149	−22	231	209	169	229	1.4	−15	−229	15
1953	−244	123	−121	−25	207	182	143	204	1.2	−13	−177	−14
1954	−210	115	−95	−27	227	200	55	160	0.9	−13	−174	27
1955	−315	42	−273	−27	149	122	43	−108	−0.6	−15	34	89
1956	50	26	76	−30	203	173	2	251	1.2	−13	−250	12
1957	−29	121	92	−32	223	191	−5	278	1.3	−13	−313	48
1958	34	119	153	−34	261	227	4	384	1.7	−10	−411	37
1959	−116	118	2	−37	233	196	–	198	0.8	−5	−68	−125
1960	−404	39	−365	−35	201	166	−6	−205	−0.8	−6	−7	218
1961	−144	51	−93	−35	223	188	−9	86	0.3	−12	23	−97
1962	−104	50	−54	−37	301	264	−14	196	0.7	−12	−195	11
1963	−123	4	−119	−38	364	326	−37	170	0.6	−16	−30	−124
1964	−551	−34	−585	−33	365	332	−74	−327	−1.0	−17	392	−48
1965	−263	−66	−329	−34	405	371	−75	−33	−0.1	−18	49	2
1966	−111	44	−67	−39	358	319	−91	161	0.4	−19	22	−164
1967	−601	157	−444	−39	354	315	−118	−247	−0.6	−25	179	93
1968	−708	341	−367	−48	303	255	−119	−231	−0.5	−26	688	−431
1969	−214	392	178	−47	468	421	−109	490	1.0	−23	−794	327
1970	−18	455	437	−56	527	471	−89	819	1.6	−22	−818	21
1971	205	590	795	−63	481	418	−90	1 123	2.0	−23	−1 330	230
1972	−736	665	−71	−52	407	355	−142	142	0.2	−35	477	−584
1973	−2 573	803	−1 770	−68	1 074	1 006	−336	−1 100	−1.5	−39	1 031	108
1974	−5 241	1 118	−4 123	−92	1 184	1 092	−302	−3 333	−4.0	−34	3 185	182
1975	−3 245	1 447	−1 798	−102	518	416	−313	−1 695	−1.6	−36	1 569	162
1976	−3 930	2 532	−1 398	−140	1 100	960	−534	−972	−0.8	−12	507	477
1977	−2 271	3 306	1 035	−152	−280	−432	−889	−286	−0.2	11	−3 286	3 561
1978	−1 534	3 777	2 243	−140	138	−2	−1 420	821	0.5	−79	−2 655	1 913
1979	−3 326	4 076	750	−130	155	25	−1 777	−1 002	−0.5	−103	864	241
1980	1 329	3 829	5 158	−82	−1 683	−1 765	−1 653	1 740	0.8	−4	−2 157	421
1981	3 238	3 951	7 189	−66	−1 058	−1 124	−1 219	4 846	1.9	−79	−5 312	545
1982	1 879	3 198	5 077	−95	−1 273	−1 368	−1 476	2 233	0.8	6	−1 233	−1 006
1983	−1 618	4 076	2 458	−89	280	191	−1 391	1 258	0.4	75	−3 287	1 954
1984	−5 409	4 491	−918	−94	1 284	1 190	−1 566	−1 294	−0.4	107	−7 130	8 317
1985	−3 416	6 767	3 351	−120	−877	−997	−2 924	−570	−0.2	185	−1 657	2 042
1986	−9 617	6 403	−3 214	−156	1 850	1 694	−2 094	−3 614	−0.9	135	−122	3 601
1987	−11 698	6 813	−4 885	−174	1 091	917	−3 570	−7 538	−1.8	333	10 764	−3 559
1988	−21 553	4 450	−17 103	−64	817	753	−3 500	−19 850	−4.2	235	17 201	2 414
1989	−24 724	3 643	−21 081	−138	−654	−792	−4 448	−26 321	−5.1	270	18 001	8 050
1990	−18 707	4 337	−14 370	−110	−2 869	−2 979	−4 932	−22 281	−4.0	497	15 083	6 701
1991	−10 223	4 102	−6 121	−63	−3 244	−3 307	−1 231	−10 659	−1.8	290	5 269	5 100
1992	−13 050	5 602	−7 448	−49	177	128	−5 534	−12 854	−2.1	421	5 090	7 343
1993	−13 066	6 741	−6 325	35	−226	−191	−5 243	−11 759	−1.8	309	11 332	118
1994	−11 126	6 509	−4 617	−170	3 518	3 348	−5 369	−6 638	−1.0	33	2 126	4 479
1995	−12 023	8 957	−3 066	−296	2 460	2 164	−7 574	−8 476	−1.2	533	2 552	5 391
1996	−13 722	11 204	−2 518	93	463	556	−4 755	−6 717	−0.9	1 260	2 811	2 646
1997	−12 342	14 106	1 764	83	3 231	3 314	−5 918	−840	−0.1	958	−8 771	8 653
1998	−21 813	14 672	−7 141	−10	12 330	12 320	−8 374	−3 195	−0.4	489	9 922	−7 216
1999	−29 051	13 597	−15 454	201	1 069	1 270	−7 533	−21 717	−2.4	747	21 416	−446
2000	−32 976	13 615	−19 361	150	4 390	4 540	−10 012	−24 833	−2.6	1 703	12 604	10 526
2001	−41 212	14 423	−26 789	66	11 598	11 664	−6 759	−21 884	−2.2	1 318	17 503	3 063
2002	−47 705	16 830	−30 875	67	23 376	23 443	−9 081	−16 513	−1.6	932	7 202	8 379
2003	−48 607	19 162	−29 445	59	24 587	24 646	−10 122	−14 921	−1.3	1 466	20 507	−7 052
2004	−60 893	25 918	−34 975	71	26 525	26 596	−10 949	−19 328	−1.6	2 063	5 641	11 624
2005	−67 298	23 056	−44 242	74	29 797	29 871	−12 179	−26 550	−2.2	2 402	17 476	6 672

1 Using series YBHA: GDP at current market prices.

1.1R Summary of balance of payments
Revisions since ONS Pink Book 2005

£ million

			Current account									
	Trade in goods	Trade in services	Total goods and services	Compensati- on of employees	Investment income	Total income	Current transfers	Current balance	Current balance as % of GDP[1]	Capital account	Financial account	Net errors & omissions
	LQCT	KTMS	KTMY	KTMP	HMBM	HMBP	KTNF	HBOG	AA6H	FKMJ	HBNT	HHDH
1946	–	–	–	–	–	–	–	–	..	–	–	–
1947	–	–	–	–	–	–	–	–	..	–	–	–
1948	–	–	–	–	–	–	–	–	–	–	–	–
1949	–	–	–	–	–	–	–	–	–	–	–	–
1950	–	–	–	–	–	–	–	–	–	–	–	–
1951	–	–	–	–	–	–	–	–	–	–	–	–
1952	–	–	–	–	–	–	–	–	–	–	–	–
1953	–	–	–	–	–	–	–	–	–	–	–	–
1954	–	–	–	–	–	–	–	–	–	–	–	–
1955	–	–	–	–	–	–	–	–	–	–	–	–
1956	–	–	–	–	–	–	–	–	–	–	–	–
1957	–	–	–	–	–	–	–	–	–	–	–	–
1958	–	–	–	–	–	–	–	–	–	–	–	–
1959	–	–	–	–	–	–	–	–	–	–	–	–
1960	–	–	–	–	–	–	–	–	–	–	–	–
1961	–	–	–	–	–	–	–	–	–	–	–	–
1962	–	–	–	–	–	–	–	–	–	–	–	–
1963	–	–	–	–	–	–	–	–	–	–	–	–
1964	–	–	–	–	–	–	–	–	–	–	–	–
1965	–	–	–	–	–	–	–	–	–	–	–	–
1966	–	–	–	–	–	–	–	–	–	–	–	–
1967	–	–	–	–	–	–	–	–	–	–	–	–
1968	–	–	–	–	–	–	–	–	–	–	–	–
1969	–	–	–	–	–	–	–	–	–	–	–	–
1970	–	–	–	–	–	–	–	–	–	–	–	–
1971	–	–	–	–	–	–	–	–	–	–	–	–
1972	–	–	–	–	–	–	–	–	–	–	–	–
1973	–	–	–	–	–	–	–	–	–	–	–	–
1974	–	–	–	–	–	–	–	–	–	–	–	–
1975	–	–	–	–	–	–	–	–	–	–	–	–
1976	–	–	–	–	–	–	–	–	–	–	–	–
1977	–	–	–	–	–	–	–	–	–	–	–	–
1978	–	–	–	–	–	–	–	–	–	–	–	–
1979	–	–	–	–	–	–	–	–	–	–	–	–
1980	–	–	–	–	–	–	–	–	–	–	–	–
1981	–	–	–	–	–	–	–	–	–	–	–	–
1982	–	–	–	–	–	–	–	–	–	–	–	–
1983	–	–	–	–	–	–	–	–	–	–	–	–
1984	–	–	–	–	–	–	–	–	–	–	–	–
1985	–	–	–	–	–	–	–	–	–	–	–	–
1986	–	–	–	–	–	–	–	–	–	–	–	–
1987	–	–	–	–	–	–	–	–	–	–	–	–
1988	–	–	–	–	–	–	–	–	–	–	–	–
1989	–	–	–	–	–	–	–	–	–	–	–	–
1990	–	–	–	–	–	–	–	–	–	–	–	–
1991	–	–	–	–	–	–	–	–	–	–	–	–
1992	–	120	120	–	–	–	–	120	–	–	1	–121
1993	–	160	160	–	–	–	–	160	0.1	–	2	–162
1994	–	130	130	–	–	–	–	130	–	–	–	–130
1995	–	476	476	–	63	63	–	539	0.1	–	–2 453	1 914
1996	–	832	832	–	–225	–225	–	607	0.1	–	–1 225	618
1997	–	908	908	–	38	38	–	946	0.1	–24	998	–1 920
1998	–	803	803	–	16	16	–	819	0.1	–27	–370	–422
1999	–	–37	–37	–	2 730	2 730	–150	2 543	0.3	–26	–697	–1 820
2000	–	–111	–111	–	–43	–43	–260	–414	–	176	–1 102	1 340
2001	–564	720	156	–	293	293	–148	301	–	112	–1 807	1 394
2002	–618	1 294	676	–	–236	–236	–466	–26	–	64	–825	787
2003	–743	2 280	1 537	–	454	454	–161	1 830	0.2	170	–1 826	–174
2004	–2 279	5 729	3 450	–	433	433	–236	3 647	0.4	83	–6 397	2 667

1 Using series YBHA: GDP at current market prices.

1.2 Current account

£ million

		1984	1985	1986	1987	1988	1989	1990	1991	1992	1993	1994
Credits												
Exports of goods and services												
Exports of goods	LQAD	70 565	78 291	72 997	79 531	80 711	92 611	102 313	103 939	107 863	122 229	135 143
Exports of services	KTMQ	21 094	23 783	24 682	27 033	26 843	28 998	31 574	32 001	36 348	41 571	45 615
Total exports of goods and services	KTMW	91 659	102 074	97 679	106 564	107 554	121 609	133 887	135 940	144 211	163 800	180 758
Income												
Compensation of employees	KTMN	323	344	369	413	445	476	543	551	551	595	681
Investment income	HMBN	50 629	51 011	46 431	47 079	55 444	72 604	77 663	75 073	66 153	72 333	73 702
Total income	HMBQ	50 952	51 355	46 800	47 492	55 889	73 080	78 206	75 624	66 704	72 928	74 383
Current transfers												
Central government	FJUM	..	..	1 973	1 475	1 929	1 507	2 050	4 892	2 180	2 826	2 138
Other sectors	FJUN	..	..	4 374	4 468	4 878	5 947	7 445	9 335	10 397	9 613	9 521
Total current transfers	KTND	4 710	4 653	6 347	5 943	6 807	7 454	9 495	14 227	12 577	12 439	11 659
Total	HBOE	**147 321**	**158 082**	**150 826**	**159 999**	**170 250**	**202 143**	**221 588**	**225 791**	**223 492**	**249 167**	**266 800**
Debits												
Imports of goods and services												
Imports of goods	LQBL	75 974	81 707	82 614	91 229	102 264	117 335	121 020	114 162	120 913	135 295	146 269
Imports of services	KTMR	16 603	17 016	18 279	20 220	22 393	25 355	27 237	27 899	30 746	34 830	39 106
Total imports of goods and services	KTMX	92 577	98 723	100 893	111 449	124 657	142 690	148 257	142 061	151 659	170 125	185 375
Income												
Compensation of employees	KTMO	417	464	525	587	509	614	653	614	600	560	851
Investment income	HMBO	49 345	51 888	44 581	45 988	54 627	73 258	80 532	78 317	65 976	72 559	70 184
Total income	HMBR	49 762	52 352	45 106	46 575	55 136	73 872	81 185	78 931	66 576	73 119	71 035
Current transfers												
Central government	FJUO	..	..	1 261	1 449	2 433	2 275	2 125	3 450	3 812	4 343	4 977
Other sectors	FJUP	..	..	7 180	8 064	7 874	9 627	12 302	12 008	14 299	13 339	12 051
Total current transfers	KTNE	6 276	7 577	8 441	9 513	10 307	11 902	14 427	15 458	18 111	17 682	17 028
Total	HBOF	**148 615**	**158 652**	**154 440**	**167 537**	**190 100**	**228 464**	**243 869**	**236 450**	**236 346**	**260 926**	**273 438**
Balances												
Trade in goods and services												
Trade in goods	LQCT	−5 409	−3 416	−9 617	−11 698	−21 553	−24 724	−18 707	−10 223	−13 050	−13 066	−11 126
Trade in services	KTMS	4 491	6 767	6 403	6 813	4 450	3 643	4 337	4 102	5 602	6 741	6 509
Total trade in goods and services	KTMY	−918	3 351	−3 214	−4 885	−17 103	−21 081	−14 370	−6 121	−7 448	−6 325	−4 617
Income												
Compensation of employees	KTMP	−94	−120	−156	−174	−64	−138	−110	−63	−49	35	−170
Investment income	HMBM	1 284	−877	1 850	1 091	817	−654	−2 869	−3 244	177	−226	3 518
Total income	HMBP	1 190	−997	1 694	917	753	−792	−2 979	−3 307	128	−191	3 348
Current transfers												
Central government	FJUQ	..	..	712	26	−504	−768	−75	1 442	−1 632	−1 517	−2 839
Other sectors	FJUR	..	..	−2 806	−3 596	−2 996	−3 680	−4 857	−2 673	−3 902	−3 726	−2 530
Total current transfers	KTNF	−1 566	−2 924	−2 094	−3 570	−3 500	−4 448	−4 932	−1 231	−5 534	−5 243	−5 369
Total (Current balance)	HBOG	**−1 294**	**−570**	**−3 614**	**−7 538**	**−19 850**	**−26 321**	**−22 281**	**−10 659**	**−12 854**	**−11 759**	**−6 638**

1.2 Current account
continued

£ million

		1995	1996	1997	1998	1999	2000	2001	2002	2003	2004	2005
Credits												
Exports of goods and services												
Exports of goods	LQAD	153 577	167 196	171 923	164 056	166 166	187 936	189 093	186 524	188 320	190 877	211 175
Exports of services	KTMQ	50 574	57 962	62 096	67 978	73 616	79 666	84 047	89 987	97 077	107 817	111 123
Total exports of goods and services	KTMW	204 151	225 158	234 019	232 034	239 782	267 602	273 140	276 511	285 397	298 694	322 298
Income												
Compensation of employees	KTMN	887	911	1 007	840	960	1 032	1 087	1 121	1 116	1 171	1 211
Investment income	HMBN	87 195	91 421	95 435	103 388	101 952	134 114	139 848	123 505	124 881	141 030	185 826
Total income	HMBQ	88 082	92 332	96 442	104 228	102 912	135 146	140 935	124 626	125 997	142 201	187 037
Current transfers												
Central government	FJUM	1 730	2 828	2 173	1 767	3 542	2 465	4 991	3 663	3 968	4 000	4 071
Other sectors	FJUN	10 821	17 201	10 898	10 597	9 678	8 076	9 453	8 572	8 235	8 917	12 242
Total current transfers	KTND	12 551	20 029	13 071	12 364	13 220	10 541	14 444	12 235	12 203	12 917	16 313
Total	HBOE	**304 784**	**337 519**	**343 532**	**348 626**	**355 914**	**413 289**	**428 519**	**413 372**	**423 597**	**453 812**	**525 648**
Debits												
Imports of goods and services												
Imports of goods	LQBL	165 600	180 918	184 265	185 869	195 217	220 912	230 305	234 229	236 927	251 770	278 473
Imports of services	KTMR	41 617	46 758	47 990	53 306	60 019	66 051	69 624	73 157	77 915	81 899	88 067
Total imports of goods and services	KTMX	207 217	227 676	232 255	239 175	255 236	286 963	299 929	307 386	314 842	333 669	366 540
Income												
Compensation of employees	KTMO	1 183	818	924	850	759	882	1 021	1 054	1 057	1 100	1 137
Investment income	HMBO	84 735	90 958	92 204	91 058	100 883	129 724	128 250	100 129	100 294	114 505	156 029
Total income	HMBR	85 918	91 776	93 128	91 908	101 642	130 606	129 271	101 183	101 351	115 605	157 166
Current transfers												
Central government	FJUO	5 022	5 297	5 260	6 787	7 482	8 015	7 584	9 296	10 944	12 304	13 499
Other sectors	FJUP	15 103	19 487	13 729	13 951	13 271	12 538	13 619	12 020	11 381	11 562	14 993
Total current transfers	KTNE	20 125	24 784	18 989	20 738	20 753	20 553	21 203	21 316	22 325	23 866	28 492
Total	HBOF	**313 260**	**344 236**	**344 372**	**351 821**	**377 631**	**438 122**	**450 403**	**429 885**	**438 518**	**473 140**	**552 198**
Balances												
Trade in goods and services												
Trade in goods	LQCT	–12 023	–13 722	–12 342	–21 813	–29 051	–32 976	–41 212	–47 705	–48 607	–60 893	–67 298
Trade in services	KTMS	8 957	11 204	14 106	14 672	13 597	13 615	14 423	16 830	19 162	25 918	23 056
Total trade in goods and services	KTMY	–3 066	–2 518	1 764	–7 141	–15 454	–19 361	–26 789	–30 875	–29 445	–34 975	–44 242
Income												
Compensation of employees	KTMP	–296	93	83	–10	201	150	66	67	59	71	74
Investment income	HMBM	2 460	463	3 231	12 330	1 069	4 390	11 598	23 376	24 587	26 525	29 797
Total income	HMBP	2 164	556	3 314	12 320	1 270	4 540	11 664	23 443	24 646	26 596	29 871
Current transfers												
Central government	FJUQ	–3 292	–2 469	–3 087	–5 020	–3 940	–5 550	–2 593	–5 633	–6 976	–8 304	–9 428
Other sectors	FJUR	–4 282	–2 286	–2 831	–3 354	–3 593	–4 462	–4 166	–3 448	–3 146	–2 645	–2 751
Total current transfers	KTNF	–7 574	–4 755	–5 918	–8 374	–7 533	–10 012	–6 759	–9 081	–10 122	–10 949	–12 179
Total (Current balance)	HBOG	**–8 476**	**–6 717**	**–840**	**–3 195**	**–21 717**	**–24 833**	**–21 884**	**–16 513**	**–14 921**	**–19 328**	**–26 550**

1.3 Summary of international investment position, financial account and investment income

£ billion

		1995	1996	1997	1998	1999	2000	2001	2002	2003	2004	2005
Investment abroad												
International investment position												
Direct investment	HBWD	213.3	211.7	232.4	309.8	438.3	618.8	616.9	637.2	691.1	689.0	753.2
Portfolio investment	HHZZ	499.3	548.3	651.0	703.8	838.3	906.1	937.4	844.0	935.8	1 092.3	1 332.1
Other investment	HLXV	808.1	851.7	1 070.4	1 107.7	1 129.7	1 427.5	1 573.1	1 635.8	1 885.1	2 156.2	2 727.1
Reserve assets	LTEB	31.8	27.3	22.8	23.3	22.2	28.8	25.6	25.5	23.8	23.3	24.7
Total	HBQA	1 552.5	1 638.9	1 976.5	2 144.7	2 428.5	2 981.2	3 153.1	3 142.4	3 535.8	3 960.7	4 837.1
Financial account transactions												
Direct investment	-HJYP	31.1	23.5	37.3	73.8	125.6	155.6	42.8	35.0	40.9	53.8	56.5
Portfolio investment	-HHZC	39.3	59.8	51.9	32.1	21.4	65.6	86.6	1.0	36.3	140.9	160.7
Financial derivatives (net)	-ZPNN	-1.7	-1.0	-1.2	3.0	-2.7	-1.6	-8.4	-1.0	5.4	7.9	2.5
Other investment	-XBMM	47.5	136.7	169.4	14.9	59.6	276.0	174.1	97.2	255.9	325.6	495.3
Reserve assets	-LTCV	-0.2	-0.5	-2.4	-0.2	-0.6	3.9	-3.1	-0.5	-1.6	0.2	0.7
Total	-HBNR	116.0	218.5	255.1	123.6	203.2	499.5	292.0	131.8	336.9	528.3	715.6
Investment income												
Direct investment	HJYW	24.8	28.6	29.5	29.9	33.1	45.0	46.7	51.5	55.1	64.4	79.1
Portfolio investment	HLYX	19.7	20.2	23.8	29.3	25.9	33.0	34.9	32.5	32.5	36.7	45.3
Other investment	AIOP	41.0	41.0	40.8	43.0	41.8	55.1	57.3	38.7	36.4	39.2	60.7
Reserve assets	HHCB	1.7	1.6	1.4	1.1	1.2	1.0	1.0	0.8	0.8	0.7	0.7
Total	HMBN	87.2	91.4	95.4	103.4	102.0	134.1	139.8	123.5	124.9	141.0	185.8
Investment in the UK												
International investment position												
Direct investment	HBWI	146.2	152.6	173.7	213.6	250.2	310.4	363.5	340.6	355.5	384.4	472.0
Portfolio investment	HLXW	406.3	480.0	583.3	692.7	828.8	998.2	958.5	892.3	1 047.3	1 177.8	1 418.3
Other investment	HLYD	1 013.0	1 061.7	1 274.3	1 355.0	1 403.9	1 696.4	1 889.6	1 945.8	2 177.1	2 509.4	3 115.6
Total	HBQB	1 565.5	1 694.4	2 031.3	2 261.4	2 482.9	3 005.0	3 211.5	3 178.7	3 579.9	4 071.6	5 006.0
Financial account transactions												
Direct investment	HJYU	13.8	17.6	22.9	45.1	55.1	80.6	37.3	16.8	16.8	42.4	87.7
Portfolio investment	HHZF	37.3	43.0	26.8	20.9	114.1	164.5	48.1	51.0	95.2	87.2	127.0
Other investment	XBMN	67.4	160.7	196.7	67.6	55.5	267.0	224.0	71.2	245.4	404.3	518.3
Total	HBNS	118.6	221.3	246.4	133.5	224.6	512.1	309.5	139.0	357.4	534.0	733.1
Investment income												
Direct investment	HJYX	13.8	16.6	14.9	8.6	17.0	27.4	21.4	16.0	21.9	27.9	34.6
Portfolio investment	HLZC	20.6	23.8	26.5	29.5	31.1	31.0	34.5	32.1	31.6	36.4	44.2
Other investment	HLZN	50.3	50.6	50.8	53.0	52.8	71.3	72.3	52.1	46.8	50.2	77.3
Total	HMBO	84.7	91.0	92.2	91.1	100.9	129.7	128.3	100.1	100.3	114.5	156.0
Net investment												
International investment position												
Direct investment	HBWQ	67.1	59.0	58.6	96.2	188.1	308.4	253.5	296.6	335.6	304.6	281.2
Portfolio investment	CGNH	93.0	68.3	67.7	11.1	9.5	-92.2	-21.1	-48.2	-111.5	-85.5	-86.3
Other investment	CGNG	-204.9	-210.1	-204.0	-247.3	-274.2	-268.9	-316.5	-310.0	-292.0	-353.2	-388.5
Reserve assets	LTEB	31.8	27.3	22.8	23.3	22.2	28.8	25.6	25.5	23.8	23.3	24.7
Net investment position	HBQC	-13.0	-55.5	-54.8	-116.7	-54.4	-23.9	-58.4	-36.3	-44.1	-110.9	-168.9
Financial account transactions												
Direct investment	HJYV	-17.3	-6.0	-14.4	-28.7	-70.5	-75.0	-5.5	-18.3	-24.1	-11.4	31.2
Portfolio investment	HHZD	-2.0	-16.8	-25.2	-11.2	92.7	99.0	-38.4	50.0	59.0	-53.6	-33.7
Financial derivatives	ZPNN	1.7	1.0	1.2	-3.0	2.7	1.6	8.4	1.0	-5.4	-7.9	-2.5
Other investment	HHYR	19.9	24.1	27.3	52.8	-4.1	-9.0	49.9	-26.0	-10.5	78.7	23.1
Reserve assets	LTCV	0.2	0.5	2.4	0.2	0.6	-3.9	3.1	0.5	1.6	-0.2	-0.7
Net transactions	HBNT	2.6	2.8	-8.8	9.9	21.4	12.6	17.5	7.2	20.5	5.6	17.5
Investment income												
Direct investment	HJYE	11.0	12.0	14.6	21.3	16.1	17.6	25.3	35.5	33.2	36.5	44.6
Portfolio investment	HLZX	-0.9	-3.5	-2.7	-0.2	-5.2	2.0	0.4	0.4	0.9	0.3	1.1
Other investment	CGNA	-9.4	-9.5	-10.0	-10.0	-11.0	-16.2	-15.1	-13.3	-10.3	-11.0	-16.5
Reserve assets	HHCB	1.7	1.6	1.4	1.1	1.2	1.0	1.0	0.8	0.8	0.7	0.7
Net earnings	HMBM	2.5	0.5	3.2	12.3	1.1	4.4	11.6	23.4	24.6	26.5	29.8

Chapter 2
Trade in goods

Summary

The balance on trade in goods has shown a deficit in all but six years since 1900, with the value of imports exceeding the value of exports. The last surplus on trade in goods was recorded for 1982. In the period 1992 to 1997, the deficit settled into the range of £11 billion – £14 billion before widening in every subsequent year.

In 2005, the deficit increased to a record £67.3 billion, driven by a rise of 10.6 per cent in the value of exports (to a record £211.2 billion) and a rise, also of 10.6 per cent, in the value of imports (to a record £278.5 billion). The deficit with EU countries widened from £30.7 billion in 2004 to a record £36.0 billion in 2005, with an 8.4 per cent rise in exports and a 10.3 per cent rise in imports. The deficit with non-EU countries widened from £30.2 billion in 2004 to a record £31.3 billion in 2005, with a 13.7 per cent rise in exports and an 11.0 per cent rise in imports.

When looking at trade figures, users should be aware that both exports and imports include the impact of VAT MTIC fraud. Following a change in the pattern of MTIC fraud, interpretation of the breakdown between EU and non-EU trade is more difficult. Originally, most carousel chains only involved EU member states. More recently, there have also been carousel chains that include non-EU countries, for example, Dubai and Switzerland. However, the MTIC trade adjustments are added to the EU import estimates as it is this part of the trading chain that is not recorded. Changes to the pattern of trading associated with MTIC fraud can therefore make it difficult to analyse trade by commodity group and by country as changes in the impact of activity associated with this fraud affect both imports and exports. In particular, adjustments affect trade in capital goods and intermediate goods – these categories include mobile phones and computer components. (For more information on MTIC fraud, see the methodological notes relating to chapter 2.)

Volume changes

Export volumes increased in every year between 1981 and 2001. The growth in exports slowed during the years 1991 to 1993 reflecting a decline in economic activity abroad. There was a period of strong growth between 1994 and 1997 followed by a marked slowdown in 1998. After a slight pick up in growth in 1999 and accelerated growth in 2000, export volume growth slowed again in 2001. In 2002 and 2003, export volumes fell as world economic activity slowed, but rose in 2004 and 2005 as world economic activity grew. In addition, there was an increase in trade associated with MTIC fraud – as shown in table 2.4 – between the two years.

Import volumes have also been generally increasing since 1981. However, a downturn in the UK economy resulted in a fall in the volume of imports in 1991. Since then imports have grown strongly in each year.

In 2005, the volume of exports rose by 8.9 per cent whilst the volume of imports rose by 6.8 per cent, both to record annual levels. Exports to EU countries rose by 6.1 per cent and to non-EU countries rose by 12.5 per cent, both to record annual levels. The volume of imports from EU countries rose by 9.7 per cent and the volume of imports from non-EU countries rose by 3.6 per cent in 2005, both to record levels again.

Figure 2.1
Trade in goods

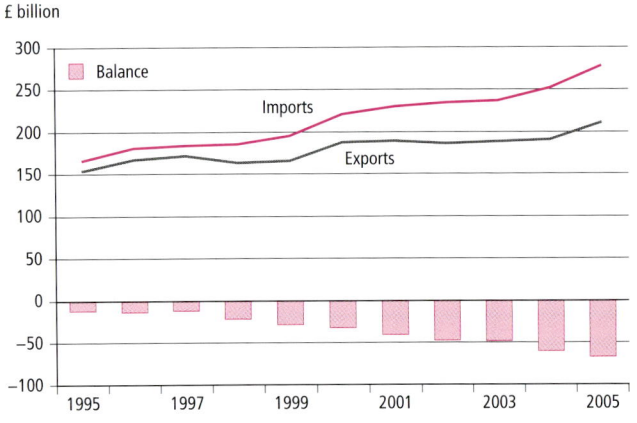

Figure 2.2
Export and import volume indices

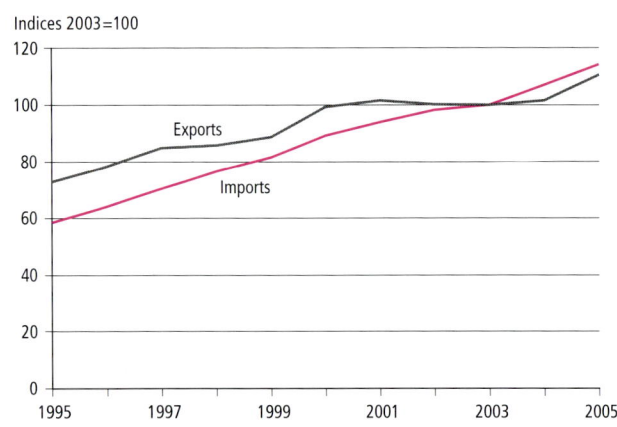

Price changes

Export and import prices rose for nine consecutive years between 1986 and 1995. The largest annual rises, 12.4 per cent for exports and 10.0 per cent for imports, occurred between 1992 and 1993 when sterling depreciated sharply following the UK's withdrawal from the Exchange Rate Mechanism (ERM). Both export and import prices fell significantly during 1997 and 1998. This reflected falls in World commodity prices and the price of crude oil feeding through into the price of manufactured goods. The price indices for crude oil increased by about 50 per cent in 1999 and by a further 70 per cent in 2000 before falling back in 2001 and 2002, only to rise again in 2003 and 2004, and to an even greater extent, in 2005.

In 2005, the overall export and import price indices both rose by around 4 per cent compared to the previous year. Excluding the oil price effect, export prices rose by 1.3 per cent and import prices by 1.9 per cent in 2005.

Figure 2.3
Export and import price indices

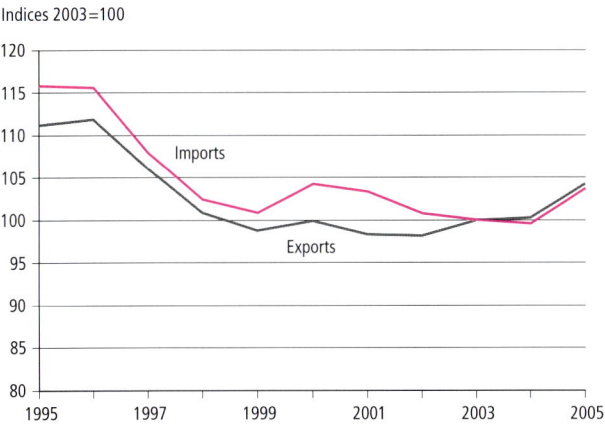

Trade in oil

While the overall balance on trade in goods has shown a deficit every year since 1982, exports of oil had consistently exceeded imports of oil in each year between 1980 and 2004 before recording a deficit in 2005. In 1985 trade in oil showed a record surplus of £8.0 billion as oil prices reached record levels. Disruptions to production in the North Sea subsequently diminished the surplus during the period 1988 to 1991. Until 1996 the annual surplus increased steadily as UK production recovered and World crude oil prices increased. Falling oil prices in 1997 and 1998 then led to a reduction in the surplus to £3.0 billion in 1998 before sharp rises in prices saw the surplus increase to £4.4 billion in 1999 and £6.5 billion in 2000 – the highest surplus since 1985. The fall in the price of crude oil reduced the oil trade surplus to £5.3 billion in 2001 and to £5.1 billion in 2002. Production difficulties from 2003 onwards led to drops in the volume of exports of crude oil and rises in the volume of crude oil imports as the existing stock of fields depleted. Coupled with rising prices, this resulted in further falls in the surplus, to £3.4 billion in 2003 and £0.9 billion in 2004 followed by the first deficit since 1979 – of £1.2 billion.

Figure 2.4
Trade in oil

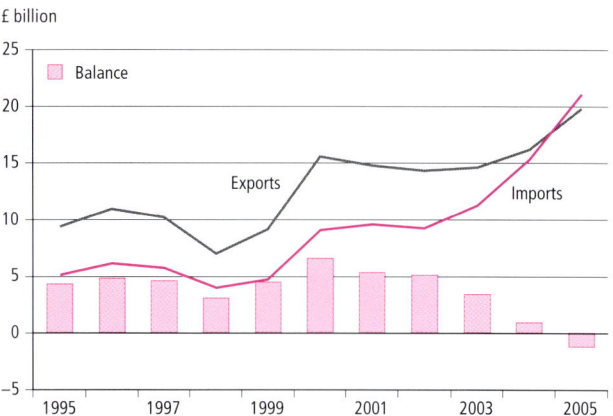

Trade in commodities other than oil

Finished manufactures accounted for more than half of both total exports and total imports in the last ten years. Their share of total exports rose from 54 per cent in 1994 to a peak of 60 per cent in 1998 before falling back steadily to 53 per cent in 2004. However, they rose to 54 per cent in 2005. Their share of total imports rose from 55 per cent in 1995 to a peak of 62 per cent in 2002 then fell back to a level of 57 per cent in 2005.

Within finished manufactures, the balance on trade in capital goods was in surplus every year between 1992 and 1999 but has moved significantly into deficit since then. The balance on trade in ships and aircraft was in surplus every year between 1992 and 1997 but then moved into deficit. The deficit peaked in 2002 and has subsequently narrowed to some extent although it rose in 2005. Trade in motor cars, other consumer goods, and intermediate goods, has been in deficit in each of the last ten years. The deficit on motor cars peaked in 2001 during a period of disruption caused by restructuring in the industry which affected production in the UK. The deficit for consumer goods other than cars has increased steadily over the last ten years.

Within semi-manufactured goods, the UK has been a net exporter of chemicals and a net importer of other semi-manufactured goods in each of the last ten years.

The balance on trade in coal, gas and electricity was in surplus from 1999 to 2003 but moved into deficit in 2004 and, to a greater extent, in 2005. This reflected higher imports of gas and electricity through the inter connectors.

In volume terms, exports of capital goods (which are affected by VAT MTIC trade) rose by 33 per cent in 2005 while exports of consumer goods other than motor cars rose by 10.3 per cent. Exports of intermediate goods rose by 8.0 per cent and of cars by 7.3 per cent. In contrast, exports of ships and aircraft fell by 5.0 per cent. Import growth, again in volume terms, was particularly strong for capital goods which rose by 27 per cent, precious stones and silver which rose by 9.3 per cent, consumer goods other than motor vehicles which rose by 9.2 per cent, and intermediate goods which rose by 4.7 per cent. Exports and imports of capital and intermediate goods are affected by changes in levels of VAT MTIC fraud related trade.

Figure 2.5
Trade in motor cars

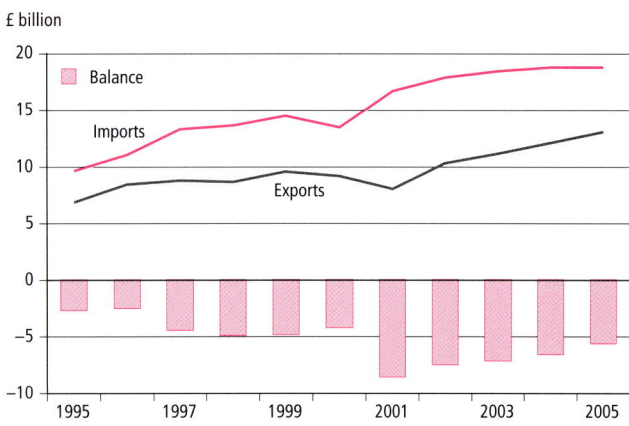

Figure 2.6
Trade in other consumer goods

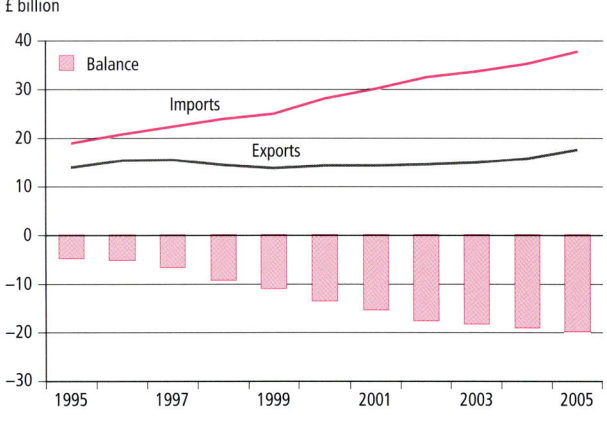

Chapter 2: Trade in goods

2.1 Trade in goods
Summary table

£ million

		SITC[1]	1995	1996	1997	1998	1999	2000	2001	2002	2003	2004	2005
Exports													
Food, beverages and tobacco	BQMV	0+1	11 192	11 328	11 103	10 216	9 947	9 908	9 630	9 993	10 879	10 578	10 650
Basic materials	ELBK	2+4	2 946	2 790	2 753	2 512	2 284	2 603	2 571	2 855	3 335	3 771	3 997
Oil													
Crude oil	BQNX	333	6 539	7 508	6 452	4 473	6 148	10 522	10 489	9 804	9 241	9 338	10 950
Oil products	BQNY	334+335	2 845	3 420	3 787	2 545	2 975	5 062	4 326	4 517	5 367	6 862	8 855
Total oil	BOKL	33	9 384	10 928	10 239	7 018	9 123	15 584	14 815	14 321	14 608	16 200	19 805
Coal, gas and electricity	BQNF	32+34+35	558	650	777	495	806	1 473	1 571	1 679	1 950	1 685	1 703
Semi-manufactured goods													
Chemicals	BQOB	5	20 999	22 166	21 901	22 102	23 071	24 992	27 514	28 386	31 373	32 008	33 280
Precious stones and silver	BQOD	667+681.1	3 117	3 609	3 555	2 833	3 633	4 744	4 709	4 728	5 138	4 909	5 545
Other	BQOC	Rest of 6	19 342	19 533	19 120	18 410	16 669	17 929	18 072	17 109	17 981	19 549	20 945
Total semi-manufactured goods	BQMX	5+6	43 458	45 308	44 576	43 345	43 373	47 665	50 295	50 223	54 492	56 466	59 770
Finished manufactured goods													
Motor cars	BQOE	781	6 898	8 450	8 789	8 710	9 585	9 178	8 046	10 297	11 183	12 107	13 077
Other consumer goods[2]	BQOF		13 893	15 391	15 554	14 448	13 840	14 280	14 360	14 606	14 997	15 782	17 605
Intermediate goods[2]	BQOG		31 311	34 320	35 881	35 637	36 659	41 130	42 089	40 025	37 370	36 672	38 638
Capital goods[2]	BQOH		27 499	30 542	32 795	33 654	33 324	37 169	37 715	34 944	31 500	29 434	37 892
Ships and aircraft	BQOI	792+793	4 611	5 536	7 513	6 125	5 730	7 261	6 978	6 508	7 143	7 301	6 917
Total finished manufactured goods	BQMQ	7+8	84 212	94 239	100 532	98 574	99 138	109 018	109 188	106 380	102 193	101 296	114 129
Commodities and transactions not classified according to kind	BOKJ	9	1 827	1 953	1 943	1 896	1 495	1 685	1 023	1 073	863	881	1 121
Total	LQAD		153 577	167 196	171 923	164 056	166 166	187 936	189 093	186 524	188 320	190 877	211 175
Imports													
Food, beverages and tobacco	BQMW	0+1	15 561	17 422	16 911	17 250	17 787	17 660	18 485	19 375	21 187	22 147	23 673
Basic materials	BQNA	2+4	6 454	6 545	6 273	5 631	5 429	6 307	6 442	5 958	6 139	6 340	6 769
Oil													
Crude oil	BQNM	333	3 093	3 810	3 414	1 967	2 106	4 825	4 878	4 752	5 705	8 191	11 212
Oil products	BQOA	334+335	1 968	2 308	2 265	2 009	2 569	4 223	4 647	4 461	5 527	7 116	9 827
Total oil	BQAQ	33	5 061	6 118	5 679	3 976	4 675	9 048	9 525	9 213	11 232	15 307	21 039
Coal, gas and electricity	BQNG	32+34+35	1 100	1 166	1 145	916	753	968	1 270	1 066	1 079	2 240	3 931
Semi-manufactured goods													
Chemicals	BQOJ	5	17 481	18 095	17 405	17 379	18 619	20 633	22 745	23 987	26 139	27 927	29 157
Precious stones and silver	BQOL	667+681.1	3 352	3 630	3 873	4 025	4 788	5 454	5 260	4 247	4 346	4 673	5 239
Other	BQOK		24 408	25 070	24 134	23 670	22 142	23 778	24 905	24 488	25 560	27 626	27 969
Total semi-manufactured goods	BQMR	5+6	45 241	46 795	45 412	45 074	45 549	49 865	52 910	52 722	56 045	60 226	62 365
Finished manufactured goods													
Motor cars	BQOM	781	9 601	10 978	13 254	13 618	14 433	13 403	16 619	17 800	18 374	18 723	18 740
Other consumer goods[2]	BQON		18 835	20 662	22 237	23 792	24 905	28 011	29 953	32 414	33 477	35 096	37 684
Intermediate goods[2]	BQOO		33 211	38 077	36 506	37 091	41 538	48 455	46 085	44 829	40 893	42 124	45 692
Capital goods[2]	BQOP		25 560	27 434	29 222	30 190	32 256	37 944	38 463	39 473	38 251	40 221	48 969
Ships and aircraft	BQOQ	792+793	3 359	3 956	5 867	6 526	6 093	7 405	9 289	9 929	8 646	7 539	7 770
Total finished manufactured goods	BQMY	7+8	90 566	101 107	107 086	111 217	119 225	135 218	140 409	144 445	139 641	143 703	158 855
Commodities and transactions not classified according to kind	BQAO	9	1 617	1 765	1 759	1 805	1 799	1 846	1 264	1 450	1 604	1 807	1 841
Total	LQBL		165 600	180 918	184 265	185 869	195 217	220 912	230 305	234 229	236 927	251 770	278 473

1 Standard International Trade Classification, Revision 3.

2 Derived from the *Classification by Broad Economic Categories defined in terms of SITC, Revision 3*, published by the United Nations.

2.1 Trade in goods
Summary table
continued

£ million

		SITC[1]	1995	1996	1997	1998	1999	2000	2001	2002	2003	2004	2005
Balances													
Food, beverages and tobacco	BQOS	0+1	−4 369	−6 094	−5 808	−7 034	−7 840	−7 752	−8 855	−9 382	−10 308	−11 569	−13 023
Basic materials	BQOR	2+4	−3 508	−3 755	−3 520	−3 119	−3 145	−3 704	−3 871	−3 103	−2 804	−2 569	−2 772
Oil													
Crude oil	BQMG	333	3 446	3 698	3 038	2 506	4 042	5 697	5 611	5 052	3 536	1 147	−262
Oil products	BQMH	334+335	877	1 112	1 522	536	406	839	−321	56	−160	−254	−972
Total oil	BQNE	33	4 323	4 810	4 560	3 042	4 448	6 536	5 290	5 108	3 376	893	−1 234
Coal, gas and electricity	BQNH	32+34+35	−542	−516	−368	−421	53	505	301	613	871	−555	−2 228
Semi-manufactured goods													
Chemicals	BQMI	5	3 518	4 071	4 496	4 723	4 452	4 359	4 769	4 399	5 234	4 081	4 123
Precious stones and silver	BQMK	667+681.1	−235	−21	−318	−1 192	−1 155	−710	−551	481	792	236	306
Other	BQMJ	Rest of 6	−5 066	−5 537	−5 014	−5 260	−5 473	−5 849	−6 833	−7 379	−7 579	−8 077	−7 024
Total semi-manufactured goods	BQOT	5+6	−1 783	−1 487	−836	−1 729	−2 176	−2 200	−2 615	−2 499	−1 553	−3 760	−2 595
Finished manufactured goods													
Motor cars	BQML	781	−2 703	−2 528	−4 465	−4 908	−4 848	−4 225	−8 573	−7 503	−7 191	−6 616	−5 663
Other consumer goods[2]	BQMM		−4 942	−5 271	−6 683	−9 344	−11 065	−13 731	−15 593	−17 808	−18 480	−19 314	−20 079
Intermediate goods[2]	BQMN		−1 900	−3 757	−625	−1 454	−4 879	−7 325	−3 996	−4 804	−3 523	−5 452	−7 054
Capital goods[2]	BQMO		1 939	3 108	3 573	3 464	1 068	−775	−748	−4 529	−6 751	−10 787	−11 077
Ships and aircraft	BQMP	792+793	1 252	1 580	1 646	−401	−363	−144	−2 311	−3 421	−1 503	−238	−853
Total finished manufactured goods	BQOV	7+8	−6 354	−6 868	−6 554	−12 643	−20 087	−26 200	−31 221	−38 065	−37 448	−42 407	−44 726
Commodities and transactions not classified according to kind	BQOU	9	210	188	184	91	−304	−161	−241	−377	−741	−926	−720
Total	LQCT		−12 023	−13 722	−12 342	−21 813	−29 051	−32 976	−41 212	−47 705	−48 607	−60 893	−67 298

1 Standard International Trade Classification, Revision 3.

2 Derived from the *Classification by Broad Economic Categories defined in terms of SITC, Revision 3*, published by the United Nations.

2.2 Trade in goods: volume indices

2003=100

		SITC[1]	1995	1996	1997	1998	1999	2000	2001	2002	2003	2004	2005
Exports													
Food, beverages and tobacco	BQPP	0+1	96	96	101	98	96	97	92	96	100	99	98
Basic materials	BQPQ	2+4	75	75	79	81	76	84	81	90	100	105	107
Oil													
Crude oil	BOGH	333	123	117	109	112	105	108	119	116	100	89	73
Oil products	BOGO	334+335	66	70	81	69	73	106	96	88	100	107	125
Total oil	BONC	33	98	96	96	93	91	106	110	106	100	95	92
Coal, gas and electricity	BOGP	32+34+35	29	28	36	32	47	73	82	90	100	74	71
Semi-manufactured goods													
Chemicals	BQLB	5	58	61	65	67	73	80	89	93	100	103	106
Precious stones and silver	BQLD	667+681.1	41	47	50	41	52	64	69	85	100	105	116
Other	BQLC	Rest of 6	100	101	104	104	98	104	104	99	100	107	110
Total semi-manufactured goods	BQPR	5+6	69	72	75	75	79	86	91	94	100	104	108
Finished manufactured goods													
Motor cars	BQLE	781	60	73	77	77	85	88	77	96	100	109	117
Other consumer goods[2]	BQLF		83	93	99	95	93	96	96	99	100	107	118
Intermediate goods[2]	BQLG		74	81	89	92	98	113	116	108	100	100	108
Capital goods[2]	BQLH		68	77	86	94	98	116	121	112	100	98	130
Ships and aircraft	BQLI	792+793	51	61	83	69	66	84	79	74	100	100	95
Total finished manufactured goods	BQPS	7+8	70	78	87	89	93	106	107	104	100	101	116
Total	BPBP		72.8	78.4	84.9	85.8	88.6	99.3	101.5	100.3	100.0	101.5	110.5
Imports													
Food, beverages and tobacco	BQPT	0+1	64	70	73	81	86	87	90	94	100	107	110
Basic materials	BQPU	2+4	97	102	103	98	96	105	109	101	100	100	102
Oil													
Crude oil	BQPV	333	88	89	92	84	61	80	90	91	100	123	115
Oil products	BQPW	334+335	102	103	107	108	101	97	117	87	100	114	111
Total oil	ELAM	33	94	95	99	94	78	87	102	89	100	118	113
Coal, gas and electricity	BQPX	32+34+35	75	80	88	81	85	107	111	116	100	177	266
Semi-manufactured goods													
Chemicals	BQLQ	5	54	59	62	65	73	78	87	95	100	107	106
Precious stones and silver	BQLS	667+681.1	56	60	68	69	81	88	93	90	100	118	129
Other	BQLR	Rest of 6	78	82	85	90	89	91	94	96	100	104	101
Total semi-manufactured goods	BQPY	5+6	65	69	73	77	81	85	91	95	100	106	106
Finished manufactured goods													
Motor cars	BQLT	781	49	55	69	71	75	72	93	97	100	104	105
Other consumer goods[2]	BQLU		54	58	64	71	75	81	86	96	100	109	119
Intermediate goods[2]	BQLV		60	70	73	82	94	107	104	105	100	106	111
Capital goods[2]	BQLW		45	49	58	67	73	87	91	100	100	108	137
Ships and aircraft	BQLX	792+793	46	52	80	89	82	91	103	112	100	90	88
Total finished manufactured goods	BQPZ	7+8	52	58	66	74	81	90	95	101	100	106	118
Total	BQBJ		58.5	64.1	70.4	76.4	81.5	89.1	93.8	98.2	100.0	106.9	114.2

1 Standard International Trade Classification, Revision 3.

2 Derived from the *Classification by Broad Economic Categories defined in terms of SITC, Revision 3,* published by the United Nations.

2.3 Trade in goods: price indices

2003=100

		SITC[1]	1995	1996	1997	1998	1999	2000	2001	2002	2003	2004	2005
Exports													
Food, beverages and tobacco	BPAI	0+1	103	105	98	95	95	94	96	96	100	99	100
Basic materials	BPAW	2+4	107	103	98	94	90	93	94	95	100	108	115
Oil													
Crude oil	BQAC	333	58	70	64	43	63	107	96	92	100	115	166
Oil products	BQAD	334+335	76	87	83	69	76	90	84	95	100	120	132
Total oil	BQAL	33	62	74	68	50	66	101	92	93	100	117	153
Coal, gas and electricity	BQAF	32+34+35	101	120	111	81	89	105	99	96	100	119	132
Semi-manufactured goods													
Chemicals	BQLJ	5	115	116	108	106	101	100	98	97	100	99	101
Precious stones and silver	BQLL	667+681.1	150	148	140	136	138	145	133	109	100	92	93
Other	BQLK	Rest of 6	109	108	102	99	95	96	96	96	100	102	105
Total semi-manufactured goods	BQAA	5+6	116	116	109	106	101	102	100	97	100	99	102
Finished manufactured goods													
Motor cars	BQPM	781	103	104	102	101	101	93	94	95	100	99	101
Other consumer goods[2]	BQLM		116	111	105	101	98	99	99	98	100	98	100
Intermediate goods[2]	BQLN		113	113	107	103	99	96	96	98	100	98	99
Capital goods[2]	BQLO		129	128	122	116	109	103	100	100	100	97	96
Ships and aircraft	BQLP	792+793	122	122	123	121	119	117	120	120	100	102	100
Total finished manufactured goods	BQAB	7+8	118	117	112	107	103	100	99	100	100	98	99
Total	BQKR		111.2	111.9	106.1	100.9	98.8	99.9	98.3	98.2	100.0	100.3	104.3
Imports													
Food, beverages and tobacco	ELAN	0+1	112	113	107	99	97	95	96	97	100	98	102
Basic materials	ELAO	2+4	105	103	99	95	92	97	96	96	100	103	108
Oil													
Crude oil	ELAS	333	61	74	64	41	60	107	96	92	100	116	168
Oil products	ELAT	334+335	35	41	38	34	47	79	71	92	100	114	162
Total oil	ELBB	33	47	56	50	37	53	91	82	92	100	115	165
Coal, gas and electricity	ELAU	32+34+35	142	141	127	110	88	94	109	93	100	150	196
Semi-manufactured goods													
Chemicals	BQLY	5	120	117	107	101	96	99	100	97	100	100	106
Precious stones and silver	BQMA	667+681.1	141	143	135	135	135	143	132	108	100	92	94
Other	BQLZ	Rest of 6	121	118	109	103	97	102	103	100	100	105	110
Total semi-manufactured goods	ELAQ	5+6	122	120	110	104	100	104	104	99	100	102	107
Finished manufactured goods													
Motor cars	BQMB	781	107	109	106	104	104	101	97	100	100	98	98
Other consumer goods[2]	BQMC		106	106	103	100	99	102	105	101	100	97	97
Intermediate goods[2]	BQMD		129	127	117	112	109	111	108	104	100	97	100
Capital goods[2]	BQME		137	134	123	118	115	113	110	105	100	97	92
Ships and aircraft	BQMF	792+793	84	88	85	85	86	94	103	102	100	96	101
Total finished manufactured goods	ELAR	7+8	120	119	112	108	106	107	106	103	100	97	97
Total	BQKS		115.8	115.6	107.9	102.4	100.8	104.2	103.3	100.7	100.0	99.5	103.7

1 Standard International Trade Classification, Revision 3.

2 Derived from the *Classification by Broad Economic Categories defined in terms of SITC, Revision 3*, published by the United Nations.

2.4 Adjustments to trade in goods on a balance of payments basis

£ million

		1995	1996	1997	1998	1999	2000	2001	2002	2003	2004	2005
Exports												
Overseas trade statistics (f.o.b.)	HGAA	154 971	169 569	173 082	165 859	168 221	189 665	190 806	187 763	189 038	191 018	211 560
Coverage adjustments												
Second-hand ships	HBYK	208	204	193	219	154	105	137	187	141	251	248
Repairs to ships and aircraft	EPAQ	12	12	12	12	12	12	12	12	12	12	12
Goods not changing ownership	HCLJ	−1 710	−1 972	−2 351	−2 565	−2 291	−2 343	−2 761	−2 788	−2 744	−2 012	−2 753
Goods procured in ports	KTPB	593	659	623	564	645	865	869	881	982	1 129	1 415
Industrial gold	DEJO	34	31	22	46	33	33	44	66	76	37	7
Other	BQPO	53	53	56	55	56	57	57	55	57	59	60
Total coverage adjustments	EHHH	−810	−1 013	−1 445	−1 671	−1 391	−1 271	−1 642	−1 587	−1 476	−524	−1 011
Other adjustments	EPAR	−584	−1 360	286	−131	−664	−460	−71	348	758	383	626
Total	LQAD	**153 577**	**167 196**	**171 923**	**164 056**	**166 166**	**187 936**	**189 093**	**186 524**	**188 320**	**190 877**	**211 175**
Imports												
Overseas trade statistics (c.i.f.)	HGAD	169 609	186 153	189 107	192 025	199 926	224 413	229 510	228 608	236 934	253 148	271 841
Coverage adjustments												
Second-hand ships	HBTY	235	232	160	185	281	112	166	113	248	223	224
Ships delivered abroad	CGER	186	96	165	217	127	540	577	586	572	302	499
Repairs to ships and aircraft	EPBA	69	9	33	35	15	11	9	9	30	54	56
Goods not changing ownership	HBYS	−1 710	−1 972	−2 351	−2 565	−2 291	−2 343	−2 761	−2 788	−2 744	−2 012	−2 753
Goods procured in ports	KTPC	590	703	709	744	780	1 035	1 218	1 438	1 865	2 240	2 380
Industrial gold	DEJP	205	209	194	135	149	164	145	163	236	230	196
Smuggling - alcohol	QHCP	101	272	270	331	266	279	43	25	29	29	..
Smuggling - tobacco	QHCT	121	328	441	693	990	1 072	1 033	1 063	1 140	1 136	..
Other	EHHI	27	25	136	28	13	21	10	8	8	7	..
Total coverage adjustments	EHHJ	−176	−98	−163	−197	330	891	440	617	1 384	2 209	1 780
Valuation adjustments												
Freight	BPGF	−3 628	−3 945	−4 171	−4 362	−4 660	−5 106	−5 423	−5 450	−5 465	−5 494	−5 534
Insurance	ENAG	−496	−522	−556	−548	−594	−654	−662	−662	−704	−733	−768
Total	HCLT	−4 124	−4 467	−4 727	−4 910	−5 254	−5 760	−6 085	−6 112	−6 169	−6 227	−6 302
Other adjustments												
Impact of MTIC fraud	BQHF	–	–	–	–	1 678	2 794	7 060	11 495	4 486	2 689	11 098
Other adjustments	EPBB	291	−670	48	−1 051	−1 462	−1 428	−620	−379	292	−49	56
Total other adjustments	CLAK	291	−670	48	−1 051	216	1 366	6 440	11 116	4 778	2 640	11 154
Total	LQBL	**165 600**	**180 918**	**184 265**	**185 869**	**195 217**	**220 912**	**230 305**	**234 229**	**236 927**	**251 770**	**278 473**

Chapter 3
Trade in services

Summary

A surplus has been recorded for trade in services in every year since 1966. There was a decrease in the surplus in 2005, from £25.9 billion in 2004 to £23.1 billion in the latest year. During, 2005 exports of services increased by 3.1 per cent whilst imports of services grew by 7.5 per cent (compared to growths of 11.1 per cent and 5.1 per cent respectively in 2004). Of the 11 main product groupings, eight showed surpluses and three (transportation, travel and government services) showed deficits. The decrease in the surplus was mainly due to a decrease in the insurance surplus, reflecting the payment of claims associated with Hurricane Katrina.

Figure 3.1
Trade in services

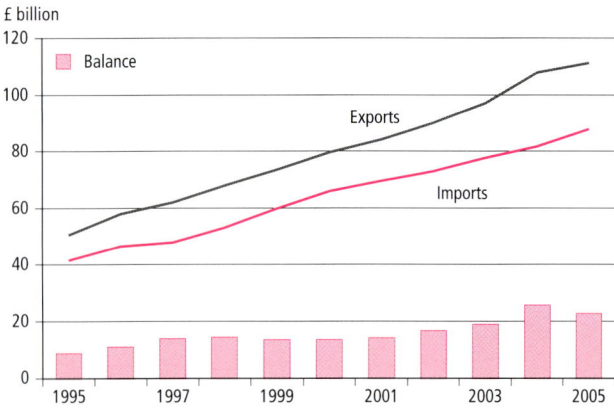

Transportation

Transportation services relate mainly to freight services on exports and imports of goods, and provision of passenger services. They are presented by mode of transport: sea, air and other.

Sea transportation was close to balance in 1995, but has recorded deficits every year since, until 2004 which shows a surplus of £1.1 billion. This surplus has grown to £1.8 billion in 2005, reflecting an increase in exports of freight services provided by UK shipping operators. The move from deficit to surplus can be explained by the continuing increase in the size of the UK fleet following the introduction of tonnage tax in July 2000.

The UK has recorded a deficit on air transport services in every year since the mid 1980s. The deficit increased from £2.9 billion in 2004 to £3.5 billion in 2005.

Figure 3.2
Trade in sea and air transport services
Exports less imports

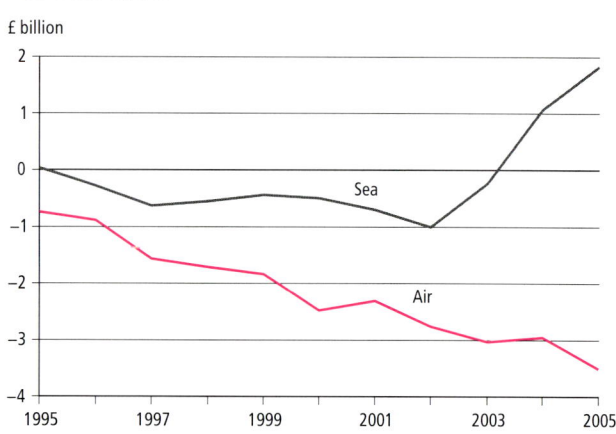

Travel

In 2005 travel expenditure by non-residents visiting the UK accounted for 15 per cent of total exports of services, while expenditure by UK residents travelling abroad accounted for 37 per cent of total imports of services.

Figure 3.3
Trade in travel services

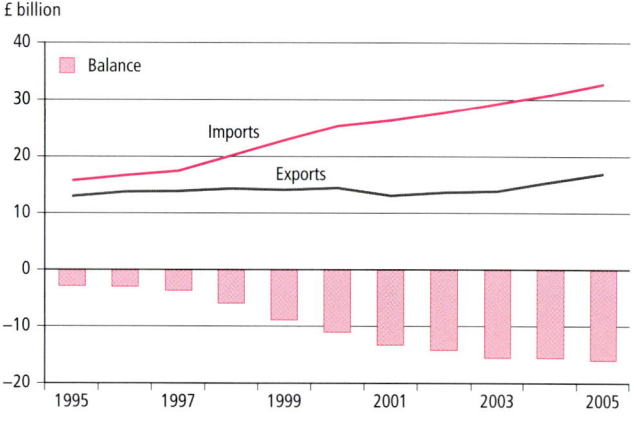

The travel deficit has grown significantly since the late 1980s. The £15.9 billion deficit in 2005 was the highest on record, up from £15.5 billion in 2004. Exports of travel services to non-resident visitors to the UK increased by 9.4 per cent in 2005 to £16.9 billion, while imports by UK residents travelling abroad grew by 6.3 per cent to £32.8 billion.

Financial services

Exports and imports of financial services from banks, fund managers, securities dealers etcetera have been presented separately since the 2001 Pink Book. As stated in the introduction to the Pink Book, Financial Services contain new estimates of banks' spread earnings on trading activity in derivatives and securities. These revisions have been introduced from 1995 onwards. The overall financial services balance rose from £16.3 billion in 2004 to £18.4 billion in 2005. This rise was mainly due to increases in exports of financial services by UK banks whose income from spread earnings and commissions and fees grew by £1.8 billion in the period.

Figure 3.4

Trade in financial services

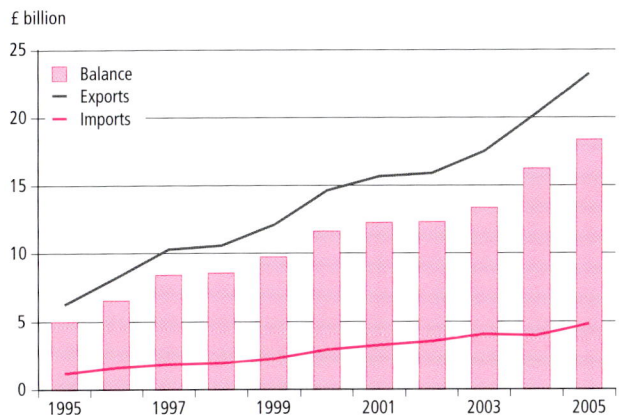

Other business services

Other business services covers a broad range of services including operational leasing, trade related services such as merchanting, and consultancy services such as advertising, engineering and legal services. Data for other business services are only available consistent with BPM5 definitions from 1991. Between 1991 and 2005 both exports and imports of other business services have increased by more than 400 per cent. The balance fell by 6.3 per cent in 2005 to £14.8 billion: exports increased by £0.4 billion to £30.8 billion, whilst imports rose by £1.4 billion, to £16.0 billion, mainly driven by imports of business services by securities dealers.

Figure 3.5

Trade in other business services

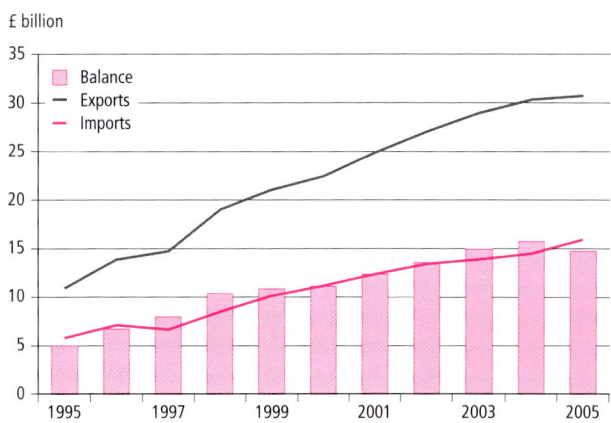

Computer and information services

Both exports and imports of computer services have shown strong growth since the late 1990s. However, the value of exports decreased by 7 per cent between 2004 and 2005, whilst the value of imports increased by 9 per cent over the same period, reducing the surplus on computer services to £2.9 billion. Both exports and imports of information services showed a decrease between 2004 and 2005, with the surplus reducing over the period from £1.0 billion in 2004 to £0.8 billion in 2005. Overall, the balance on total computer and information services decreased by 14.7 per cent in 2005, to £3.7 billion.

Figure 3.6

Trade in computer and information services

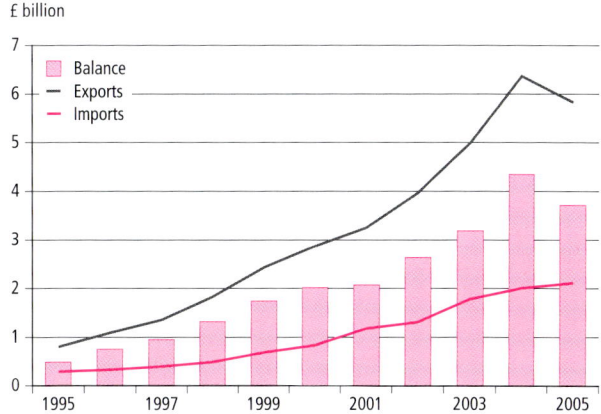

3.1 Trade in services
Summary table

£ million

		1995	1996	1997	1998	1999	2000	2001	2002	2003	2004	2005
Exports												
Transportation	FJOD	10 170	10 915	11 199	11 522	11 764	12 675	12 768	12 522	13 629	16 373	17 974
Travel	FJPF	12 990	13 691	13 805	14 302	14 060	14 446	13 110	13 595	13 876	15 414	16 868
Communications	FJPH	1 009	1 057	1 196	1 209	1 564	1 864	2 034	2 219	2 479	2 933	3 036
Construction	FJPI	130	173	266	332	275	220	174	195	245	303	522
Insurance	FJPJ	2 344	2 656	3 191	2 851	3 280	2 680	3 667	5 601	5 427	4 965	1 578
Financial	FJPK	6 263	8 223	10 309	10 561	12 112	14 620	15 644	15 910	17 498	20 281	23 260
Computer and information	FJPL	795	1 090	1 357	1 826	2 433	2 865	3 253	3 954	4 986	6 373	5 832
Royalties and license fees	FJPM	3 854	4 253	4 148	4 270	5 092	5 389	5 673	5 786	6 174	6 704	7 313
Other business	FJPN	10 906	13 853	14 713	19 013	21 017	22 395	24 844	27 026	28 937	30 305	30 738
Personal, cultural and recreational	FJPR	690	774	820	880	962	1 305	1 358	1 601	1 892	2 145	1 966
Government	FJPU	1 423	1 277	1 092	1 132	1 057	1 207	1 522	1 578	1 934	2 021	2 036
Total	**KTMQ**	**50 574**	**57 962**	**62 096**	**67 978**	**73 616**	**79 666**	**84 047**	**89 987**	**97 077**	**107 817**	**111 123**
Imports												
Transportation	FJPV	10 733	11 916	13 291	13 799	14 180	15 972	16 282	16 922	17 416	18 671	20 101
Travel	APQA	15 793	16 642	17 443	20 201	22 930	25 385	26 376	27 697	29 355	30 873	32 806
Communications	FJQZ	1 328	1 340	1 381	1 582	1 805	1 867	1 993	2 040	2 158	2 372	2 664
Construction	FJRA	95	120	168	115	98	55	107	104	120	142	455
Insurance	FJRB	495	567	594	577	575	721	762	758	778	830	880
Financial	FJRE	1 239	1 629	1 859	1 968	2 308	2 959	3 314	3 553	4 073	3 982	4 866
Computer and information	FJRF	293	333	405	494	691	838	1 175	1 316	1 792	2 012	2 110
Royalties and license fees	FJRG	3 295	4 042	3 747	4 015	4 285	4 379	4 494	4 609	4 810	5 007	4 986
Other business	FJRH	5 855	7 128	6 695	8 557	10 143	11 206	12 424	13 464	13 928	14 547	15 973
Personal, cultural and recreational	FJRL	493	556	546	489	608	779	724	797	855	904	788
Government	FJRO	1 998	2 485	1 861	1 509	2 396	1 890	1 973	1 897	2 630	2 559	2 438
Total	**KTMR**	**41 617**	**46 758**	**47 990**	**53 306**	**60 019**	**66 051**	**69 624**	**73 157**	**77 915**	**81 899**	**88 067**
Balances												
Transportation	FJRP	−563	−1 001	−2 092	−2 277	−2 416	−3 297	−3 514	−4 400	−3 787	−2 298	−2 127
Travel	FJSR	−2 803	−2 951	−3 638	−5 899	−8 870	−10 939	−13 266	−14 102	−15 479	−15 459	−15 938
Communications	FJST	−319	−283	−185	−293	−241	−3	41	179	321	561	372
Construction	FJSU	35	53	98	217	177	165	67	91	125	161	67
Insurance	FJSV	1 849	2 089	2 597	2 274	2 705	1 959	2 905	4 843	4 649	4 135	698
Financial	FJTA	5 024	6 594	8 450	8 593	9 804	11 661	12 330	12 357	13 425	16 299	18 394
Computer and information	FJTB	502	757	952	1 332	1 742	2 027	2 078	2 638	3 194	4 361	3 722
Royalties and license fees	FJTC	559	211	401	255	807	1 010	1 179	1 177	1 364	1 697	2 327
Other business	FJTD	5 051	6 725	8 018	10 456	10 874	11 189	12 420	13 562	15 009	15 758	14 765
Personal, cultural and recreational	FJTH	197	218	274	391	354	526	634	804	1 037	1 241	1 178
Government	FJUL	−575	−1 208	−769	−377	−1 339	−683	−451	−319	−696	−538	−402
Total	**KTMS**	**8 957**	**11 204**	**14 106**	**14 672**	**13 597**	**13 615**	**14 423**	**16 830**	**19 162**	**25 918**	**23 056**

3.2 Transportation

£ million

		1995	1996	1997	1998	1999	2000	2001	2002	2003	2004	2005
Exports												
Sea transport												
Passenger												
Passenger revenue	FJAL	693	705	697	462	463	630	488	569	993	846	692
Time charter receipts	FJAM	8	9	9	–	9	8	–	11	–	36	52
Total passenger	FJOF	701	714	706	462	472	638	488	580	993	882	744
Freight												
Dry cargo												
Freight on UK exports	HECV	421	409	416	322	375	400	406	481	525	444	541
Freight on cross-trades	HDVI	1 354	1 345	1 614	1 602	1 511	1 453	1 609	1 844	2 069	3 380	4 118
Time charter receipts	FJAO	125	125	138	109	90	140	106	118	196	640	943
Wet cargo												
Freight on UK exports	HEIX	64	71	68	60	59	98	82	96	126	173	199
Freight on cross-trades	HECX	488	550	536	442	350	458	497	420	742	1 305	1 423
Time charter receipts	FJAP	139	120	68	70	87	104	336	162	247	472	639
Total Freight	FJOG	2 591	2 620	2 840	2 605	2 472	2 653	3 036	3 121	3 905	6 414	7 863
Disbursements in the UK	FJAR	946	950	981	1 139	1 063	1 042	1 086	1 008	952	801	800
Total sea transport	FJOE	4 238	4 284	4 527	4 206	4 007	4 333	4 610	4 709	5 850	8 097	9 407
Air transport												
Passenger revenue	FJOJ	3 721	4 200	4 040	4 242	4 402	4 690	4 455	4 162	3 856	3 907	4 072
Freight on UK exports and cross trades	FJOK	361	428	407	408	380	428	365	350	368	394	397
Other												
Disbursements in the UK	FJAX	983	1 024	1 177	1 565	1 765	1 994	2 167	1 991	2 111	2 302	2 461
Other revenue	HBWB	176	221	242	236	294	303	258	247	240	267	182
Total other	FJOL	1 159	1 245	1 419	1 801	2 059	2 297	2 425	2 238	2 351	2 569	2 643
Total air transport	FJOI	5 241	5 873	5 866	6 451	6 841	7 415	7 245	6 750	6 575	6 870	7 112
Other transport												
Rail												
Passenger	FJOS	71	77	80	108	132	109	113	90	91	112	136
Freight	FJOT	10	11	8	16	17	20	16	12	15	16	17
Total rail	FJOR	81	88	88	124	149	129	129	102	106	128	153
Road												
Passenger	FJOW	–	–	–	–	–	–	–	–	–	–	–
Freight	FJOX	570	634	682	703	730	750	728	905	1 042	1 222	1 246
Total road	FJOV	570	634	682	703	730	750	728	905	1 042	1 222	1 246
Pipeline transport	FJPD	40	36	36	38	37	48	56	56	56	56	56
Total other transport	FJOM	691	758	806	865	916	927	913	1 063	1 204	1 406	1 455
Total	FJOD	**10 170**	**10 915**	**11 199**	**11 522**	**11 764**	**12 675**	**12 768**	**12 522**	**13 629**	**16 373**	**17 974**

3.2 Transportation
continued

£ million

		1995	1996	1997	1998	1999	2000	2001	2002	2003	2004	2005
Imports												
Sea transport												
Passenger												
Passenger expenditure	FJBP	457	396	486	494	429	413	450	486	476	476	508
Time charter payments	FJBQ	65	70	92	22	24	24	19	19	15	25	10
Total passenger	FJPX	522	466	578	516	453	437	469	505	491	501	518
Freight												
Dry cargo												
Freight on UK imports	HCJO	1 811	1 846	2 008	2 063	2 202	2 531	2 552	2 698	3 023	2 851	2 711
Time charter payments	FJBS	135	145	190	217	122	149	316	236	221	667	563
Wet cargo												
Freight on UK imports	HCNJ	194	221	315	282	415	280	305	330	355	417	424
Time charter payments	FJBT	186	243	161	181	89	172	176	140	184	359	452
Freight on UK coastal routes	HFAA	116	132	135	135	135	172	202	199	190	177	176
Total Freight	FJPY	2 442	2 587	2 809	2 878	2 963	3 304	3 551	3 603	3 973	4 471	4 326
Other												
Disbursements - dry cargo	FJBU	1 134	1 375	1 670	1 291	953	1 036	1 231	1 528	1 508	1 917	2 547
Disbursements - wet cargo	FJBW	107	134	104	78	76	55	54	81	118	138	185
Total other	FJPZ	1 241	1 509	1 774	1 369	1 029	1 091	1 285	1 609	1 626	2 055	2 732
Total sea transport	FJPW	4 205	4 562	5 161	4 763	4 445	4 832	5 305	5 717	6 090	7 027	7 576
Air transport												
Passenger expenditure	FJQB	3 115	3 505	3 863	4 197	4 650	5 192	5 255	5 559	5 949	6 443	7 121
Freight	FJQC	401	481	543	583	685	740	822	818	768	681	688
Disbursements abroad	FJCA	2 459	2 764	3 015	3 372	3 336	3 951	3 468	3 132	2 880	2 692	2 799
Total air transport	FJQA	5 975	6 750	7 421	8 152	8 671	9 883	9 545	9 509	9 597	9 816	10 608
Other transport												
Rail												
Passenger	FJQK	54	85	98	121	154	167	168	172	151	167	165
Freight	FJQL	13	13	10	21	26	37	43	44	46	47	44
Total rail	FJQJ	67	98	108	142	180	204	211	216	197	214	209
Road												
Passenger	FJQO	–	–	–	–	–	–	–	–	–	–	–
Freight	FJQP	422	457	550	694	836	1 001	1 169	1 428	1 480	1 562	1 656
Total road	FJQN	422	457	550	694	836	1 001	1 169	1 428	1 480	1 562	1 656
Pipeline transport	FJQV	64	49	51	48	48	52	52	52	52	52	52
Total other transport	FJQE	553	604	709	884	1 064	1 257	1 432	1 696	1 729	1 828	1 917
Total	FJPV	**10 733**	**11 916**	**13 291**	**13 799**	**14 180**	**15 972**	**16 282**	**16 922**	**17 416**	**18 671**	**20 101**

3.2 Transportation
continued

£ million

		1995	1996	1997	1998	1999	2000	2001	2002	2003	2004	2005
Balances												
Sea transport												
Passenger	FJRR	179	248	128	−54	19	201	19	75	502	381	226
Freight												
Dry cargo	FJNJ	−46	−112	−30	−247	−348	−687	−747	−491	−454	946	2 328
Wet cargo	FJNM	311	277	196	109	−8	208	434	208	576	1 174	1 385
Other	FJVC	−116	−132	−135	−135	−135	−172	−202	−199	−190	−177	−176
Total Freight	FJRS	149	33	31	−273	−491	−651	−515	−482	−68	1 943	3 537
Other												
Dry cargo	FJVF	−1 134	−1 375	−1 670	−1 291	−953	−1 036	−1 231	−1 528	−1 508	−1 917	−2 547
Wet Cargo	FJVG	−107	−134	−104	−78	−76	−55	−54	−81	−118	−138	−185
Other	FJVI	946	950	981	1 139	1 063	1 042	1 086	1 008	952	801	800
Total other	FJRT	−295	−559	−793	−230	34	−49	−199	−601	−674	−1 254	−1 932
Total sea transport	FJRQ	33	−278	−634	−557	−438	−499	−695	−1 008	−240	1 070	1 831
of which												
Ships owned or chartered-in by UK residents	FLMZ	*1 665*	*1 367*	*1 329*	*1 278*	*1 680*	*1 855*	*1 728*	*1 697*	*2 852*	*4 190*	*4 850*
Ships operated by non-residents	FLNF	*−1 632*	*−1 645*	*−1 963*	*−1 835*	*−2 118*	*−2 354*	*−2 423*	*−2 705*	*−3 092*	*−3 120*	*−3 019*
Air transport												
Passenger	FJRV	606	695	177	45	−248	−502	−800	−1 397	−2 093	−2 536	−3 049
Freight	FJRW	−40	−53	−136	−175	−305	−312	−457	−468	−400	−287	−291
Other	FJRX	−1 300	−1 519	−1 596	−1 571	−1 277	−1 654	−1 043	−894	−529	−123	−156
Total air transport	FJRU	−734	−877	−1 555	−1 701	−1 830	−2 468	−2 300	−2 759	−3 022	−2 946	−3 496
Other transport												
Rail												
Passenger	FJSE	17	−8	−18	−13	−22	−58	−55	−82	−60	−55	−29
Freight	FJSF	−3	−2	−2	−5	−9	−17	−27	−32	−31	−31	−27
Total rail	FJSD	14	−10	−20	−18	−31	−75	−82	−114	−91	−86	−56
Road												
Passenger	FJSI	–	–	–	–	–	–	–	–	–	–	–
Freight	FJSJ	148	177	132	9	−106	−251	−441	−523	−438	−340	−410
Total road	FJSH	148	177	132	9	−106	−251	−441	−523	−438	−340	−410
Pipeline transport	FJSP	−24	−13	−15	−10	−11	−4	4	4	4	4	4
Total other transport	FJRY	138	154	97	−19	−148	−330	−519	−633	−525	−422	−462
Total	FJRP	**−563**	**−1 001**	**−2 092**	**−2 277**	**−2 416**	**−3 297**	**−3 514**	**−4 400**	**−3 787**	**−2 298**	**−2 127**

3.3 Travel

£ million

		1995	1996	1997	1998	1999	2000	2001	2002	2003	2004	2005
Exports												
Business												
Expenditure by seasonal & border workers[1]	FJCQ	52	60	53	132	114	147	163	219	169	203	222
Other	FJNO	3 240	3 246	3 533	3 857	3 998	4 084	3 615	3 618	3 478	3 735	4 095
Total business travel	FJPG	3 292	3 306	3 586	3 989	4 112	4 231	3 778	3 837	3 647	3 938	4 317
Personal												
Health related[2]	FJCX	53	105	112	79	93	66	83	64	144	68	69
Education related	FJDD	2 237	2 512	2 492	2 696	2 534	2 484	2 723	2 592	2 881	3 072	3 331
Other	FJDG	7 408	7 768	7 615	7 538	7 321	7 665	6 526	7 102	7 204	8 336	9 151
Total personal travel	FJTU	9 698	10 385	10 219	10 313	9 948	10 215	9 332	9 758	10 229	11 476	12 551
Total	FJPF	**12 990**	**13 691**	**13 805**	**14 302**	**14 060**	**14 446**	**13 110**	**13 595**	**13 876**	**15 414**	**16 868**
Imports												
Business												
Expenditure by seasonal & border workers[1]	FJDO	71	55	56	118	197	192	215	102	225	159	193
Other	FJNP	3 044	3 435	3 451	4 231	4 352	4 811	4 479	4 336	4 135	4 243	4 695
Total business travel	FJQY	3 115	3 490	3 507	4 349	4 549	5 003	4 694	4 438	4 360	4 402	4 888
Personal												
Health related[2]	FJDT	4	3	11	3	10	19	16	12	33	45	59
Education related	FJDV	106	118	111	133	180	99	108	110	102	117	167
Other	APPW	12 568	13 031	13 814	15 716	18 191	20 264	21 558	23 137	24 860	26 309	27 692
Total personal travel	APQW	12 678	13 152	13 936	15 852	18 381	20 382	21 682	23 259	24 995	26 471	27 918
Total	APQA	15 793	16 642	17 443	20 201	22 930	25 385	26 376	27 697	29 355	30 873	32 806
Balances												
Business												
Expenditure by seasonal & border workers[1]	FJCR	−19	5	−3	14	−83	−45	−52	117	−56	44	29
Other	FJCW	196	−189	82	−374	−354	−727	−864	−718	−657	−508	−600
Total business travel	FJSS	177	−184	79	−360	−437	−772	−916	−601	−713	−464	−571
Personal												
Health related[2]	FJCY	49	102	101	76	83	47	67	52	111	23	10
Education related	FJDE	2 131	2 394	2 381	2 563	2 354	2 385	2 615	2 482	2 779	2 955	3 164
Other	FJDH	−5 160	−5 263	−6 199	−8 178	−10 870	−12 599	−15 032	−16 035	−17 656	−17 973	−18 541
Total personal travel	FJTW	−2 980	−2 767	−3 717	−5 539	−8 433	−10 167	−12 350	−13 501	−14 766	−14 995	−15 367
Total	FJSR	**−2 803**	**−2 951**	**−3 638**	**−5 899**	**−8 870**	**−10 939**	**−13 266**	**−14 102**	**−15 479**	**−15 459**	**−15 938**

1 There are no firm data for expenditure by seasonal & border workers before 1994, but for continuity some estimates have been included in other business travel.
2 There are no firm data for health related travel before 1994, but for continuity broad estimates have been included in other personal travel.

3.4 Communications services

£ million

		1995	1996	1997	1998	1999	2000	2001	2002	2003	2004	2005
Exports												
Postal and courier services												
Postal services	FJTN	109	85	93	88	109	118	97	110	112	124	121
Courier services	FJTO	24	23	15	13	52	29	80	67	111	120	139
Total postal and courier services	FJED	133	108	108	101	161	147	177	177	223	244	260
Telecommunications services	FJAS	876	949	1 088	1 188	1 403	1 717	1 857	2 042	2 256	2 689	2 776
Total	FJPH	**1 009**	**1 057**	**1 196**	**1 289**	**1 564**	**1 864**	**2 034**	**2 219**	**2 479**	**2 933**	**3 036**
Imports												
Postal and courier services												
Postal services	FJTP	223	217	200	218	239	260	200	200	225	181	169
Courier services	FJTQ	19	19	14	39	48	18	55	58	90	94	108
Total postal and courier services	FJEI	242	236	214	257	287	278	255	258	315	275	277
Telecommunications services	FJAT	1 086	1 104	1 167	1 325	1 518	1 589	1 738	1 782	1 843	2 097	2 387
Total	FJQZ	**1 328**	**1 340**	**1 381**	**1 582**	**1 805**	**1 867**	**1 993**	**2 040**	**2 158**	**2 372**	**2 664**
Balances												
Postal and courier services												
Postal services	FJTR	−114	−132	−107	−130	−130	−142	−103	−90	−113	−57	−48
Courier services	FJTS	5	4	1	−26	4	11	25	9	21	26	31
Total postal and courier services	FJEE	−109	−128	−106	−156	−126	−131	−78	−81	−92	−31	−17
Telecommunications services	FJAQ	−210	−155	−79	−137	−115	128	119	260	413	592	389
Total	FJST	**−319**	**−283**	**−185**	**−293**	**−241**	**−3**	**41**	**179**	**321**	**561**	**372**

3.5 Insurance services

£ million

		1995	1996	1997	1998	1999	2000	2001	2002	2003	2004	2005
Exports												
Life insurance and pension funds	FJEU	238	415	494	838	1 557	1 417	2 174	797	8	−712	−538
Freight insurance	FJJL	2	31	82	76	47	41	49	80	129	90	68
Other direct insurance[1]	FJEW	562	839	925	439	653	412	579	2 164	1 935	3 350	208
Reinsurance	FJEX	409	339	718	331	−49	−296	1 011	1 473	2 241	1 023	601
Auxiliary insurance services (insurance brokers)	FJEY	1 133	1 032	972	1 167	1 072	1 106	1 012	1 087	1 114	1 214	1 239
Total[2]	FJPJ	**2 344**	**2 656**	**3 191**	**2 851**	**3 280**	**2 680**	**3 667**	**5 601**	**5 427**	**4 965**	**1 578**
Imports												
Life insurance and pension funds	FJRC	–	–	–	–	–	–	–	–	–	–	–
Freight insurance	FJRD	495	567	594	577	575	721	762	758	778	830	880
Other direct insurance	FJFC	–	–	–	–	–	–	–	–	–	–	–
Reinsurance	FJFD	–	–	–	–	–	–	–	–	–	–	–
Auxiliary insurance services	FJFE	–	–	–	–	–	–	–	–	–	–	–
Total	FJRB	**495**	**567**	**594**	**577**	**575**	**721**	**762**	**758**	**778**	**830**	**880**
Balances												
Life insurance and pension funds	FJSW	238	415	494	838	1 557	1 417	2 174	797	8	−712	−538
Freight insurance	FJSX	−493	−536	−512	−501	−528	−680	−713	−678	−649	−740	−812
Other direct insurance	FJJM	562	839	925	439	653	412	−579	2 164	1 935	3 350	208
Reinsurance	FJJN	409	339	718	331	−49	−296	1 011	1 473	2 241	1 023	601
Auxiliary insurance services	FJJO	1 133	1 032	972	1 167	1 072	1 106	1 012	1 087	1 114	1 214	1 239
Total	FJSV	**1 849**	**2 089**	**2 597**	**2 274**	**2 705**	**1 959**	**2 905**	**4 843**	**4 649**	**4 135**	**698**

1 Other direct insurance by UK insurance companies includes facultative reinsurance on marine, aviation and transport business.
2 Exports of insurance services are net of expenditure abroad by UK insurance companies.

3.6 Financial services

£ million

		1995	1996	1997	1998	1999	2000	2001	2002	2003	2004	2005
Exports												
Monetary financial institutions (banks)												
Commissions and fees	APUP	1 178	1 269	1 778	2 108	2 506	3 041	2 986	3 215	2 677	3 458	4 198
Spread earnings	APVA	1 460	1 732	1 961	1 737	1 628	1 809	2 370	2 922	4 536	4 904	5 914
Total monetary financial institutions (banks)	ZXTE	2 638	3 001	3 739	3 845	4 134	4 850	5 356	6 137	7 213	8 362	10 112
Fund managers	FNMM	457	743	904	849	866	868	853	1 045	1 528	1 925	2 387
Securities dealers												
Commissions and fees	CDFI	1 649	2 103	2 761	2 831	3 996	5 632	5 211	4 290	3 922	4 316	4 628
Spread earnings	QZCM	690	934	1 253	1 233	1 209	1 033	1 492	1 168	1 316	1 666	1 925
Total securities dealers	ZXTF	2 339	3 037	4 014	4 064	5 205	6 665	6 703	5 458	5 238	5 982	6 553
Baltic Exchange	APRJ	315	280	340	320	320	336	377	357	398	577	777
Other	ZSHJ	514	1 162	1 312	1 483	1 587	1 901	2 355	2 913	3 121	3 435	3 431
Total	FJPK	6 263	8 223	10 309	10 561	12 112	14 620	15 644	15 910	17 498	20 281	23 260
Imports												
Monetary financial institutions (banks)	APVW	412	463	573	549	733	1 003	1 157	1 475	1 701	1 556	1 693
Fund managers	FNMS	32	150	155	171	143	160	229	219	336	420	426
Securities dealers[1]	RWMG	287	411	506	689	829	1 199	1 296	1 009	795	862	1 244
Baltic Exchange	APSZ	24	20	24	23	27	39	27	35	18	26	42
Other	ZXTG	484	585	601	536	576	558	605	815	1 223	1 118	1 461
Total	FJRE	1 239	1 629	1 859	1 968	2 308	2 959	3 314	3 553	4 073	3 982	4 866
Balances												
Monetary financial institutions	ZXLV	2 226	2 538	3 166	3 296	3 401	3 847	4 199	4 662	5 512	6 806	8 419
Fund managers	ZXLW	425	593	749	678	723	708	624	826	1 192	1 505	1 961
Securities dealers	ZXLX	2 052	2 626	3 508	3 375	4 376	5 466	5 407	4 449	4 443	5 120	5 309
Baltic Exchange	ZXLY	291	260	316	297	293	297	350	322	380	551	735
Other	ZXLZ	30	577	711	947	1 011	1 343	1 750	2 098	1 898	2 317	1 970
Total	FJTA	5 024	6 594	8 450	8 593	9 804	11 661	12 330	12 357	13 425	16 299	18 394

1 For securities dealers, the move to a gross presentation means that imports of non-financial services are moved to the other business services accounts (see table 3.9).

3.7 Computer and information services

£ million

		1995	1996	1997	1998	1999	2000	2001	2002	2003	2004	2005
Exports												
Computer services	FJCN	695	956	1 183	1 640	2 056	2 478	2 725	3 328	3 705	4 977	4 637
Information services	FJCO	100	134	174	186	377	387	528	626	1 281	1 396	1 195
Total	FJPL	**795**	**1 090**	**1 357**	**1 826**	**2 433**	**2 865**	**3 253**	**3 954**	**4 986**	**6 373**	**5 832**
Imports												
Computer services	FJDL	253	283	339	473	593	745	859	1 122	1 478	1 600	1 746
Information services	FJDM	40	50	66	21	98	93	316	194	314	412	364
Total	FJRF	**293**	**333**	**405**	**494**	**691**	**838**	**1 175**	**1 316**	**1 792**	**2 012**	**2 110**
Balances												
Computer Services	FJJP	442	673	844	1 167	1 463	1 733	1 866	2 206	2 227	3 377	2 891
Information services	FJJQ	60	84	108	165	279	294	212	432	967	984	831
Total	FJTB	**502**	**757**	**952**	**1 332**	**1 742**	**2 027**	**2 078**	**2 638**	**3 194**	**4 361**	**3 722**

3.8 Royalties and license fees

£ million

		1995	1996	1997	1998	1999	2000	2001	2002	2003	2004	2005
Exports												
Film and television	FJFO	744	879	705	775	868	934	982	880	911	890	994
Other royalties and license fees	FFVJ	3 110	3 374	3 443	3 495	4 224	4 455	4 691	4 906	5 263	5 814	6 319
Total	FJPM	**3 854**	**4 253**	**4 148**	**4 270**	**5 092**	**5 389**	**5 673**	**5 786**	**6 174**	**6 704**	**7 313**
Imports												
Film and television	FJFQ	763	829	863	882	932	1 020	1 176	1 315	1 449	1 533	1 561
Other royalties and license fees	FFVP	2 532	3 213	2 884	3 133	3 353	3 359	3 318	3 294	3 361	3 474	3 425
Total	FJRG	**3 295**	**4 042**	**3 747**	**4 015**	**4 285**	**4 379**	**4 494**	**4 609**	**4 810**	**5 007**	**4 986**
Balances												
Film and television	FFVV	–19	50	–158	–107	–64	–86	–194	–435	–538	–643	–567
Other royalties and license fees	FFWB	578	161	559	362	871	1 096	1 373	1 612	1 902	2 340	2 894
Total	FJTC	**559**	**211**	**401**	**255**	**807**	**1 010**	**1 179**	**1 177**	**1 364**	**1 697**	**2 327**

3.9 Other business services

£ million

		1995	1996	1997	1998	1999	2000	2001	2002	2003	2004	2005
Exports												
Merchanting and other trade related services												
Merchanting	FJFS	508	481	314	569	868	626	782	699	573	549	518
Other trade related services	FJFX	547	709	662	732	1 504	1 759	1 881	1 720	1 899	1 698	1 967
Total merchanting and other trade related services	FJPO	1 055	1 190	976	1 301	2 372	2 385	2 663	2 419	2 472	2 247	2 485
Operational leasing services	FJPP	121	129	113	40	92	299	248	190	239	342	334
Miscellaneous business, professional and technical services												
Legal, accounting and management consulting												
Law society	FJGE	537	565	675	824	760	1 171	1 339	1 465	1 335	1 470	1 615
Commercial bar association	FJCP	33	41	47	61	62	61	77	85	95	86	116
Other legal services[1]	FJGD	..	161	202	275	349	288	363	481	600	435	436
Accounting	FJBX	156	178	258	477	603	662	642	728	733	892	1 000
Business management and management consulting	FJNV	610	668	933	952	1 101	1 083	1 069	2 545	3 127	3 288	3 310
of which Recruitment and training	TVLQ	..	..	..	..	..	..	..	*354*	*359*	*350*	*365*
Advertising and market research	FJGP	633	717	1 022	1 174	1 150	1 432	1 622	1 703	2 155	1 965	2 111
Research and development	FJDP	986	1 311	1 616	2 300	2 801	2 421	2 933	2 899	3 467	4 463	4 355
Architectural, engineering and other technical services												
North Sea oil and gas[2]	FJCV	358	331	..	..	..	..	..	..	..	..	..
Architectural	FJGT	52	51	83	67	82	76	153	71	106	110	87
Engineering	FJGU	1 843	2 243	2 491	2 987	2 676	2 441	3 239	3 049	3 475	3 501	3 102
Surveying	FJGV	37	37	31	41	45	68	66	62	57	137	225
Other Technical	FJGW	612	807	798	1 083	1 027	1 113	1 220	1 931	1 629	1 698	1 579
Agricultural, mining and on-site processing services	FJHC	21	26	21	52	47	54	41	31	202	234	182
Other miscellaneous business services	FJHH	3 351	4 716	4 576	6 398	6 749	7 561	7 688	7 748	7 140	7 295	7 729
of which Other business services exported by UK banks	APVQ	*564*	*622*	*505*	*1 008*	*1 325*	*1 414*	*1 277*	*1 490*	*2 118*	*1 892*	*2 134*
Services between affiliated enterprises, n.i.e.	FJHF	501	682	871	981	1 101	1 280	1 481	1 619	2 105	2 142	2 072
Total miscellaneous business, professional, and technical services	FJPQ	9 730	12 534	13 624	17 672	18 553	19 711	21 933	24 417	26 226	27 716	27 919
Total	FJPN	**10 906**	**13 853**	**14 713**	**19 013**	**21 017**	**22 395**	**24 844**	**27 026**	**28 937**	**30 305**	**30 738**
Imports												
Merchanting and other trade related services												
Merchanting	FJHN	88	110	44	65	38	71	55	148	35	81	80
Other trade related services	FJHR	517	652	444	633	884	965	952	854	752	1 122	577
Total merchanting and other trade related services	FJRI	605	762	488	698	922	1 036	1 007	1 002	787	1 203	657
Operational leasing services	FJRJ	163	194	196	193	226	560	457	450	456	784	816
Miscellaneous business, professional and technical services												
Legal, accounting and management consulting												
Legal[1]	FJHX	24	173	209	249	307	490	380	486	453	416	391
Accounting	FJVJ	105	128	98	108	119	213	228	251	300	324	311
Business management and management consulting	FJNW	263	310	327	371	387	456	569	1 428	1 924	2 328	2 550
of which Recruitment and training	TVLV	..	..	..	..	..	..	..	*237*	*264*	*357*	*410*
Advertising and market research	FJID	443	493	460	581	719	789	841	860	946	842	813
Research and development	FJDQ	639	767	657	753	781	723	661	644	1 148	1 806	1 775
Architectural, engineering and other technical services												
North Sea oil and gas[2]	FJDR	567	442	..	..	..	..	..	..	..	..	..
Architectural	FJIF	5	6	7	12	12	13	35	25	50	11	4
Engineering	FJIG	546	629	909	1 228	977	724	1 075	868	1 107	1 325	1 354
Surveying	FJIH	18	17	36	26	15	55	31	29	24	48	85
Other Technical	FJII	188	232	358	435	410	429	431	463	368	384	532
Agricultural, mining and on-site processing services	FJIN	6	7	7	27	50	71	142	77	53	63	73
Other miscellaneous business services	FJIP	1 954	2 562	2 486	3 157	4 448	4 839	5 498	5 741	5 181	3 815	5 318
of which Other business sevices imported by UK banks	APWA	*381*	*421*	*184*	*509*	*794*	*520*	*448*	*619*	*760*	*497*	*591*
Other business services imported by Security dealers	RWMH	*592*	*865*	*787*	*986*	*1 511*	*2 294*	*2 027*	*1 358*	*1 149*	*1 462*	*3 036*
Services between affiliated enterprises, n.i.e.	FJHG	329	406	457	719	770	808	1 069	1 140	1 131	1 198	1 294
Total miscellaneous business, professional and technical services	FJRK	5 087	6 172	6 011	7 666	8 995	9 610	10 960	12 012	12 685	12 560	14 500
Total	FJRH	**5 855**	**7 128**	**6 695**	**8 557**	**10 143**	**11 206**	**12 424**	**13 464**	**13 928**	**14 547**	**15 973**

1 Other legal services are included indistinguishably within other miscellaneous business services for years before 1996.
2 North Sea oil and gas services are included indistinguishably within engineering services for years after 1996.

3.9 Other business services
continued

£ million

		1995	1996	1997	1998	1999	2000	2001	2002	2003	2004	2005
Balances												
Merchanting and other trade related services												
Merchanting	FJFT	420	371	270	504	830	555	727	551	538	468	438
Other trade related services	FJFY	30	57	218	99	620	794	929	866	1 147	576	1 390
Total merchanting and other trade related services	FJTE	450	428	488	603	1 450	1 349	1 656	1 417	1 685	1 044	1 828
Operational leasing services	FJTF	−42	−65	−83	−153	−134	−261	−209	−260	−217	−442	−482
Miscellaneous business, professional and technical services												
Legal, accounting and management consulting												
Legal	FJGG	546	594	715	911	864	1 030	1 399	1 545	1 577	1 575	1 776
Accounting	FJGI	51	50	160	369	484	449	414	477	433	568	689
Business management and management consulting	FJGK	347	358	606	581	714	627	500	1 117	1 203	960	760
Advertising and market research	FJGQ	190	224	562	593	431	643	781	843	1 209	1 123	1 298
Research and development	FJGS	347	544	959	1 547	2 020	1 698	2 272	2 255	2 319	2 657	2 580
Architectural, engineering and other technical services	FJGY	1 578	2 143	2 093	2 477	2 416	2 477	3 106	3 728	3 718	3 678	3 018
Agricultural, mining and on-site processing services	FJHD	15	19	14	25	−3	−17	−101	−46	149	171	109
Services between affiliated enterprises, n.i.e.	FJHL	172	276	414	262	331	472	412	479	974	944	778
Other	FJHI	1 397	2 154	2 090	3 241	2 301	2 722	2 190	2 007	1 959	3 480	2 411
Total miscellaneous business, professional, and technical services	FJTG	4 643	6 362	7 613	10 006	9 558	10 101	10 973	12 405	13 541	15 156	13 419
Total	FJTD	5 051	6 725	8 018	10 456	10 874	11 189	12 420	13 562	15 009	15 758	14 765

3.10 Personal, cultural and recreational services

£ million

		1995	1996	1997	1998	1999	2000	2001	2002	2003	2004	2005
Exports												
Audiovisual and related services												
Film and television	FKJO	422	395	461	480	531	726	737	856	1 077	1 274	1 111
Other	FFWH	77	101	152	167	189	252	172	184	204	286	259
Total audiovisual and related services	FJPS	499	496	613	647	720	978	909	1 040	1 281	1 560	1 370
Other personal, cultural and recreational services	FJPT	191	278	207	233	242	327	449	561	611	585	596
Total	FJPR	**690**	**774**	**820**	**880**	**962**	**1 305**	**1 358**	**1 601**	**1 892**	**2 145**	**1 966**
Imports												
Audiovisual and related services												
Film and television	FKJX	400	441	450	411	496	532	512	615	463	587	554
Other	FFWN	20	25	22	35	40	55	46	39	59	89	43
Total audiovisual and related services	FJRM	420	466	472	446	536	587	558	654	522	676	597
Other personal, cultural and recreational services	FJRN	73	90	74	43	72	192	166	143	333	228	191
Total	FJRL	**493**	**556**	**546**	**489**	**608**	**779**	**724**	**797**	**855**	**904**	**788**
Balances												
Audiovisual and related services	FJTI	79	30	141	201	184	391	351	386	759	884	773
Other personal, cultural and recreational services	FJTJ	118	188	133	190	170	135	283	418	278	357	405
Total	FJTH	**197**	**218**	**274**	**391**	**354**	**526**	**634**	**804**	**1 037**	**1 241**	**1 178**

3.11 Government services

£ million

		1995	1996	1997	1998	1999	2000	2001	2002	2003	2004	2005
Exports												
Expenditure by foreign embassies and consulates in the UK	FJUK	367	393	357	371	385	385	389	393	397	401	405
Military units and agencies												
Expenditure by US forces in UK	FJKB	364	328	250	293	247	271	262	262	264	264	264
Other military receipts by UK government	HCOJ	108	91	56	40	21	58	48	67	248	312	286
Total military units and agencies	FJIX	472	419	306	333	268	329	310	329	512	576	550
Other												
EU institutions	FKIE	301	241	240	216	213	226	525	487	494	543	561
Other receipts	HCQO	283	224	189	212	191	267	298	369	531	501	520
Total other	FJJA	584	465	429	428	404	493	823	856	1 025	1 044	1 081
Total	FJPU	**1 423**	**1 277**	**1 092**	**1 132**	**1 057**	**1 207**	**1 522**	**1 578**	**1 934**	**2 021**	**2 036**
Imports												
Expenditure abroad by UK embassies and consulates	FJUJ	194	259	208	177	219	106	142	215	190	177	169
Expenditure abroad by UK military units and agencies	FJJD	1 632	2 030	1 418	1 116	1 972	1 584	1 629	1 494	2 144	1 892	1 830
Civil non-EU services	FJJF	172	196	235	216	205	200	202	188	296	490	439
Total	FJRO	**1 998**	**2 485**	**1 861**	**1 509**	**2 396**	**1 890**	**1 973**	**1 897**	**2 630**	**2 559**	**2 438**
Balances												
Embassies and consulates	FJIW	173	134	149	194	166	279	247	178	207	224	236
Military units and agencies	FJIY	−1 160	−1 611	−1 112	−783	−1 704	−1 255	−1 319	−1 165	−1 632	−1 316	−1 280
Other	FJJB	412	269	194	212	199	293	621	668	729	554	642
Total	FJUL	**−575**	**−1 208**	**−769**	**−377**	**−1 339**	**−683**	**−451**	**−319**	**−696**	**−538**	**−402**

Chapter 4
Income

Summary

The balance on investment income has been in surplus for all years since 1994. The investment income surplus grew strongly between 2000 and 2002, largely due to a doubling in the net earnings on direct investment. Since 2002, the surplus has continued to rise, if more slowly, reaching a record £29.8 billion in 2005.

In the decade to 2001, earnings on both investment abroad and investment in the UK nearly doubled. In 2002 however, both fell sharply – credits down 12 per cent and debits down 22 per cent. This was largely due to cuts in official interest rates, both abroad and in the UK, post 11 September (2001) and throughout 2002, and subsequent falls in interest receipts and payments on loans and deposits. Since 2002, income has risen significantly and by 2005 both investment income credits and debits were up around 50 per cent on 2002. This mainly reflects stronger profits on foreign direct investment and significant levels of investment over the period.

2005, compared with 23 per cent in 1995. Other investment income, which is mostly earnings from loans and deposits, now only accounts for 33 per cent of total earnings, down from 47 per cent in 1995. Similarly, growth in foreign earnings on investment in the UK over the last 10 years has predominantly been in direct and portfolio investment, although other investment remains the largest component, accounting for approximately half of all investment income paid.

The investment income surplus has grown strongly since 1999, reaching a record £29.8 billion in 2005. By component, direct investment has recorded a surplus in every year since 1986. Within portfolio investment, a net surplus on interest receipts and payments on debt securities has largely been offset by net dividend payments on equity securities. Other investment has recorded a net deficit in every year since 1988, reaching £16.2 billion in 2000 before falling to £10.3 billion in 2003. It started to rise again in 2004 due to rising global interest rates and the fact that the UK has an excess of other investment liabilities over assets, and in 2005 was a record £16.5 billion.

Figure 4.1
Income

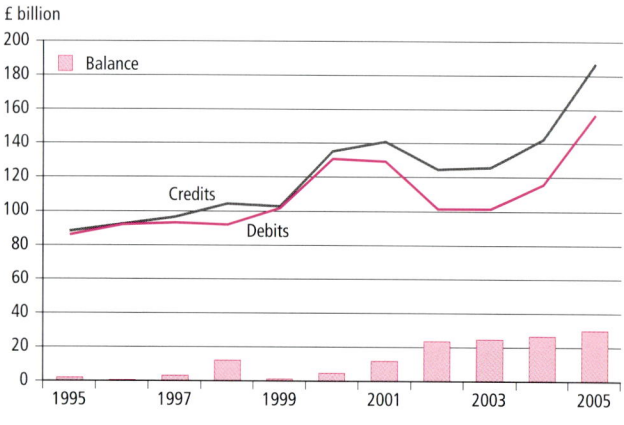

Figure 4.2
Investment income
Credits less debits

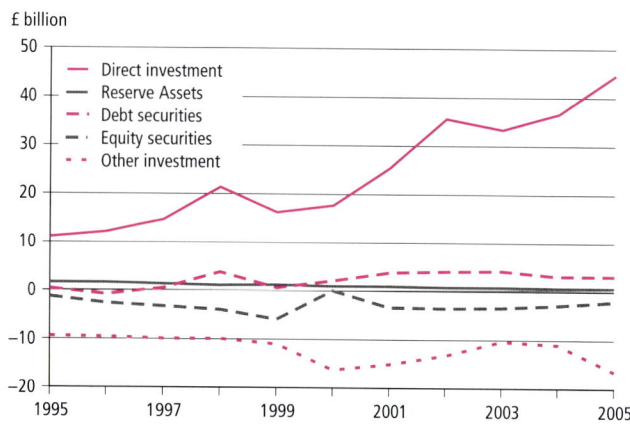

Earnings on direct investment abroad have been the largest component of investment income credits in recent years, accounting for over 40 per cent of total earnings, compared to only 28 per cent in 1995. The boom in UK merger and acquisition activity in the late 1990s and 2000 and subsequent growth in earnings from abroad has been the main driver of this change. Earnings on portfolio investment abroad have been broadly rising in line with total investment income, accounting for 24 per cent of total earnings from abroad in

By sector, net earnings of UK monetary financial institutions (banks and building societies), fell by £2.0 billion to £10.7 billion in 2005. This fall was driven by a fall in the balance on other investment, where the deficit increased from £3.0 billion to £6.3 billion. Partly offsetting this, the surplus on income from portfolio investment rose by £1.3 billion. Over the same period non-financial corporations' net earnings on direct investment increased from £29.1 billion to £37.2 billion.

The balance on compensation of employees has shown a small surplus in nine of the past ten years.

Direct investment

Direct investment income credits have exceeded debits in every year since 1986, and the surplus increased to a record £44.6 billion in 2005. Earnings from direct investment abroad increased over 20 per cent in 2005, to a record £79.1 billion, primarily due to higher foreign earnings of both financial and non-financial corporations. Foreign earnings on direct investment in the UK increased to a record £34.6 billion, resulting from increased earnings of foreign-owned monetary financial institutions and private non-financial corporations. Foreign earnings on direct investment in the UK tend to be more erratic than earnings on direct investment abroad, partly because of their concentration in the financial sector. Foreign-owned banks and other financial corporations often locate in the UK to be close to the financial markets in London and their profits have previously reflected the difficult trading conditions, that is, in 1998 and, to a lesser extent, 2002.

Figure 4.3

Direct investment income

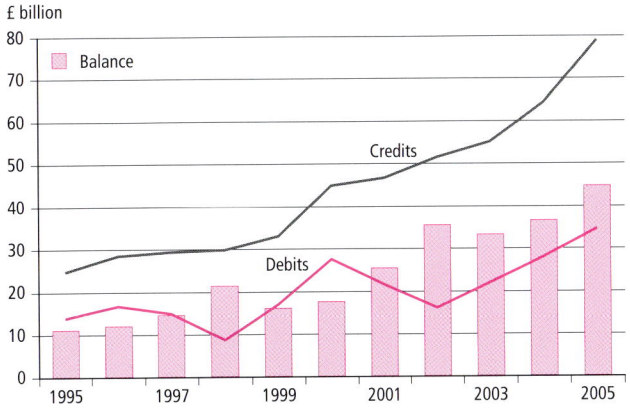

Portfolio investment

Until 2000, the UK generally recorded a deficit on portfolio investment, with a net surplus on debt securities being more than offset by a net deficit on equities. Since 2000, higher UK earnings on holdings of foreign equity, partly reflecting increased UK investment into foreign equities, have helped move the overall portfolio investment balance from deficit to surplus. By instrument, the UK has paid out more dividends on UK equity securities owned by non-residents than have been received on foreign securities owned by UK residents, in all years since 1987. In contrast, the UK has recorded a surplus on debt securities in all but one of the last ten years, with a surplus on earnings from bonds and notes only partly offset by a deficit on money market instruments. UK banks have almost trebled their net earnings on portfolio investment in the last 10 years, moving from a surplus of £5.0 billion in 1995 to a surplus of £14.1 billion in 2005. UK banks traditionally tended to hold debt rather than equity securities but in recent years UK banks have steadily increased their levels of investment in foreign equity securities, which has resulted in a similar rise in dividend receipts. UK banks' dividend receipts were £3.4 billion in 2005 compared with just £0.3 billion in 1995 and £0.9 billion in 2000. UK banks' interest receipts on foreign debt securities rose to £22.6 billion in 2005, up 22 per cent on 2004, due to increased investment in those instruments and higher interest rates. UK insurance companies, pension funds and other financial intermediaries (securities dealers, unit and investment trusts) mainly hold equity rather than debt. On the debits side, foreign earnings from UK equity have been about £13–£14 billion since 2001. Strong foreign investment into UK debt securities has led to a sharp rise in interest paid, reaching a record £29.3 billion in 2005.

Figure 4.4

Portfolio investment income

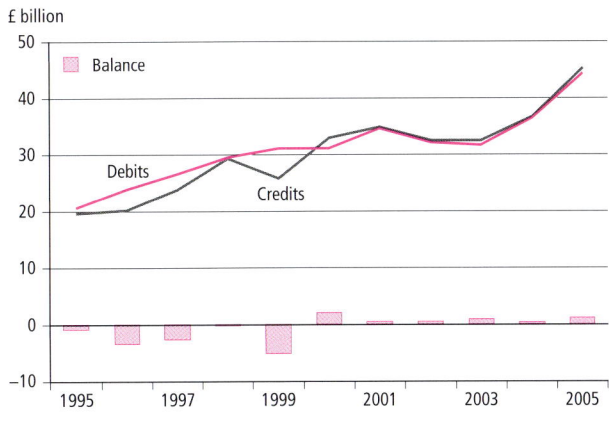

Other investment

Movements in the other investment balance are mainly driven by interest rate changes, which impact on interest paid and received on loans and deposits. As the UK has an excess of other investment liabilities over assets, there is generally a deficit on other investment income, with rising interest rates leading to a rising deficit and falling interest rates to a falling deficit. So, falling global interest rates from 2001 through to 2003 led to the other investment deficit declining from £15.1 billion to £10.3 billion over that period. In recent years global interest rates have been on the up and the deficit on other investment widened to £16.5 billion in 2005.

Earnings on deposits and loans abroad by UK banks accounted for nearly 80 per cent of total other investment credits in 2005. The vast majority of these earnings are made from foreign currency, reflecting the international nature of banking in the United Kingdom (as many of the banks trading with the rest of the world are actually branches or subsidiaries of foreign banks).

Figure 4.5

Other investment income

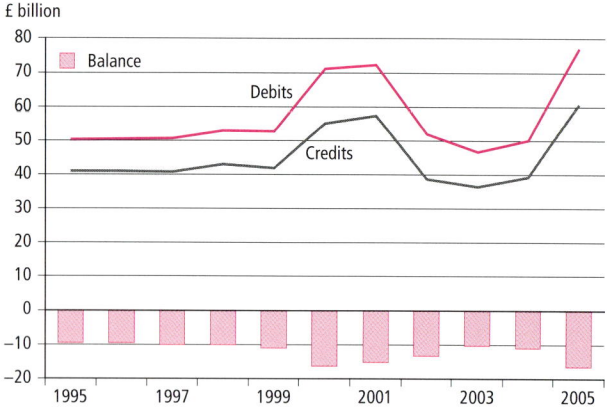

Sectoral breakdown of investment income

UK banks remain the single biggest investing sector, earning around 44 per cent of total UK investment income credits and paying out 46 per cent of debits in 2005. Banks have earned an investment income surplus in every year since 1997, reaching a record £13.0 billion in 2003. That surplus declined to £10.7 billion in 2005 reflecting increased net interest payments on foreign currency deposits as a result of global interest rate rises. When considering the banking sector's overall contribution to the UK's balance of payments, it is important to also include bank's financial service fees and commissions earned from foreign clients – a net £2.5 billion in 2005.

Central government has recorded a net annual deficit of about £3 to £5 billion in recent years (mostly interest payments on Gilts), whilst other sectors – predominantly private non-financial corporations and non-monetary financial institutions – have historically recorded net surpluses. In 2005, these other sectors recorded a surplus of £24.6 billion, largely due to strong direct investment earnings from abroad by private non-financial corporations.

Figure 4.6

Investment income of banks
Credits less debits

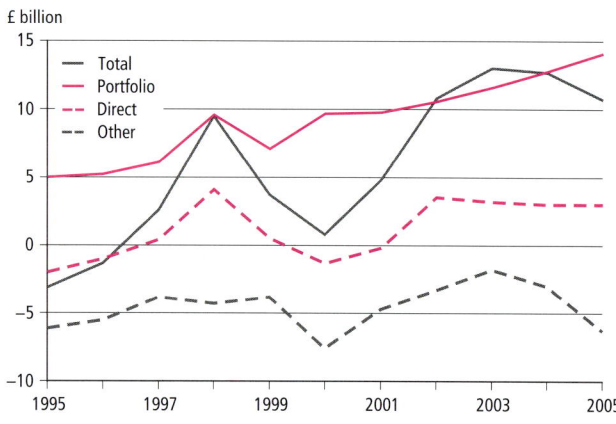

4.1 Income
Summary table

£ million

		1995	1996	1997	1998	1999	2000	2001	2002	2003	2004	2005
Credits												
Compensation of employees	KTMN	887	911	1 007	840	960	1 032	1 087	1 121	1 116	1 171	1 211
Investment income												
Earnings on direct investment abroad	HJYW	24 831	28 584	29 470	29 919	33 144	45 042	46 741	51 473	55 093	64 442	79 146
Earnings on portfolio investment abroad												
Earnings on equity securities	HCPL	4 451	4 768	5 449	6 061	7 773	9 872	9 861	10 530	10 385	11 220	12 792
Earnings on debt securities	HLYW	15 274	15 479	18 377	23 237	18 095	23 101	25 021	21 954	22 165	25 489	32 483
Total portfolio investment	HLYX	19 725	20 247	23 826	29 298	25 868	32 973	34 882	32 484	32 550	36 709	45 275
Earnings on other investment abroad	AIOP	40 953	41 039	40 767	43 039	41 779	55 114	57 264	38 728	36 447	39 174	60 746
Earnings on reserve assets	HHCB	1 686	1 551	1 372	1 132	1 161	985	961	820	791	705	659
Total investment income	HMBN	87 195	91 421	95 435	103 388	101 952	134 114	139 848	123 505	124 881	141 030	185 826
Total	HMBQ	88 082	92 332	96 442	104 228	102 912	135 146	140 935	124 626	125 997	142 201	187 037
Debits												
Compensation of employees	KTMO	1 183	818	924	850	759	882	1 021	1 054	1 057	1 100	1 137
Investment income												
Foreign earnings on direct investment in the UK	HJYX	13 819	16 630	14 916	8 585	17 003	27 435	21 437	16 016	21 919	27 936	34 574
Foreign earnings on portfolio investment in the UK												
Earnings on equity securities	ZMRB	5 612	7 359	8 601	9 930	13 542	9 899	13 189	14 054	13 674	14 002	14 829
Earnings on debt securities	HLZB	14 988	16 405	17 937	19 523	17 533	21 111	21 286	18 002	17 930	22 379	29 335
Total portfolio investment	HLZC	20 600	23 764	26 538	29 453	31 075	31 010	34 475	32 056	31 604	36 381	44 164
Earnings on other investment in the UK	HLZN	50 316	50 564	50 750	53 020	52 805	71 279	72 338	52 057	46 771	50 188	77 291
Total investment income	HMBO	84 735	90 958	92 204	91 058	100 883	129 724	128 250	100 129	100 294	114 505	156 029
Total	HMBR	85 918	91 776	93 128	91 908	101 642	130 606	129 271	101 183	101 351	115 605	157 166
Balances (Net earnings)												
Compensation of employees	KTMP	−296	93	83	−10	201	150	66	67	59	71	74
Investment income												
Direct investment	HJYE	11 012	11 954	14 554	21 334	16 141	17 607	25 304	35 457	33 174	36 506	44 572
Portfolio investment												
Earnings on equity securities	HLZO	−1 161	−2 591	−3 152	−3 869	−5 769	−27	−3 328	−3 524	−3 289	−2 782	−2 037
Earnings on debt securities	HLZP	286	−926	440	3 714	562	1 990	3 735	3 952	4 235	3 110	3 148
Total portfolio investment	HLZX	−875	−3 517	−2 712	−155	−5 207	1 963	407	428	946	328	1 111
Other investment	CGNA	−9 363	−9 525	−9 983	−9 981	−11 026	−16 165	−15 074	−13 329	−10 324	−11 014	−16 545
Reserve assets	HHCB	1 686	1 551	1 372	1 132	1 161	985	961	820	791	705	659
Total investment income	HMBM	2 460	463	3 231	12 330	1 069	4 390	11 598	23 376	24 587	26 525	29 797
Total	HMBP	2 164	556	3 314	12 320	1 270	4 540	11 664	23 443	24 646	26 596	29 871

4.2 Investment income
Sector analysis

£ million

		1995	1996	1997	1998	1999	2000	2001	2002	2003	2004	2005
Credits (Earnings of UK residents on investment abroad)												
Monetary financial institutions												
Banks	CGNB	45 949	48 587	49 302	56 219	52 883	69 599	72 606	54 837	54 348	58 339	81 781
Building societies	GJXE	81	119	103	134	176	292	333	337	276	282	324
Total monetary financial institutions	CGND	46 030	48 706	49 405	56 353	53 059	69 891	72 939	55 174	54 624	58 621	82 105
Central government	CGNY	1 695	1 561	1 380	1 267	1 165	989	965	823	795	707	667
Public corporations	CGNP	118	111	356	410	329	364	438	371	389	802	1 290
Other sectors	CGNW	39 352	41 043	44 294	45 358	47 399	62 870	65 506	67 137	69 073	80 900	101 764
Total	HMBN	**87 195**	**91 421**	**95 435**	**103 388**	**101 952**	**134 114**	**139 848**	**123 505**	**124 881**	**141 030**	**185 826**
Debits (Foreign earnings on investment in UK)												
Monetary financial institutions (banks and building societies)	CGPN	49 103	50 010	46 814	46 896	49 330	69 074	68 102	44 420	41 660	45 894	71 402
Central government	CGNZ	5 276	5 557	5 797	5 826	5 027	4 592	4 280	3 914	4 181	4 874	6 142
Local authorities	CGOB	38	30	21	16	12	7	4	2	–	–	–
Public corporations	CGOD	43	34	28	20	–	–	–	–	–	–	–
Other sectors	CGSE	30 275	35 327	39 544	38 300	46 514	56 051	55 864	51 793	54 453	63 737	78 485
Total	HMBO	**84 735**	**90 958**	**92 204**	**91 058**	**100 883**	**129 724**	**128 250**	**100 129**	**100 294**	**114 505**	**156 029**
Balances (Net earnings)												
Monetary financial institutions (banks and building societies)	CGSO	–3 073	–1 304	2 591	9 457	3 729	817	4 837	10 754	12 964	12 727	10 703
Central government	CGOE	–3 581	–3 996	–4 417	–4 559	–3 862	–3 603	–3 315	–3 091	–3 386	–4 167	–5 475
Local authorities	-CGOB	–38	–30	–21	–16	–12	–7	–4	–2	–	–	–
Public corporations	CGOF	75	77	328	390	329	364	438	371	389	802	1 290
Other sectors	CGTX	9 077	5 716	4 750	7 058	885	6 819	9 642	15 344	14 620	17 163	23 279
Total	HMBM	**2 460**	**463**	**3 231**	**12 330**	**1 069**	**4 390**	**11 598**	**23 376**	**24 587**	**26 525**	**29 797**

4.3 Earnings on direct investment

£ million

		1995	1996	1997	1998	1999	2000	2001	2002	2003	2004	2005
Credits (Earnings of UK residents on direct investment abroad)												
Earnings on equity												
Dividends and distributed branch profits												
Dividends	CNZN	8 808	8 833	11 791	12 246	8 795	14 679	14 412	15 255	29 482	27 994	31 050
Distributed branch profits	HDNG	1 387	1 670	1 468	1 158	1 278	2 231	2 552	2 381	2 723	3 273	6 074
Total dividends and distributed branch profits	HMAE	10 195	10 503	13 259	13 404	10 073	16 910	16 964	17 636	32 205	31 267	37 124
Reinvested earnings	-HDNY	14 378	17 271	16 112	14 071	21 392	25 178	27 220	32 209	21 456	32 430	40 597
Earnings on property investment	HHBW	107	124	85	89	264	358	433	380	399	439	496
Total earnings on equity	HMAK	24 680	27 898	29 456	27 564	31 729	42 446	44 617	50 225	54 060	64 136	78 217
Earnings on other capital [1]	HDNQ	151	686	14	2 355	1 415	2 596	2 124	1 248	1 033	306	929
Total	HJYW	**24 831**	**28 584**	**29 470**	**29 919**	**33 144**	**45 042**	**46 741**	**51 473**	**55 093**	**64 442**	**79 146**
Debits (Foreign earnings on direct investment in the UK)												
Earnings on equity												
Dividends and distributed branch profits												
Dividends	BCEA	5 131	5 895	6 146	6 945	8 198	9 472	14 418	7 638	8 170	10 142	12 243
Distributed branch profits	CYFD	1 354	1 531	787	−2 534	323	2 713	2 251	−1 079	56	1 999	2 576
Total dividends and distributed branch profits	HMAH	6 485	7 426	6 933	4 411	8 521	12 185	16 669	6 559	8 226	12 141	14 819
Reinvested earnings	CYFV	5 254	7 873	6 386	1 522	4 607	10 788	−992	3 647	7 429	9 320	11 095
Earnings on property investment	HESG	213	219	234	259	1 167	1 258	1 398	1 507	1 614	1 663	1 796
Total earnings on equity	HMAG	11 952	15 518	13 553	6 192	14 295	24 231	17 075	11 713	17 269	23 124	27 710
Earnings on other capital [1]	CYFN	1 867	1 112	1 363	2 393	2 708	3 204	4 362	4 303	4 650	4 812	6 864
Total	HJYX	**13 819**	**16 630**	**14 916**	**8 585**	**17 003**	**27 435**	**21 437**	**16 016**	**21 919**	**27 936**	**34 574**
Balances (Net earnings)												
Earnings on equity												
Dividends and distributed branch profits												
Dividends	LTMA	3 677	2 938	5 645	5 301	597	5 207	−6	7 617	21 312	17 852	18 807
Distributed branch profits	LTMB	33	139	681	3 692	955	−482	301	3 460	2 667	1 274	3 498
Total dividends and distributed branch profits	HHZA	3 710	3 077	6 326	8 993	1 552	4 725	295	11 077	23 979	19 126	22 305
Reinvested earnings	LTMC	9 124	9 398	9 726	12 549	16 785	14 390	28 212	28 562	14 027	23 110	29 502
Earnings on property investment	LTMD	−106	−95	−149	−170	−903	−900	−965	−1 127	−1 215	−1 224	−1 300
Total earnings on equity	HHYY	12 728	12 380	15 903	21 372	17 434	18 215	27 542	38 512	36 791	41 012	50 507
Earnings on other capital [1]	HMAM	−1 716	−426	−1 349	−38	−1 293	−608	−2 238	−3 055	−3 617	−4 506	−5 935
Total	HJYE	**11 012**	**11 954**	**14 554**	**21 334**	**16 141**	**17 607**	**25 304**	**35 457**	**33 174**	**36 506**	**44 572**

1 Earnings on other capital consists of interest accrued to/from direct investors from/to associated enterprises abroad.

4.4 Earnings on direct investment
Sector analysis

£ million

		1995	1996	1997	1998	1999	2000	2001	2002	2003	2004	2005
Credits (Earnings of UK residents on investment abroad)												
Monetary financial institutions (banks)	HCVU	137	1 409	1 407	1 682	2 613	3 639	4 596	4 464	5 700	5 806	8 036
Insurance companies	CNZD	1 608	1 270	1 600	793	1 488	930	790	628	2 326	3 555	2 259
Other financial intermediaries	HCWW	1 888	2 540	2 547	2 209	2 990	3 017	2 924	3 232	4 293	5 132	6 878
Private non-financial corporations	HCUS	21 109	23 266	23 823	25 136	25 944	37 335	38 269	42 949	42 499	49 663	61 649
Public corporations	HDMG	14	9	12	14	17	17	40	54	87	54	54
Household sector [1]	HHLI	75	90	81	85	92	104	122	146	188	232	270
Total	HJYW	**24 831**	**28 584**	**29 470**	**29 919**	**33 144**	**45 042**	**46 741**	**51 473**	**55 093**	**64 442**	**79 146**
Debits (Foreign earnings on direct investment in UK)												
Monetary financial institutions (banks)	GPAZ	2 134	2 379	1 037	−2 433	2 109	4 979	4 795	1 008	2 537	2 844	5 076
Insurance companies	HDPK	379	881	1 138	1 333	4	612	−955	179	898	1 842	2 081
Other financial intermediaries												
Securities dealers	HDQX	269	799	375	−643	1 124	1 495	1 272	1 337	449	1 476	1 353
Other	HFBT	127	204	237	415	361	780	593	829	1 754	1 237	1 565
Total other financial intermediaries	HFCY	396	1 003	612	−228	1 485	2 275	1 865	2 166	2 203	2 713	2 918
Private non-financial corporations	BCEB	10 910	12 367	12 129	9 913	13 405	19 569	15 732	12 663	16 281	20 537	24 499
Total	HJYX	**13 819**	**16 630**	**14 916**	**8 585**	**17 003**	**27 435**	**21 437**	**16 016**	**21 919**	**27 936**	**34 574**
Balances (Net earnings)												
Monetary financial institutions (banks)	LTME	−1 997	−970	370	4 115	504	−1 340	−199	3 456	3 163	2 962	2 960
Insurance companies	LTMF	1 229	389	462	−540	1 484	318	1 745	449	1 428	1 713	178
Other financial intermediaries	LTMG	1 492	1 537	1 935	2 437	1 505	742	1 059	1 066	2 090	2 419	3 960
Private non-financial corporations	LTMH	10 199	10 899	11 694	15 223	12 539	17 766	22 537	30 286	26 218	29 126	37 150
Public corporations	HDMG	14	9	12	14	17	17	40	54	87	54	54
Households	HHLI	75	90	81	85	92	104	122	146	188	232	270
Total	HJYE	**11 012**	**11 954**	**14 554**	**21 334**	**16 141**	**17 607**	**25 304**	**35 457**	**33 174**	**36 506**	**44 572**

1 The household sector includes non-profit institutions serving households.

4.5 Earnings on portfolio investment

£ million

		1995	1996	1997	1998	1999	2000	2001	2002	2003	2004	2005
Credits (Earnings of UK residents on portfolio investment abroad)												
Earnings on equity securities (shares) by:												
Monetary financial Institutions (banks)	HHRX	298	414	411	521	609	865	1 261	1 473	2 299	2 676	3 410
Central Government	LOEN	–	–	–	–	–	–	–	–	–	–	6
Insurance companies and pension funds												
Insurance companies	CGOM	1 494	1 490	1 511	1 715	1 939	2 237	2 289	2 455	2 075	2 191	2 358
Pension funds[1]	HPDL	1 538	1 544	1 388	2 023	2 132	1 861	1 543	1 598	1 456	1 827	2 189
Total insurance companies and pension funds	CGOX	3 032	3 034	2 899	3 738	4 071	4 098	3 832	4 053	3 531	4 018	4 547
Other financial intermediaries	CGOY	990	1 173	1 952	1 610	2 914	4 677	4 465	4 690	4 275	4 202	4 352
Private non-financial corporations	EGMS	6	5	8	9	10	41	124	127	110	126	135
Household sector[2]	HEOG	125	142	179	183	169	191	179	187	170	198	342
Total earnings on equity securities	HCPL	4 451	4 768	5 449	6 061	7 773	9 872	9 861	10 530	10 385	11 220	12 792
Earnings on debt securities												
Earnings on bonds and notes by:												
Monetary financial institutions												
Banks	HHRY	9 429	10 283	11 934	13 369	11 153	15 538	16 066	15 259	15 036	17 161	20 952
Building societies	GJXE	81	119	103	134	176	292	333	337	276	282	324
Total monetary financial institutions	HTCQ	9 510	10 402	12 037	13 503	11 329	15 830	16 399	15 596	15 312	17 443	21 276
Insurance companies and pension funds												
Insurance companies	CGON	733	718	770	1 122	1 075	1 121	1 370	1 718	1 998	1 770	2 233
Pension funds[1]	HPDM	315	361	317	415	509	517	565	621	703	866	1 170
Total insurance companies and pension funds	CGOZ	1 048	1 079	1 087	1 537	1 584	1 638	1 935	2 339	2 701	2 636	3 403
Other financial intermediaries	CGPA	2 259	2 679	3 759	3 759	2 807	2 762	3 468	2 071	2 206	2 808	4 845
Private non-financial corporations	EGNF	168	156	218	61	54	43	108	111	117	210	213
Household sector[2]	HEOH	1 431	396	336	312	266	286	260	240	255	238	269
Total earnings on bonds and notes	HCPK	14 416	14 712	17 437	19 172	16 040	20 559	22 170	20 357	20 591	23 335	30 006
Earnings on money market instruments by:												
Monetary financial institutions (banks)	HBMX	659	543	700	3 933	1 908	2 292	2 569	1 233	984	1 451	1 665
Central government	LSPA	–	–	–	–	–	–	18	26	19	9	2
Other financial intermediaries	NHQV	74	70	113	49	73	131	118	130	205	249	281
Private non-financial corporations	HGBX	125	154	127	83	74	119	146	208	366	445	529
Total earnings on money market instruments	HCHG	858	767	940	4 065	2 055	2 542	2 851	1 597	1 574	2 154	2 477
Total earnings on debt securities	HLYW	15 274	15 479	18 377	23 237	18 095	23 101	25 021	21 954	22 165	25 489	32 483
Total	HLYX	**19 725**	**20 247**	**23 826**	**29 298**	**25 868**	**32 973**	**34 882**	**32 484**	**32 550**	**36 709**	**45 275**

1 The pension funds data only covers self-administered funds, see glossary.
2 The household sector includes non-profit institutions serving households.

4.5 Earnings on portfolio investment
continued

£ million

		1995	1996	1997	1998	1999	2000	2001	2002	2003	2004	2005
Debits (Foreign earnings on portfolio investment in the UK)												
Earnings on equity securities (shares) issued by:												
Monetary financial institutions (banks and building societies)	HBQJ	336	441	516	305	296	115	131	109	112	114	121
Other sectors[1]	HBQK	5 276	6 918	8 085	9 625	13 246	9 784	13 058	13 945	13 562	13 888	14 708
Total foreign earnings on UK equity securities	ZMRB	5 612	7 359	8 601	9 930	13 542	9 899	13 189	14 054	13 674	14 002	14 829
Earnings on debt securities												
Earnings on bonds and notes												
Issues by central government												
UK foreign currency bonds and notes	ZMRA	866	817	667	339	311	339	265	128	20	37	38
Earnings on British government stocks by:												
Foreign central banks (exchange reserves)	HESK	1 389	1 339	1 237	1 392	1 244	1 318	1 182	1 133	1 064	1 124	1 202
Other foreign residents	HCEV	2 836	3 232	3 779	4 014	3 418	2 918	2 743	2 562	3 019	3 539	4 715
Total foreign earnings on British government stocks	HENI	4 225	4 571	5 016	5 406	4 662	4 236	3 925	3 695	4 083	4 663	5 917
Total issues by central government	HBQU	5 091	5 388	5 683	5 745	4 973	4 575	4 190	3 823	4 103	4 700	5 955
Local authorities' bonds	HHGH	–	–	–	–	–	–	–	–	–	–	–
Public corporations' bonds	HESY	–	–	–	–	–	–	–	–	–	–	–
Issues by monetary financial institutions (banks and building societies)												
Bonds	HGUV	1 304	1 451	1 577	1 540	1 620	1 976	1 897	1 945	2 101	2 696	3 289
European medium term notes and other medium-term paper:												
Issued by UK banks	HCEY	745	897	1 025	1 071	1 035	1 138	1 350	1 418	1 788	2 587	3 417
Issued by UK building societies	HCFB	290	234	163	80	54	109	100	86	53	75	103
Total medium-term paper	HGMM	1 035	1 131	1 188	1 151	1 089	1 247	1 450	1 504	1 841	2 662	3 520
Total issues by monetary financial institutions	HBOT	2 339	2 582	2 765	2 691	2 709	3 223	3 347	3 449	3 942	5 358	6 809
Issues by other sectors[1]	HGUW	4 057	4 517	4 907	4 793	5 042	6 151	5 909	6 054	6 538	8 393	10 236
Total foreign earnings on UK bonds and notes	HLZA	11 487	12 487	13 355	13 229	12 724	13 949	13 446	13 326	14 583	18 451	23 000
Earnings on money market instruments												
Earnings on treasury bills (issued by central government)												
Sterling treasury bills	XAMR	55	64	29	49	38	3	13	20	24	126	145
Euro treasury bills	HHNV	106	85	67	18	3	–	–	–	–	–	–
Total earnings on treasury bills	HHZU	161	149	96	67	41	3	13	20	24	126	145
Earnings on certificates of deposit (Issued by monetary financial institutions)												
Issued by UK banks	HCEB	2 335	2 612	3 199	4 371	3 075	4 910	6 049	3 473	2 324	2 437	3 853
Issued by UK building societies	HGUY	47	37	26	19	21	35	20	17	40	70	60
Total earnings on certificates of deposit	HCEE	2 382	2 649	3 225	4 390	3 096	4 945	6 069	3 490	2 364	2 507	3 913
Earnings on commercial paper												
Issued by monetary financial institutions												
Issued by UK banks	HCEC	177	275	447	928	586	803	813	572	570	755	1 246
Issued by UK building societies	HHBC	206	205	140	51	100	161	110	36	42	86	197
Total earnings on mfi issued commercial paper	HCEF	383	480	587	979	686	964	923	608	612	841	1 443
Issued by other sectors[1]	HHZT	575	640	674	858	986	1 250	835	558	347	454	834
Total earnings on commercial paper	HHBO	958	1 120	1 261	1 837	1 672	2 214	1 758	1 166	959	1 295	2 277
Total foreign earnings on UK Money Market Instruments	HLYZ	3 501	3 918	4 582	6 294	4 809	7 162	7 840	4 676	3 347	3 928	6 335
Total foreign earnings on UK debt securities	HLZB	14 988	16 405	17 937	19 523	17 533	21 111	21 286	18 002	17 930	22 379	29 335
Total	HLZC	**20 600**	**23 764**	**26 538**	**29 453**	**31 075**	**31 010**	**34 475**	**32 056**	**31 604**	**36 381**	**44 164**

1 These series relate to non-governmental sectors other than monetary financial institutions.

4.5 Earnings on portfolio investment
continued

£ million

		1995	1996	1997	1998	1999	2000	2001	2002	2003	2004	2005
Balances (net earnings)												
Earnings on equity securities (shares)	HLZO	–1 161	–2 591	–3 152	–3 869	–5 769	–27	–3 328	–3 524	–3 289	–2 782	–2 037
Earnings on debt securities												
Earnings on bonds and notes	HLZQ	2 929	2 225	4 082	5 943	3 316	6 610	8 724	7 031	6 008	4 884	7 006
Earnings on money market instruments	HLZR	–2 643	–3 151	–3 642	–2 229	–2 754	–4 620	–4 989	–3 079	–1 773	–1 774	–3 858
Total foreign earnings on UK debt securities	HLZP	286	–926	440	3 714	562	1 990	3 735	3 952	4 235	3 110	3 148
Total	HLZX	**–875**	**–3 517**	**–2 712**	**–155**	**–5 207**	**1 963**	**407**	**428**	**946**	**328**	**1 111**

1 These series relate to non-governmental sectors other than monetary financial institutions.

4.6 Earnings on portfolio investment
Sector analysis

£ million

		1995	1996	1997	1998	1999	2000	2001	2002	2003	2004	2005
Credits (Earnings of UK residents on portfolio investment abroad)												
Earnings from portfolio investment abroad by UK:												
Monetary financial institutions												
Banks	AINB	10 386	11 240	13 045	17 823	13 670	18 695	19 896	17 965	18 319	21 288	26 027
Building societies	GJXE	81	119	103	134	176	292	333	337	276	282	324
Total monetary financial institutions	AIND	10 467	11 359	13 148	17 957	13 846	18 987	20 229	18 302	18 595	21 570	26 351
Central government	LOEO	–	–	–	–	–	–	18	26	19	9	8
Insurance companies and pension funds	AINE	4 080	4 113	3 986	5 275	5 655	5 736	5 767	6 392	6 232	6 654	7 950
Other financial intermediaries	AINF	3 323	3 922	5 824	5 418	5 794	7 570	8 051	6 891	6 686	7 259	9 478
Private non-financial corporations	AINI	299	315	353	153	138	203	378	446	593	781	877
Household sector[1]	AINK	1 556	538	515	495	435	477	439	427	425	436	611
Total	HLYX	19 725	20 247	23 826	29 298	25 868	32 973	34 882	32 484	32 550	36 709	45 275
Debits (Foreign earnings on portfolio investment in the UK)												
Foreign earnings from portfolio investment in UK:												
Monetary financial institutions (banks and building societies)	HBXI	5 440	6 152	7 093	8 365	6 787	9 247	10 470	7 656	7 030	8 820	12 286
Central government	HBXM	5 252	5 537	5 779	5 812	5 014	4 578	4 203	3 843	4 127	4 826	6 100
Local authorities	HHGH	–	–	–	–	–	–	–	–	–	–	–
Public corporations	HESY	–	–	–	–	–	–	–	–	–	–	–
Other sectors	HBXR	9 908	12 075	13 666	15 276	19 274	17 185	19 802	20 557	20 447	22 735	25 778
Total	HLZC	20 600	23 764	26 538	29 453	31 075	31 010	34 475	32 056	31 604	36 381	44 164
Balances (Net earnings)												
Monetary financial institutions	LTMI	5 027	5 207	6 055	9 592	7 059	9 740	9 759	10 646	11 565	12 750	14 065
Central government	ZPOF	–5 252	–5 537	–5 779	–5 812	–5 014	–4 578	–4 185	–3 817	–4 108	–4 817	–6 092
Local authorities	-HHGH	–	–	–	–	–	–	–	–	–	–	–
Public corporations	-HESY	–	–	–	–	–	–	–	–	–	–	–
Other sectors	LTMJ	–650	–3 187	–2 988	–3 935	–7 252	–3 199	–5 167	–6 401	–6 511	–7 605	–6 862
Total	HLZX	–875	–3 517	–2 712	–155	–5 207	1 963	407	428	946	328	1 111

1 The household sector includes non-profit institutions serving households.

4.7 Earnings on other investment

£ million

		1995	1996	1997	1998	1999	2000	2001	2002	2003	2004	2005
Credits (Earnings of UK residents on other investment abroad)												
Earnings on trade credit												
Central government	XBGJ	–	–	–	–	–	–	–	–	–	–	–
Other sectors[1]	HGQD	146	138	157	177	–	–	–	–	–	–	–
Total earnings on trade credit	AIOM	146	138	157	177	–	–	–	–	–	–	–
Earnings on loans												
Long-term												
Bank loans under ECGD guarantee	AINM	786	708	721	664	594	508	378	235	205	198	235
Inter-government loans by the UK	XBGI	9	10	8	9	4	4	4	3	4	2	2
Loans by Commonwealth Development Corporation (public corporations)	HGEN	104	102	110	123	115	101	74	74	74	74	74
Loans by the Export Credit Guarantee Department	CY95	..	..	234	273	197	246	324	243	228	674	1 162
Loans by specialist leasing companies[1]	HBXC	–	–	–	–	–	–	–	–	–	–	–
Total long-term loans	AIOO	899	820	1 073	1 069	910	859	780	555	511	948	1 473
Short-term loans	VTUN	81	66	68	54	37	36	36	36	36	36	36
Total earnings on loans	CGKJ	980	886	1 141	1 123	947	895	816	591	547	984	1 509
Earnings on deposits												
By UK monetary financial institutions (banks)												
Sterling deposits	CGEJ	3 598	3 995	5 518	6 842	6 842	7 639	7 249	5 761	6 203	8 160	9 104
Foreign currency deposits	HCAT	31 032	31 231	28 606	29 205	29 164	39 118	40 487	26 412	23 921	22 887	38 379
Total deposits by UK banks	CGGT	34 630	35 226	34 124	36 047	36 006	46 757	47 736	32 173	30 124	31 047	47 483
Deposits by securities dealers	HGTD	1 104	1 272	1 080	789	854	1 376	2 908	1 733	1 762	1 904	3 338
Deposits by other UK residents[1]	CGJK	3 801	3 202	3 927	4 425	3 584	5 538	5 315	3 906	3 709	4 910	8 083
Total earnings on deposits abroad	CGJQ	39 535	39 700	39 131	41 261	40 444	53 671	55 959	37 812	35 595	37 861	58 904
Earnings on other assets (Non-governmental sectors other than monetary financial institutions)												
Trusts and annuities	HHLF	292	315	338	352	388	548	489	325	305	329	333
Miscellaneous central government receipts	HPPK	–	–	–	126	–	–	–	–	–	–	–
Total earnings on other assets	CGKM	292	315	338	478	388	548	489	325	305	329	333
Total	AIOP	40 953	41 039	40 767	43 039	41 779	55 114	57 264	38 728	36 447	39 174	60 746

1 These series relate to non-governmental sectors other than monetary financial institutions (and securities dealers).

4.7 Earnings on other investment
continued

£ million

		1995	1996	1997	1998	1999	2000	2001	2002	2003	2004	2005
Debits (Foreign earnings on other investment in the UK)												
Earnings on trade credit												
Public corporations	XBGW	–	–	–	–	–	–	–	–	–	–	–
Other sectors [1]	HHLW	150	152	143	140	–	–	–	–	–	–	–
Total earnings on trade credit	CGMA	150	152	143	140	–	–	–	–	–	–	–
Earnings on loans [2]												
Loans to:												
Central government	CGLF	24	20	18	14	13	14	77	71	54	48	42
Local authorities	CGLG	38	30	21	16	12	7	4	2	–	–	–
Public corporations	CGLH	43	34	28	20	–	–	–	–	–	–	–
Securities dealers	CGLI	3 273	3 354	5 293	5 120	4 762	7 502	9 525	6 722	6 853	6 557	9 613
Other [1]	CGMD	4 464	4 477	5 428	5 546	6 285	7 880	8 605	8 271	6 528	8 256	12 654
Total earnings on loans	CGNO	7 842	7 915	10 788	10 716	11 072	15 403	18 211	15 066	13 435	14 861	22 309
Earnings on deposits [2] (Monetary financial institutions)												
Deposits with UK banks												
Sterling deposits	HCEG	5 787	6 118	6 492	8 044	7 566	9 437	9 100	7 095	6 898	9 834	11 621
Foreign currency deposits	HCEH	35 260	34 931	31 837	32 653	32 644	45 101	43 508	28 477	25 016	24 173	42 139
Total deposits with UK banks	HCEQ	41 047	41 049	38 329	40 697	40 210	54 538	52 608	35 572	31 914	34 007	53 760
Deposits with UK building societies	HHLS	482	430	355	267	224	310	229	184	179	223	280
Total earnings on deposits	HMAS	41 529	41 479	38 684	40 964	40 434	54 848	52 837	35 756	32 093	34 230	54 040
Earnings on other liabilities (Non-governmental sectors other than monetary financial institutions)												
Imputed income to foreign households from UK insurance companies technical reserves	HBWS	795	1 018	1 135	1 200	1 299	1 028	1 290	1 235	1 243	1 097	942
Other liabilities	CGME	–	–	–	–	–	–	–	–	–	–	–
Total earnings on other liabilities	CGMH	795	1 018	1 135	1 200	1 299	1 028	1 290	1 235	1 243	1 097	942
Total	HLZN	50 316	50 564	50 750	53 020	52 805	71 279	72 338	52 057	46 771	50 188	77 291
Balances (Net earnings)												
Trade credit	LTMK	–4	–14	14	37	–	–	–	–	–	–	–
Loans	LTML	–6 862	–7 029	–9 647	–9 593	–10 125	–14 508	–17 395	–14 475	–12 888	–13 877	–20 800
Currency and deposits	LTMM	–1 994	–1 779	447	297	10	–1 177	3 122	2 056	3 502	3 631	4 864
Other investment	LTMN	–503	–703	–797	–722	–911	–480	–801	–910	–938	–768	–609
Total	CGNA	–9 363	–9 525	–9 983	–9 981	–11 026	–16 165	–15 074	–13 329	–10 324	–11 014	–16 545

1 These series relate to non-governmental sectors other than monetary financial institutions.
2 It is not possible to separate out earnings on foreign loans to UK banks from earnings on foreign deposits with UK banks. Earnings on such loans are therefore included indistinguishably within earnings on deposits.

4.8 Earnings on other investment
Sector analysis

£ million

		1995	1996	1997	1998	1999	2000	2001	2002	2003	2004	2005
Credits (Earnings of UK residents on other investment abroad)												
Earnings from other investment by UK:												
Monetary financial institutions (banks)	CGMM	35 426	35 938	34 850	36 714	36 600	47 265	48 114	32 408	30 329	31 245	47 718
Central government	CGMN	9	10	8	135	4	4	4	3	4	2	2
Public corporations	ZPOP	104	102	344	396	312	347	398	317	302	748	1 236
Other sectors	CGMR	5 414	4 989	5 565	5 794	4 863	7 498	8 748	6 000	5 812	7 179	11 790
Total	AIOP	**40 953**	**41 039**	**40 767**	**43 039**	**41 779**	**55 114**	**57 264**	**38 728**	**36 447**	**39 174**	**60 746**
Debits (Foreign earnings on other investment in the UK)												
Foreign earnings from other investment in UK:												
Monetary financial institutions												
Banks	HCEQ	41 047	41 049	38 329	40 697	40 210	54 538	52 608	35 572	31 914	34 007	53 760
Building societies	HHLS	482	430	355	267	224	310	229	184	179	223	280
Total monetary financial institutions	HMAS	41 529	41 479	38 684	40 964	40 434	54 848	52 837	35 756	32 093	34 230	54 040
Central government	CGLF	24	20	18	14	13	14	77	71	54	48	42
Local authorities	CGLG	38	30	21	16	12	7	4	2	–	–	–
Public corporations	CGMV	43	34	28	20	–	–	–	–	–	–	–
Other sectors	CGMZ	8 682	9 001	11 999	12 006	12 346	16 410	19 420	16 228	14 624	15 910	23 209
Total	HLZN	**50 316**	**50 564**	**50 750**	**53 020**	**52 805**	**71 279**	**72 338**	**52 057**	**46 771**	**50 188**	**77 291**
Balances (Net earnings)												
Monetary financial institutions	LTMO	–6 103	–5 541	–3 834	–4 250	–3 834	–7 583	–4 723	–3 348	–1 764	–2 985	–6 322
Central government	LTMP	–15	–10	–10	121	–9	–10	–73	–68	–50	–46	–40
Local authorities	-CGLG	–38	–30	–21	–16	–12	–7	–4	–2	–	–	–
Public corporations	LTMQ	61	68	316	376	312	347	398	317	302	748	1 236
Other sectors	LTMR	–3 268	–4 012	–6 434	–6 212	–7 483	–8 912	–10 672	–10 228	–8 812	–8 731	–11 419
Total	CGNA	**–9 363**	**–9 525**	**–9 983**	**–9 981**	**–11 026**	**–16 165**	**–15 074**	**–13 329**	**–10 324**	**–11 014**	**–16 545**

Chapter 5
Current transfers

Summary

The current transfers deficit almost doubled between 1993 and 2000, growing from £5.2 billion in 1993 to £10.0 billion in 2000. After decreasing to £6.8 billion in 2001, the deficit increased again in each of the four subsequent years, to £12.2 billion at the end of 2005 – the highest cash figure on record.

The deficit on the government sector widened, from a deficit of £8.3 billion in 2004 to a deficit of £9.4 billion in 2005. Over the same period the deficit for other sectors increased by £0.1 billion to £2.8 billion. Overall receipts from EU institutions rose by £0.8 billion in 2005, to £7.8 billion, while payments increased from £11.5 billion in 2004 to £13.1 billion in 2005.

Figure 5.1
Current transfers

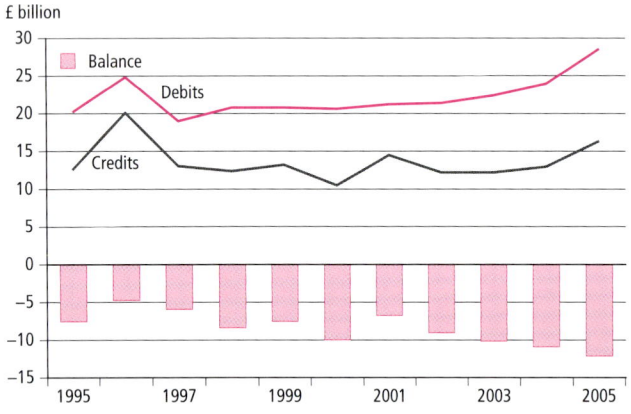

Central government transfers

Central government transfers include: taxes and social contributions received from non-resident workers and businesses; current transfers with international organisations (for example, EU Institutions); bilateral aid; social security payments abroad; military grants; and miscellaneous transfers. On the credits side, there was little movement, with the total increasing by £0.1 billion to £4.1 billion for 2005. Debits increased by £1.2 billion between 2004 and 2005, mainly driven by GNP fourth resource contributions to EU institutions, which rose by £1.0 billion to £8.6 billion in 2005, the first full year in which the EU budget covered an enlarged EU.

Figure 5.2
Transfers by central government

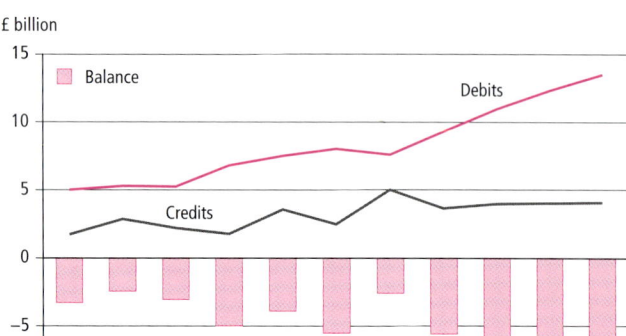

Other sector transfers

Non-government transfers include those EU transfers where the UK government simply acts as the agent for the final beneficiary (for example, social fund and agricultural guidance fund receipts) or original payer (for example, VAT based contributions). Other sectors transfers also include: taxes on income and wealth paid by UK workers and outward direct investors to foreign governments; insurance premiums and claims; and other transfers (workers remittances, and other private transfers such as gifts). Other sectors credits rose by

Figure 5.3
Transfers by other sectors

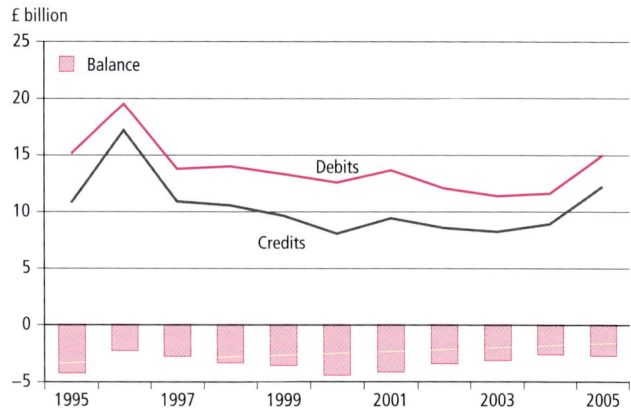

£3.3 billion between 2004 and 2005, whilst other sectors debits increased by £3.4 billion over the same period. The increases in both credits and debits in 2005 were both driven by offsetting rises in net non-life insurance premiums paid to and claims paid by UK companies as a result of payments of claims relating to Hurricane Katrina.

EU institutions

Transfers with EU institutions constitute the largest single component within current transfers. They showed a deficit in every year from 1986 to 2005; the lowest deficit recorded over the last ten years is £2.1 billion (in 1996) and the highest deficit £5.4 billion (in 2000). The deficit increased by £0.8 billion between 2004 and 2005, to £5.3 billion.

Figure 5.4

Transfers with other EU institutions

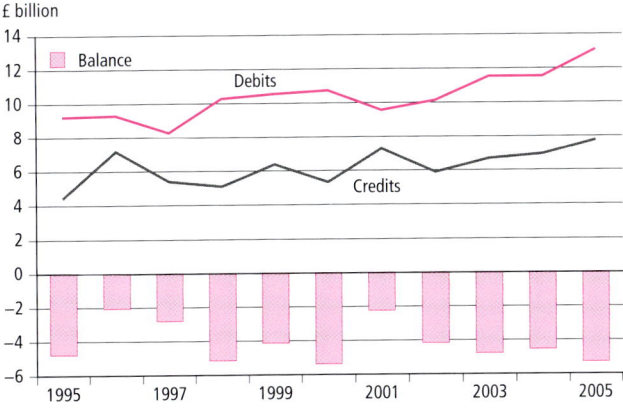

5.1 Current transfers

£ million

		1995	1996	1997	1998	1999	2000	2001	2002	2003	2004	2005
Credits												
Central government												
Current taxes on income, wealth etc.	FJKI	472	376	402	354	337	357	398	527	375	374	394
Other taxes on production	FJKH	–	–	–	–	–	–	–	–	–	–	–
Other subsidies on production	FJBC	–	–	–	–	–	–	–	–	–	–	–
Social contributions	FJBH	25	28	32	29	29	24	25	24	23	22	22
Social benefits	FJBL	–	–	–	–	–	–	–	–	–	–	–
EU Institutions:												
(a) Abatement	FKKL	1 208	2 411	1 733	1 377	3 171	2 084	4 560	3 099	3 560	3 592	3 655
(b) Other EU receipts	FKIJ	25	13	6	7	5	–	8	13	10	12	–
Miscellaneous receipts[1]	FKIK	–	–	–	–	–	–	–	–	–	–	–
Total central government	FJUM	1 730	2 828	2 173	1 767	3 542	2 465	4 991	3 663	3 968	4 000	4 071
Other sectors												
Current taxes on income, wealth etc.	FJBJ	–	–	–	–	–	–	–	–	–	–	–
Other taxes on production	FJGC	–	–	–	–	–	–	–	–	–	–	–
Other subsidies on production	FJBA	–	–	–	–	–	–	–	–	–	–	–
Social contributions	FJAB	70	80	75	70	60	31	34	53	21	13	–8
EU Institutions:												
(a) Agricultural Guarantee Fund	EBGL	2 392	3 931	3 063	2 935	2 781	2 571	2 336	2 381	2 691	2 909	3 216
(b) Social Fund	HDIZ	755	804	615	783	434	659	370	412	427	433	900
(c) ECSC Grant	FJKP	39	29	5	1	–	–	1	–	–	2	–
Net non-life insurance premiums[2]	NQQP	4 993	9 763	4 423	4 168	3 663	2 144	3 998	3 009	2 364	2 846	5 311
Non-life insurance claims[3]	FJFA	–	5	5	7	10	18	25	19	19	47	108
Other receipts of households[4]	FKIL	2 572	2 589	2 712	2 633	2 730	2 653	2 689	2 698	2 713	2 667	2 715
Total other sectors	FJUN	10 821	17 201	10 898	10 597	9 678	8 076	9 453	8 572	8 235	8 917	12 242
Total	KTND	**12 551**	**20 029**	**13 071**	**12 364**	**13 220**	**10 541**	**14 444**	**12 235**	**12 203**	**12 917**	**16 313**
Of which: Receipts from EU institutions	FKIM	4 419	7 188	5 422	5 103	6 391	5 314	7 275	5 905	6 688	6 948	7 771

1 Includes contributions by other countries towards the UK's cost of the 1991 Gulf conflict.
2 Premiums paid to UK insurance companies.
3 Claims paid to UK residents by foreign insurance companies.
4 Includes estimates for workers' remittances and for non-profit institutions serving households.

5.1 Current transfers
continued

£ million

		1995	1996	1997	1998	1999	2000	2001	2002	2003	2004	2005
Debits												
Central government												
Current taxes on income, wealth etc.	FJKK	–	–	–	–	–	–	–	–	–	–	–
Other taxes on production	FJKN	–	–	–	–	–	–	–	–	–	–	–
Other subsidies on production	FJCE	–	–	–	–	–	–	–	–	–	–	–
Social contributions	FJCH	–	–	–	–	–	–	–	–	–	–	–
Social security benefits	FJCK	972	1 029	1 102	1 162	1 183	1 218	1 292	1 388	1 452	1 575	1 631
Contributions to international organisations												
EU Institutions:												
(a) GNP: 4th Resource	HCSO	1 639	2 488	2 655	3 516	4 403	4 243	3 859	5 259	6 622	7 565	8 597
(b) GNP adjustments	HCSM	187	–34	–197	404	229	136	–1	76	150	–16	135
(c) Inter governmental agreements	HCBW	–	–	–	–	–	–	–	–	–	–	–
(d) Other	FKIN	8	8	31	–1	11	6	24	10	18	–3	106
Other organisations:												
(a) Military	HDKF	116	112	168	139	118	157	195	192	152	169	111
(b) Multilateral economic assistance	HCHJ	358	273	268	314	245	503	434	539	367	546	333
(c) Other	HCKL	835	633	429	402	479	691	647	459	488	557	602
Bilateral aid:												
(a) Non-project grants	FJKT	249	214	131	142	133	175	185	206	268	303	318
(b) Technical cooperation	FJKU	604	543	644	692	651	859	904	1 038	1 320	1 478	1 544
Military grants	HDJO	54	31	29	17	30	27	45	129	107	130	122
Total central government	FJUO	5 022	5 297	5 260	6 787	7 482	8 015	7 584	9 296	10 944	12 304	13 499
Other sectors												
Current taxes on income, wealth etc.	FJCI	557	610	638	454	682	775	523	644	444	615	717
Other taxes on production	FJLB	–	–	–	–	–	–	–	–	–	–	–
Other subsidies on production	FJCC	–	–	–	–	–	–	–	–	–	–	–
Social contributions	FJBG	–	–	–	–	–	–	–	–	–	–	–
Social benefits	FJCM	70	80	75	70	60	31	34	53	21	13	–8
EU Institutions:												
(a) Customs duties and agricultural levies	QYRD	2 458	2 318	2 291	2 076	2 024	2 086	2 069	1 919	1 937	2 145	2 220
(b) Sugar levies	GTBA	55	26	91	42	46	44	31	25	18	25	24
(c) VAT based contributions	HCML	4 635	4 441	3 646	3 758	3 920	4 104	3 624	2 720	2 775	1 764	1 980
(d) VAT adjustments	FSVL	210	30	–249	470	–109	100	–49	88	–35	25	19
(e) ECSC Production levy	GTBB	–	–	–	–	–	–	–	–	–	–	–
Net non-life insurance premiums[1]	FJDB	–	5	5	7	10	18	25	19	19	47	108
Non-life insurance claims[2]	NQQR	4 993	9 763	4 423	4 168	3 663	2 144	3 998	3 009	2 364	2 846	5 311
Other payments by households[3]	FKIQ	2 125	2 214	2 809	2 906	2 975	3 236	3 364	3 543	3 838	4 082	4 622
Total other sectors	FJUP	15 103	19 487	13 729	13 951	13 271	12 538	13 619	12 020	11 381	11 562	14 993
Total	KTNE	**20 125**	**24 784**	**18 989**	**20 738**	**20 753**	**20 553**	**21 203**	**21 316**	**22 325**	**23 866**	**28 492**
Of which: Payments to EU institutions	FKIR	9 192	9 277	8 268	10 265	10 524	10 719	9 557	10 097	11 485	11 505	13 081

1 Premiums paid by UK residents to foreign insurance companies.
2 Claims paid by UK insurance companies to non-residents.
3 Includes estimates for workers' remittances and for non-profit institutions serving households.

5.1 Current transfers
continued

£ million

		1995	1996	1997	1998	1999	2000	2001	2002	2003	2004	2005
Balances												
Central government												
Current taxes on income, wealth etc.	FJKJ	472	376	402	354	337	357	398	527	375	374	394
Other taxes on production	FJIZ	–	–	–	–	–	–	–	–	–	–	–
Other subsidies on production	FJBD	–	–	–	–	–	–	–	–	–	–	–
Social contributions	FJBI	25	28	32	29	29	24	25	24	23	22	22
Social benefits	FJBM	–972	–1 029	–1 102	–1 162	–1 183	–1 218	–1 292	–1 388	–1 452	–1 575	–1 631
Other current transfers[1]	FJKW	–2 817	–1 844	–2 419	–4 241	–3 123	–4 713	–1 724	–4 796	–5 922	–7 125	–8 213
Total central government	FJUQ	–3 292	–2 469	–3 087	–5 020	–3 940	–5 550	–2 593	–5 633	–6 976	–8 304	–9 428
Other sectors												
Current taxes on income, wealth etc.	FJHU	–557	–610	–638	–454	–682	–775	–523	–644	–444	–615	–717
Other taxes on production	FJHT	–	–	–	–	–	–	–	–	–	–	–
Other subsidies on production	FJHV	–	–	–	–	–	–	–	–	–	–	–
Social contributions	FJHJ	70	80	75	70	60	31	34	53	21	13	–8
Social benefits	FJJG	685	724	540	713	374	628	336	359	406	420	908
Other current transfers[1]	FJLT	–4 480	–2 480	–2 808	–3 683	–3 345	–4 346	–4 013	–3 216	–3 129	–2 463	–2 934
Total other sectors	FJUR	–4 282	–2 286	–2 831	–3 354	–3 593	–4 462	–4 166	–3 448	–3 146	–2 645	–2 751
Total	KTNF	**–7 574**	**–4 755**	**–5 918**	**–8 374**	**–7 533**	**–10 012**	**–6 759**	**–9 081**	**–10 122**	**–10 949**	**–12 179**
Of which: EU institutions	FKIS	–4 773	–2 089	–2 846	–5 162	–4 133	–5 405	–2 282	–4 192	–4 797	–4 557	–5 310

1 Includes an estimate for workers' remittances.

form
Part 2: Capital account, financial account and International investment position

Chapter 6
Capital account

Summary

The capital account has remained in surplus for over 20 years. A surplus of £2.4 billion was recorded in 2005; the £0.3 billion increase compared to 2004 was mainly due to a rise in European Regional Development Fund receipts from the EU.

Figure 6.1

Capital account

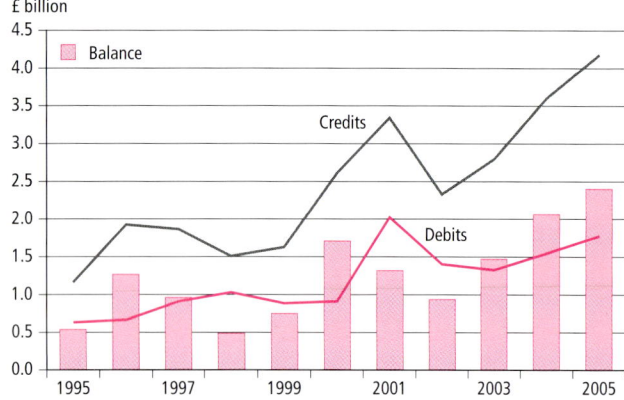

6.1 Capital account

£ million

		1995	1996	1997	1998	1999	2000	2001	2002	2003	2004	2005
Credits												
Capital transfers												
Central government												
Debt forgiveness	FJUU	–	–	–	–	–	–	–	–	–	–	–
Other capital transfers	FJLY	–	–	–	–	–	–	–	–	–	–	–
Total central government	FJMD	–	–	–	–	–	–	–	–	–	–	–
Other sectors												
Migrants' transfers	FJMG	678	703	754	967	1 144	1 371	2 267	1 864	1 951	2 298	2 491
Debt forgiveness	FJNC	–	–	–	–	–	–	–	–	–	–	–
Other capital transfers												
EU Institutions:												
Regional development fund	FKIT	437	620	812	357	285	989	543	296	622	1 062	1 393
Agricultural guidance fund	FJXL	48	30	57	56	47	82	26	–	2	49	80
Other capital transfers	EBGO	–	524	178	43	–	–	322	–	–	–	–
Total EU institutions	FKIV	485	1 174	1 047	456	332	1 071	891	296	624	1 111	1 473
Total other sectors	FJMU	1 163	1 877	1 801	1 423	1 476	2 442	3 158	2 160	2 575	3 409	3 964
Total capital transfers	FJMX	1 163	1 877	1 801	1 423	1 476	2 442	3 158	2 160	2 575	3 409	3 964
Sales of non-produced, non-financial assets	FJUX	–	49	68	89	152	165	177	172	218	193	214
Total	FKMH	**1 163**	**1 926**	**1 869**	**1 512**	**1 628**	**2 607**	**3 335**	**2 332**	**2 793**	**3 602**	**4 178**
Debits												
Capital transfers												
Central government												
Debt forgiveness	FJUV	28	23	24	146	22	22	18	15	16	13	16
Other capital transfers (project grants)	FJMB	149	143	169	182	171	225	237	263	345	390	408
Total central government	FJME	177	166	193	328	193	247	255	278	361	403	424
Other sectors												
Migrants' transfers	FJMH	453	465	592	531	499	461	1 300	582	547	515	551
Debt forgiveness												
Monetary financial institutions[1]	FJNF	–	–	–	–	–	–	–	–	–	–	–
Public corporations[2]	HMLY	–	–	24	27	49	55	188	236	130	109	119
Total debt forgiveness	IZZZ	–	–	24	27	49	55	188	236	130	109	119
Other capital transfers	FJMS	–	–	–	–	–	–	–	–	–	–	–
Total other sectors	FJMV	453	465	616	558	548	516	1 488	818	677	624	670
Total capital transfers	FJMY	630	631	809	886	741	763	1 743	1 096	1 038	1 027	1 094
Purchases of non-produced, non-financial assets	FJUY	–	35	102	137	140	141	274	304	289	512	682
Total	FKMI	**630**	**666**	**911**	**1 023**	**881**	**904**	**2 017**	**1 400**	**1 327**	**1 539**	**1 776**
Balances												
Capital transfers												
Central government												
Debt forgiveness	FJUW	−28	−23	−24	−146	−22	−22	−18	−15	−16	−13	−16
Other capital transfers	FJMC	−149	−143	−169	−182	−171	−225	−237	−263	−345	−390	−408
Total central government	FJMF	−177	−166	−193	−328	−193	−247	−255	−278	−361	−403	−424
Other sectors												
Migrants' transfers	FJMI	225	238	162	436	645	910	967	1 282	1 404	1 783	1 940
Debt forgiveness	FJNG	–	–	−24	−27	−49	−55	−188	−236	−130	−109	−119
Other capital transfers	FJMT	485	1 174	1 047	456	332	1 071	891	296	624	1 111	1 473
Total other sectors	FJMW	710	1 412	1 185	865	928	1 926	1 670	1 342	1 898	2 785	3 294
Total capital transfers	FJMZ	533	1 246	992	537	735	1 679	1 415	1 064	1 537	2 382	2 870
Non-produced, non-financial assets	NHSG	–	14	−34	−48	12	24	−97	−132	−71	−319	−468
Total	FKMJ	**533**	**1 260**	**958**	**489**	**747**	**1 703**	**1 318**	**932**	**1 466**	**2 063**	**2 402**

1 This series also appears in the Financial Account (see Table 7.7).
2 This series also appears in the Financial Account (see Table 7.7) as series HMLW.

Chapter 7
Financial account

Summary

Investment abroad and into the UK both increased dramatically from the mid-1990s, reflecting the increased globalisation of the world economy. From 1998 to 2000 the increase was driven by global merger and acquisition activity. Since then, portfolio investment and banking activity have dominated cross-border investment.

In recent years the United Kingdom has needed to borrow from abroad to finance a continuing current account deficit, which has resulted in inward investment (UK liabilities) exceeding outward investment (UK assets).

In 2005, direct investment in the UK exceeded direct investment abroad for the first time since 1990; this was mainly due to the restructuring of The 'Shell' Transport and Trading Company Plc and Royal Dutch Petroleum Company into Royal Dutch Shell, which is treated as a Dutch company for balance of payments purposes. The historical pattern of portfolio investment has also been to record net investment abroad, although this pattern has been distorted by the attractiveness of UK debt securities to foreign investors in recent years. In 1999 and 2000 high investment in UK equity resulted from substantial UK direct investment acquisitions in foreign telecom and pharmaceutical companies, which were funded by the issue of UK shares to foreign shareholders; this is recorded as portfolio investment in the UK. Other investment is the largest and most volatile form of investment. The amounts recorded in the gross flows of loans and deposits are as much

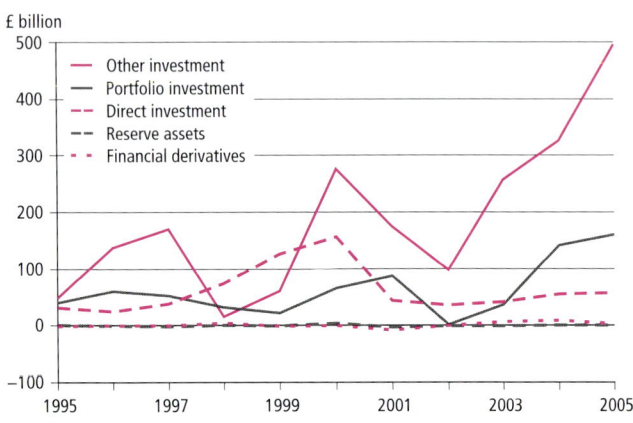

Figure 7.2

UK investment abroad
Credits less debits

a consequence of how the transaction is carried out between resident and non-resident banks, as overall market conditions.

UK investment abroad reached a record £715.6 billion in 2005, with record outward portfolio and other investment. Direct investment abroad increased to its highest level since 2000, but is still substantially lower than the levels seen during the mergers and acquisition boom of the late 1990s. Inward investment showed a similar pattern to outward investment, reaching a record £733.1 billion in 2005. Direct and other investment in the UK were at record levels, while portfolio investment in the UK was at its highest since 2000.

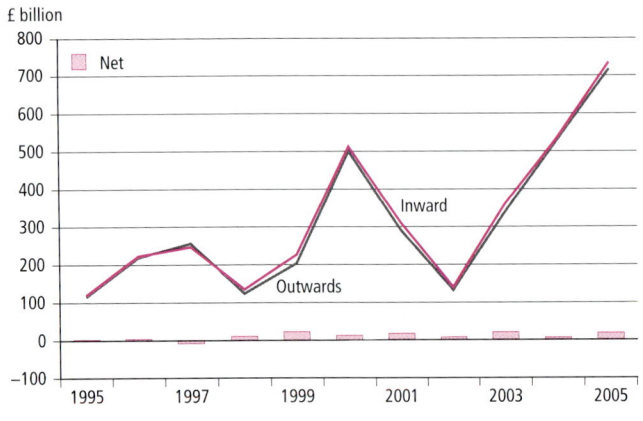

Figure 7.1

Financial account

Figure 7.3

Foreign investment in the UK

Direct investment

Outward direct investment peaked at £155.6 billion in 2000, reflecting booming merger and acquisition activity – the largest outward acquisitions were the investment in Mannesmann AG by Vodafone Airtouch for a reported £100 billion and the purchase of Atlantic Richfield Company by BP Amoco plc for a reported £18 billion. Outward direct investment then declined to £35.0 billion in 2002, before rising in each of the following years to reach £56.6 billion in 2005. Outward merger and acquisition activity increased for the second consecutive year with 365 acquisitions and 110 disposals but activity was still well down on the 1999 high of 590 acquisitions and 198 disposals. These transactions are reflected in the equity capital component of direct investment abroad. The major component of outward investment in recent years has been reinvested earnings rather than equity capital.

Until 2004, inward direct investment showed a pattern similar to outward investment, with direct investment in the UK reaching £80.6 billion in 2000, followed by lower levels of investment due to the slowdown in global merger and acquisition activity. In 2004 and 2005 however, there have been a number of inward acquisitions, including the purchase of Abbey National by Banco Santander and the Shell restructuring in respective years. The latter being one of the main factors behind the record inward direct investment in 2005 of £87.7 billion.

Figure 7.4
Direct investment

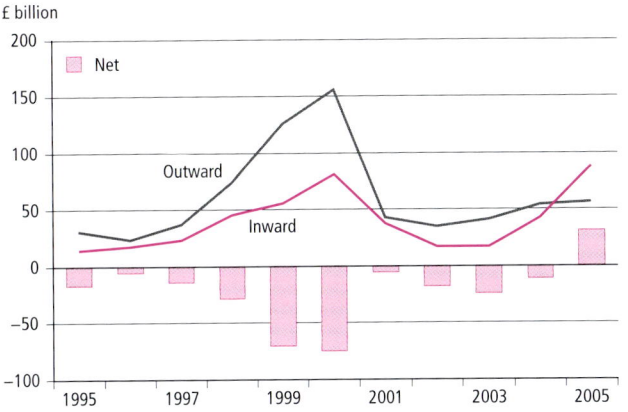

Portfolio investment

Portfolio investment abroad has shown net investment in every year since 1994, reaching a record £160.7 billion in 2005. Generally investment in foreign debt exceeds investment in foreign equities. In 2005, there was record investment in both foreign equities and foreign debt, with banks buying £36.5 billion of foreign equities and £64.7 billion of foreign debt securities. Net disposals of foreign equity securities occurred in three years coinciding with financial shocks: the UK's exit from

Figure 7.5
Portfolio investment

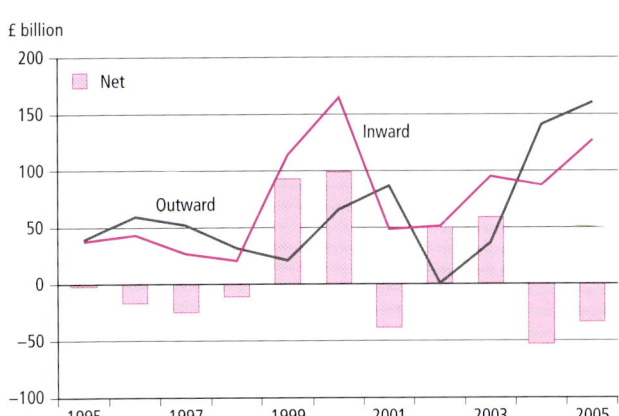

the Exchange Rate Mechanism in 1992; the South-East Asia crisis in 1997; and the collapse in equity markets in 2002. There has been inward portfolio investment in every year data are available. In the early 1990s, the majority of inward investment was in bonds and notes. This switched to UK issued equity in the late 1990s as the counterpart to the outward direct investment occurring then. Since 2002, there has been strong net investment in UK debt securities. The attractiveness of UK debt to foreign investors may reflect higher interest rates in the UK compared to other major economies, and a switch from dollar to sterling issued debt due to the fall in the value of the dollar over this period.

Other investment

In most recent years, loans and deposits by UK banks constitute the major component of other investment. Loans and deposits by UK banks are carried out predominantly in foreign currency, so will be partly influenced by relative exchange rates and interest rates as well as the global financial conditions generally. In 2005, UK banks made deposits in foreign currency of

Figure 7.6
Other investment

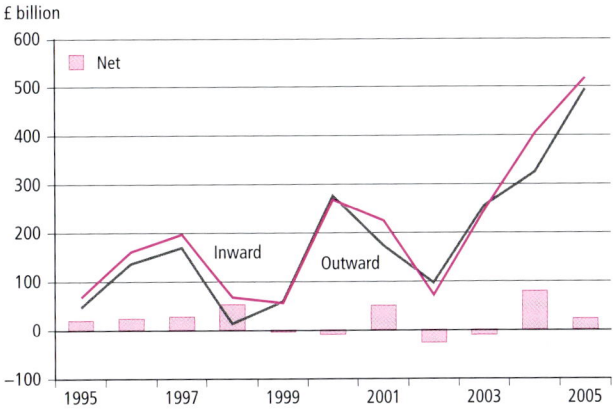

£125.2 billion and made loans in foreign currency of £115.0 billion. Other investment in the UK is largely composed of deposits with UK banks and short-term loans to securities dealers and other sectors (principally private non-financial corporations). In 2005, net foreign currency deposits with UK banks amounted to £232.6 billion and short-term loans to securities dealers amounted to £223.8 billion.

Sectoral breakdown of the financial account

In 2005, UK banks' reported net investment abroad of £87.6 billion, the first time they have reported net outward investment since 1998. This turnaround was mainly due to increased loans to and deposits with non-residents. On the other hand, other UK sectors (mainly other financial intermediaries and private non-financial corporations) showed net inward investment in 2005 of £75.5 billion, the first net inward investment since 2000. This was due to inward direct investment activity and an increase in short-term loans.

7.1 Financial account
Summary table

£ million

		1995	1996	1997	1998	1999	2000	2001	2002	2003	2004	2005
UK investment abroad												
(UK assets = net debits)												
Direct investment abroad												
Equity capital	-HJYM	10 491	6 070	20 358	47 640	101 897	147 679	16 890	26 155	20 639	18 469	16 672
Reinvested earnings	-HDNY	14 378	17 271	16 112	14 071	21 392	25 178	27 220	32 209	21 456	32 430	40 597
Other capital transactions	-HMAB	6 242	175	832	12 075	2 313	–17 275	–1 283	–23 323	–1 206	2 932	–730
Total direct investment abroad	-HJYP	31 111	23 516	37 302	73 786	125 602	155 582	42 827	35 041	40 889	53 831	56 539
Portfolio investment abroad												
Equity securities	-HBVI	8 386	10 361	–4 202	2 713	14 455	20 521	44 464	–3 189	19 684	56 647	64 766
Debt securities	-XBMW	30 888	49 399	56 143	29 360	6 935	45 042	42 087	4 200	16 583	84 206	95 944
Total portfolio investment abroad	-HHZC	39 274	59 760	51 941	32 073	21 390	65 563	86 551	1 011	36 267	140 853	160 710
Financial derivatives (net)	-ZPNN	–1 667	–963	–1 156	3 043	–2 685	–1 553	–8 417	–1 001	5 401	7 875	2 451
Other investment abroad	-XBMM	47 498	136 669	169 420	14 887	59 557	276 028	174 086	97 185	255 863	325 588	495 261
Reserve assets	-LTCV	–200	–510	–2 380	–164	–639	3 915	–3 085	–459	–1 559	196	656
Total	-HBNR	**116 016**	**218 472**	**255 127**	**123 625**	**203 225**	**499 535**	**291 962**	**131 777**	**336 861**	**528 343**	**715 617**
Investment in the UK												
(UK liabilities = net credits)												
Direct investment in the UK												
Equity capital	HJYR	12 756	8 156	11 599	27 895	46 709	59 811	20 954	11 809	4 464	24 400	76 531
Reinvested earnings	CYFV	5 254	7 873	6 386	1 522	4 607	10 788	–992	3 647	7 429	9 320	11 095
Other capital transactions	HMAD	–4 179	1 535	4 915	15 637	3 750	9 967	17 386	1 326	4 883	8 696	99
Total direct investment in the UK	HJYU	13 831	17 564	22 900	45 054	55 066	80 566	37 348	16 782	16 776	42 416	87 725
Portfolio investment in the UK												
Equity securities	XBLW	5 111	6 073	4 793	37 785	72 127	113 593	23 065	3 164	9 738	–8 345	2 670
Debt securities	XBLX	32 204	36 923	21 993	–16 932	41 979	50 950	25 083	47 846	85 484	95 592	124 351
Total portfolio investment in the UK	HHZF	37 315	42 996	26 786	20 853	114 106	164 543	48 148	51 010	95 222	87 247	127 021
Other investment in the UK	XBMN	67 422	160 723	196 670	67 640	55 469	267 030	223 969	71 187	245 370	404 321	518 347
Total	HBNS	**118 568**	**221 283**	**246 356**	**133 547**	**224 641**	**512 139**	**309 465**	**138 979**	**357 368**	**533 984**	**733 093**
Net transactions												
(net credits *less* net debits)												
Direct investment												
Equity capital	HBWN	2 265	2 086	–8 759	–19 745	–55 188	–87 868	4 064	–14 346	–16 175	5 931	59 859
Reinvested earnings	HBWT	–9 124	–9 398	–9 726	–12 549	–16 785	–14 390	–28 212	–28 562	–14 027	–23 110	–29 502
Other capital transactions	HBWU	–10 421	1 360	4 083	3 562	1 437	27 242	18 669	24 649	6 089	5 764	829
Total net direct investment	HJYV	–17 280	–5 952	–14 402	–28 732	–70 536	–75 016	–5 479	–18 259	–24 113	–11 415	31 186
Portfolio investment												
Equity securities	HBWV	–3 275	–4 288	8 995	35 072	57 672	93 072	–21 399	6 353	–9 946	–64 992	–62 096
Debt securities	HBWX	1 316	–12 476	–34 150	–46 292	35 044	5 908	–17 004	43 646	68 901	11 386	28 407
Total net portfolio investment	HHZD	–1 959	–16 764	–25 155	–11 220	92 716	98 980	–38 403	49 999	58 955	–53 606	–33 689
Financial derivatives	ZPNN	1 667	963	1 156	–3 043	2 685	1 553	8 417	1 001	–5 401	–7 875	–2 451
Other investment	HHYR	19 924	24 054	27 250	52 753	–4 088	–8 998	49 883	–25 998	–10 493	78 733	23 086
Reserve assets	LTCV	200	510	2 380	164	639	–3 915	3 085	459	1 559	–196	–656
Total	HBNT	**2 552**	**2 811**	**–8 771**	**9 922**	**21 416**	**12 604**	**17 503**	**7 202**	**20 507**	**5 641**	**17 476**

7.2 Financial account
Summary table

£ million

		1995	1996	1997	1998	1999	2000	2001	2002	2003	2004	2005
UK investment abroad (UK assets = net debits)												
By:													
Monetary financial institutions												
Banks	-HFAM	48 911	85 819	165 973	75 467	6 598	229 288	123 341	60 188	181 157	326 077	408 260
Building societies	HEQN	529	−369	2 557	1 334	949	4 382	1 476	−903	−1 786	706	38
Total monetary financial institutions	-HFAQ	49 440	85 450	168 530	76 801	7 547	233 670	124 817	59 285	179 371	326 783	408 298
Central government	-HFAN	269	1 387	−2 263	−245	−311	4 251	−2 620	610	−2 325	511	988
Public corporations	-HFAO	163	92	91	−7	304	582	62	−164	−571	−180	−572
Other sectors	-HFAP	66 144	131 543	88 769	47 076	195 685	261 032	169 703	72 046	160 386	201 229	306 903
Total	-HBNR	**116 016**	**218 472**	**255 127**	**123 625**	**203 225**	**499 535**	**291 962**	**131 777**	**336 861**	**528 343**	**715 617**
Investment in the UK (UK liabilities = net credits)												
In:												
Monetary financial institutions (banks and building societies)	CGUL	48 296	94 858	167 280	31 242	36 026	243 355	151 682	116 120	201 950	360 036	320 758
Central government	HFAR	719	5 301	−4 315	439	−4 434	−244	−96	−4 534	13 657	13 539	29 724
Local authorities	HFAS	−51	13	−53	−87	−106	−188	22	26	204	633	200
Public corporations	HFAT	−151	−14	−206	−5	–	–	–	–	–	–	–
Other sectors	GGCJ	69 755	121 125	83 650	101 958	193 155	269 216	157 857	27 367	141 557	159 776	382 411
Total	HBNS	**118 568**	**221 283**	**246 356**	**133 547**	**224 641**	**512 139**	**309 465**	**138 979**	**357 368**	**533 984**	**733 093**
Net transactions (net credits *less* net debits)												
In assets and liabilities of:												
Monetary financial institutions (banks and building societies)	GGCK	−1 144	9 408	−1 250	−45 559	28 479	9 685	26 865	56 835	22 579	33 253	−87 540
Central government	HFAV	450	3 914	−2 052	684	−4 123	−4 495	2 524	−5 144	15 982	13 028	28 736
Local authorities	HFAS	−51	13	−53	−87	−106	−188	22	26	204	633	200
Public corporations	HFAW	−314	−106	−297	2	−304	−582	−62	164	571	180	572
Other sectors	GGCL	3 611	−10 418	−5 119	54 882	−2 530	8 184	−11 846	−44 679	−18 829	−41 453	75 508
Total	HBNT	**2 552**	**2 811**	**−8 771**	**9 922**	**21 416**	**12 604**	**17 503**	**7 202**	**20 507**	**5 641**	**17 476**

7.3 Direct investment

£ million

		1995	1996	1997	1998	1999	2000	2001	2002	2003	2004	2005
Direct investment abroad (UK assets = net debits)												
Equity capital												
Claims on affiliated enterprises (net acquisition of ordinary shares)												
Purchases of ordinary shares	-HDOA	11 655	12 703	25 476	60 627	114 693	181 488	40 221	35 374	26 128	26 942	22 994
Sales of ordinary shares	-HDOC	–3 840	–7 672	–4 041	–13 677	–13 620	–34 693	–25 073	–10 440	–8 479	–10 325	–8 093
Total claims on affiliated enterprises	-HJYL	7 815	5 031	21 435	46 950	101 073	146 795	15 148	24 934	17 649	16 617	14 901
Net acquisition of property	-HHVG	2 676	1 039	–1 077	690	824	884	1 742	1 221	2 990	1 852	1 771
Total equity capital	-HJYM	10 491	6 070	20 358	47 640	101 897	147 679	16 890	26 155	20 639	18 469	16 672
Reinvested earnings	-HDNY	14 378	17 271	16 112	14 071	21 392	25 178	27 220	32 209	21 456	32 430	40 597
Other capital transactions [1]												
Claims on affiliated enterprises												
Debt securities issued by affiliated enterprises												
Purchases of debt securities	-HDOD	1 175	89	529	396	636	952	2 263	513	1 598	1 903	812
Sales of debt securities	-HDOE	–	–52	–117	–315	–578	–496	–304	–1 080	–2 312	–608	–946
Other claims on affiliated enterprises												
Change in inter-company accounts	-HDOF	4 631	2 506	5 040	20 721	15 806	15 110	5 072	17 140	10 178	20 476	8 183
Change in branch indebtedness	-HDOI	669	500	1 053	1 493	–483	–3 360	5 153	–610	1 783	–874	–4 786
Total claims on affiliated enterprises	-HJYN	6 475	3 043	6 505	22 295	15 381	12 206	12 184	15 963	11 247	20 897	3 263
Liabilities to affiliated enterprises												
Change in inter-company accounts	-HDOG	10	3 292	5 299	8 453	14 340	28 278	12 880	38 774	10 568	20 103	4 097
Change in branch indebtedness	-HDOJ	223	–424	374	1 767	–1 272	1 203	587	512	1 885	–2 138	–104
Total liabilities to affiliated enterprises	-HJYO	233	2 868	5 673	10 220	13 068	29 481	13 467	39 286	12 453	17 965	3 993
Total other capital transactions	-HMAB	6 242	175	832	12 075	2 313	–17 275	–1 283	–23 323	–1 206	2 932	–730
Total	-HJYP	31 111	23 516	37 302	73 786	125 602	155 582	42 827	35 041	40 889	53 831	56 539
Direct investment in the UK (UK liabilities = net credits)												
Equity capital												
Liabilities to direct investors												
Quoted ordinary shares												
Purchases of quoted ordinary shares	CYFY	4 255	6 510	7 434	24 660	40 393	16 253	3 502	5 951	1 739	15 380	62 698
Sales of quoted ordinary shares	CYFZ	–191	–1 206	–1 293	–4 336	–10 526	–2 038	–1 185	–775	–1 200	–	–569
Unquoted ordinary shares												
Purchases of unquoted ordinary shares	CYGA	11 755	3 039	5 055	7 147	20 721	48 154	20 381	11 068	4 238	10 645	18 317
Sales of unquoted ordinary shares	CYGB	–3 287	–800	–447	–274	–4 692	–4 187	–2 535	–5 183	–708	–2 248	–4 512
Total liabilities to direct investors	HJYQ	12 532	7 543	10 749	27 197	45 896	58 182	20 163	11 061	4 069	23 777	75 934
Net acquisition of property	CGLO	224	613	850	698	813	1 629	791	748	395	623	597
Total equity capital	HJYR	12 756	8 156	11 599	27 895	46 709	59 811	20 954	11 809	4 464	24 400	76 531
Reinvested earnings	CYFV	5 254	7 873	6 386	1 522	4 607	10 788	–992	3 647	7 429	9 320	11 095
Other capital transactions [1]												
Liabilities to direct investors												
Debt securities issued by affiliated enterprises												
Purchases of debt securities	CYGC	540	3	1 516	783	558	710	1 318	598	1 844	3 531	1 992
Sales of debt securities	CYGD	–	–	–22	–183	–567	–183	–571	–377	–484	–1 146	–339
Other liabilities to direct investors												
Change in inter-company accounts	CYGH	–315	1 915	5 571	25 700	17 253	11 338	17 420	10 756	–1 264	307	–1 469
Change in branch indebtedness	CYGL	42	112	629	392	–210	869	285	403	1 738	467	724
Total liabilities to direct investors	HJYT	267	2 030	7 694	26 692	17 034	12 734	18 452	11 380	1 834	3 159	908
Claims on direct investors												
Change in inter-company accounts	CYGF	4 441	423	3 575	11 199	13 266	2 495	561	9 990	–2 112	–5 324	658
Change in branch indebtedness	CYGK	5	72	–796	–144	18	272	505	64	–937	–213	151
Total claims on direct investors	HJYS	4 446	495	2 779	11 055	13 284	2 767	1 066	10 054	–3 049	–5 537	809
Total other capital transactions	HMAD	–4 179	1 535	4 915	15 637	3 750	9 967	17 386	1 326	4 883	8 696	99
Total	HJYU	13 831	17 564	22 900	45 054	55 066	80 566	37 348	16 782	16 776	42 416	87 725

1 From Pink Book 2005 the presentation of Other capital transactions no longer mirror each other between UK assets and liabilities: both are now shown from the perspective of the direct investor.

7.3 Direct investment
continued

£ million

		1995	1996	1997	1998	1999	2000	2001	2002	2003	2004	2005
Net transactions (net credits less net debits)												
Equity capital												
Net acquisition of ordinary shares	LTMS	4 717	2 512	−10 686	−19 753	−55 177	−88 613	5 015	−13 873	−13 580	7 160	61 033
Net acquisition of property	LTMT	−2 452	−426	1 927	8	−11	745	−951	−473	−2 595	−1 229	−1 174
Total equity capital	HBWN	2 265	2 086	−8 759	−19 745	−55 188	−87 868	4 064	−14 346	−16 175	5 931	59 859
Reinvested earnings	HBWT	−9 124	−9 398	−9 726	−12 549	−16 785	−14 390	−28 212	−28 562	−14 027	−23 110	−29 502
Other capital transactions	HBWU	−10 421	1 360	4 083	3 562	1 437	27 242	18 669	24 649	6 089	5 764	829
Total	HJYV	**−17 280**	**−5 952**	**−14 402**	**−28 732**	**−70 536**	**−75 016**	**−5 479**	**−18 259**	**−24 113**	**−11 415**	**31 186**

7.4 Direct investment
Sector analysis

£ million

		1995	1996	1997	1998	1999	2000	2001	2002	2003	2004	2005
Direct investment abroad (UK assets = net debits)												
By:												
UK Monetary financial institutions (banks)	-HCWJ	1 820	1 444	169	971	1 028	3 378	4 283	2 825	1 942	16 231	9 639
Insurance companies	-CNZE	2 343	506	3 137	969	−2 135	2 166	−256	1 388	3 038	4 436	767
Other financial intermediaries	-HCXL	1 977	2 631	5 711	11 676	8 469	9 716	4 468	4 071	13 858	−1 543	3 316
Private non-financial corporations	-HCVH	22 463	17 684	29 132	59 663	117 330	139 101	32 575	25 859	18 598	33 216	41 070
Public corporations	-HDND	64	−9	68	20	280	574	201	258	−185	158	158
Household sector[1]	-AAQN	2 444	1 260	−915	487	630	647	1 556	640	3 638	1 333	1 589
Total	-HJYP	**31 111**	**23 516**	**37 302**	**73 786**	**125 602**	**155 582**	**42 827**	**35 041**	**40 889**	**53 831**	**56 539**
Direct investment in the UK (UK liabilities = net credits)												
In:												
Monetary financial institutions (banks)	GPBQ	2 092	2 458	3 494	678	1 616	4 133	3 387	1 757	2 683	11 744	3 876
Insurance companies	HDQI	−144	1 615	891	−138	1 763	2 492	1 304	312	876	1 659	541
Other financial intermediaries												
Securities dealers	HDRU	356	905	338	−1 188	836	1 919	938	706	212	1 552	2 086
Other	HFCL	1 453	87	−34	9 865	−232	5 792	8 098	3 298	5 395	−3 840	−3 590
Total other financial intermediaries	HFDR	1 809	992	304	8 677	604	7 711	9 036	4 004	5 607	−2 288	−1 504
Private non-financial corporations	BCEC	10 074	12 499	18 211	35 837	51 083	66 230	23 621	10 709	7 610	31 301	84 812
Total	HJYU	**13 831**	**17 564**	**22 900**	**45 054**	**55 066**	**80 566**	**37 348**	**16 782**	**16 776**	**42 416**	**87 725**
Net transaction (net credits less net debits)												
In assets and liabilities of:												
Monetary financial institutions	LTMU	272	1 014	3 325	−293	588	755	−896	−1 068	741	−4 487	−5 763
Insurance companies	LTMV	−2 487	1 109	−2 246	−1 107	3 898	326	1 560	−1 076	−2 162	−2 777	−226
Other financial intermediaries	LTMW	−168	−1 639	−5 407	−2 999	−7 865	−2 005	4 568	−67	−8 251	−745	−4 820
Private non-financial corporations	LTMX	−12 389	−5 185	−10 921	−23 826	−66 247	−72 871	−8 954	−15 150	−10 988	−1 915	43 742
Public corporations	HDND	−64	9	−68	−20	−280	−574	−201	−258	185	−158	−158
Household sector[1]	AAQN	−2 444	−1 260	915	−487	−630	−647	−1 556	−640	−3 638	−1 333	−1 589
Total	HJYV	**−17 280**	**−5 952**	**−14 402**	**−28 732**	**−70 536**	**−75 016**	**−5 479**	**−18 259**	**−24 113**	**−11 415**	**31 186**

1 The household sector includes non-profit institutions serving households.

7.5 Portfolio investment

£ million

		1995	1996	1997	1998	1999	2000	2001	2002	2003	2004	2005
Portfolio investment abroad (UK assets = net debits)												
Transactions in equity securities (shares) by:												
Monetary financial Institutions (banks)	-VTWC	161	4 570	–3 138	4 549	100	7 195	–1 287	–11 767	18 824	31 597	36 515
Central Government	LOEQ	–	–	–	–	–	–	–	–	–	–	20
Insurance companies and pension funds												
Insurance companies	-HBHM	3 688	3 147	1 335	1 015	3 111	–4 297	6 520	2 959	–3 354	6 116	19 087
Pension funds[1]	-HBHO	–1 884	2 828	–3 326	2 073	–518	–12 798	11 720	15 256	4 394	6 491	8 823
Total insurance companies and pension funds	-HBRD	1 804	5 975	–1 991	3 088	2 593	–17 095	18 240	18 215	1 040	12 607	27 910
Other financial intermediaries												
Securities dealers	-HGLG	3 600	–3 058	810	–7 634	5 783	13 673	24 128	–12 050	–796	9 734	–15 115
Unit and Investment Trusts	-HBHQ	3 178	2 724	919	3 567	6 468	9 968	3 913	3 329	2 121	1 174	7 305
Other	-HBRC	–290	–456	–831	–833	–1 300	–1 446	–1 077	–856	–1 563	–1 621	–1 668
Total other financial intermediaries	-HBRE	6 488	–790	898	–4 900	10 951	22 195	26 964	–9 577	–238	9 287	–9 478
Private non-financial corporations	-XBNL	–222	188	–62	84	241	9 047	444	–52	17	–380	–90
Household sector[2]	HALH	155	418	91	–108	570	–821	103	–8	41	3 536	9 889
Total transactions in equity securities	-HBVI	8 386	10 361	–4 202	2 713	14 455	20 521	44 464	–3 189	19 684	56 647	64 766
Transactions in debt securities												
Transactions in bonds and notes by:												
Monetary financial institutions												
Banks	-VTWA	23 672	19 034	17 442	43 090	11 011	34 007	37 604	3 774	–11 215	57 131	62 455
Building societies	RYWJ	496	67	691	1 417	1 099	2 464	854	–338	–1 498	767	263
Total monetary financial institutions	-HPCP	24 168	19 101	18 133	44 507	12 110	36 471	38 458	3 436	–12 713	57 898	62 718
Central Government	MDZJ	–	–	–	–	–	–	–	–	–	–	–
Insurance companies and pension funds												
Insurance companies	-HBHN	1 052	4 096	3 614	11 615	7 103	5 363	8 200	8 535	1 618	1 522	4 451
Pension funds[1]	-HBHP	732	1 650	4 696	3 581	2 933	5 875	1 267	–3 604	1 732	3 980	3 648
Total insurance companies and pension funds	-HBRF	1 784	5 746	8 310	15 196	10 036	11 238	9 467	4 931	3 350	5 502	8 099
Other financial intermediaries												
Securities dealers	CGFO	3 039	26 584	22 318	–33 645	–28 883	–1 935	–19 589	–1 114	9 912	21 829	21 546
Unit and investment trusts	-HBHR	–133	351	195	1 452	1 121	664	1 478	720	2 445	1 531	1 430
Other	-HBRG	–22	–35	45	–154	–38	–36	–57	–72	–76	–101	–125
Total other financial intermediaries	-HBRH	2 884	26 900	22 558	–32 347	–27 800	–1 307	–18 168	–466	12 281	23 259	22 851
Private non-financial corporations	-XBNM	–135	840	–2 370	553	–1 299	1 179	566	300	1 292	197	–721
Household sector[2]	HBRI	–556	–1 586	186	184	–380	256	88	88	88	88	88
Total transactions in bonds and notes	-HEPK	28 145	51 001	46 817	28 093	–7 333	47 837	30 411	8 289	4 298	86 944	93 035
Transactions in Money Market Instruments												
Transactions in commercial paper by:												
Monetary financial institutions:												
Banks	-HBXH	2 233	–3 547	7 295	4 112	9 729	–963	6 700	–3 980	7 583	–4 470	2 274
Building societies	TAIH	–339	14	254	–169	66	899	635	–564	–191	99	–25
Central government	-RUUR	–	–	–	–	–	–	458	467	–925	–1	–
Insurance companies and pension funds	-HBVK	178	292	617	–1 558	243	–106	–159	333	70	602	34
Other financial intermediaries	-HGIS	420	1 206	611	–815	504	–2 077	2 505	–602	2 579	615	–861
Private non-financial corporations	-HBRL	–2	438	279	–956	722	1 110	1 912	1 110	3 798	615	1 661
Total transactions in commercial paper	-HGLU	2 490	–1 597	9 056	614	11 264	–1 137	12 051	–3 236	12 914	–2 540	3 083
Transactions in certificates of deposit by:												
Monetary financial institutions (Building societies)	TAIF	–25	2	261	210	–71	409	37	563	39	–248	–106
Other financial intermediaries	-RZUV	278	–7	9	443	3 075	–2 067	–412	–1 416	–668	50	–68
Total transactions in certificates of deposit	HEPH	253	–5	270	653	3 004	–1 658	–375	–853	–629	–198	–174
Total transactions in Money Market Instruments	-HHZM	2 743	–1 602	9 326	1 267	14 268	–2 795	11 676	–4 089	12 285	–2 738	2 909
Total transactions in debt securities	-XBMW	30 888	49 399	56 143	29 360	6 935	45 042	42 087	4 200	16 583	84 206	95 944
Total	-HHZC	**39 274**	**59 760**	**51 941**	**32 073**	**21 390**	**65 563**	**86 551**	**1 011**	**36 267**	**140 853**	**160 710**

1 The pension funds data only covers self-administered funds, see glossary.   2 The household sector includes non-profit institutions serving households.

7.5 Portfolio investment
continued

£ million

		1995	1996	1997	1998	1999	2000	2001	2002	2003	2004	2005
Portfolio investment in the UK (UK liabilities = net credits)												
Transactions in equity securities (shares) issued by:													
Monetary financial Institutions (banks and building societies)	HBQG	471	1 477	1 939	−5 798	−2 735	−2 901	135	−1 167	588	−329	16
Other sectors[1]	HBQH	4 640	4 596	2 854	43 583	74 862	116 494	22 930	4 331	9 150	−8 016	2 654
Total transactions in equity securities	XBLW	5 111	6 073	4 793	37 785	72 127	113 593	23 065	3 164	9 738	−8 345	2 670
Transactions in debt securities												
Transactions in bonds and notes												
Issues by central government												
UK foreign currency bonds and notes	HEZP	101	−1 632	−3 058	−1 660	241		988	−3 342	−2 811	886	38	−32
Other central government bonds	HHJM	–	–	–	–	–	–	–	–	–	–	–
Transactions in British government stocks (gilts) by:												
Foreign central banks (exchange reserves)	AING	−250	261	−1 586	1 692	489	1 049	1 157	−1 245	−748	−2 339	384
Other foreign residents	VTWG	−879	7 604	2 244	1 802	−6 017	−2 338	1 512	424	11 059	14 920	30 330
Total transactions in British government stocks	HEPC	−1 129	7 865	658	3 494	−5 528	−1 289	2 669	−821	10 311	12 581	30 714
Total issues by central government	HBRX	−1 028	6 233	−2 400	1 834	−5 287	−301	−673	−3 632	11 197	12 619	30 682
Local authorities' bonds	HBQT	–	–	–	–	–	–	–	–	–	–	–
Public corporations' bonds	HCEW	−7	–	–	–	–	–	–	–	–	–	–
Issues by monetary financial Institutions (banks and building societies)												
Bonds	HBRY	3 233	2 863	3 158	−1 163	6 574	1 905	506	4 887	15 129	13 214	19 373
European medium term notes and other medium-term paper:												
Issued by UK banks	HCEZ	1 572	5 585	3 137	1 881	4 244	891	3 425	1 706	12 117	16 525	19 240
Issued by UK building societies	HCFC	−399	−315	−116	−140	252	1 814	630	69	1 754	2 222	3 498
Total	HBRV	1 173	5 270	3 021	1 741	4 496	2 705	4 055	1 775	13 871	18 747	22 738
Total monetary financial institutions	HMBD	4 406	8 133	6 179	578	11 070	4 610	4 561	6 662	29 000	31 961	42 111
Issues by other sectors[1]	HBRT	10 063	9 466	9 835	−3 622	20 465	5 928	1 574	15 213	47 202	40 776	60 314
Total transactions in bonds and notes	XBLY	13 434	23 832	13 614	−1 210	26 248	10 237	5 462	18 243	87 399	85 356	133 107
Transactions in Money Market Instruments												
Transactions in treasury bills (issued by central government)												
Sterling treasury bills	AARB	853	−663	−183	−820	637	−251	304	−180	2 150	1 973	−1 007
Euro treasury bills	HHNW	471	425	−729	−913	−227	–	–	–	–	–	–
Total treasury bills	HHZO	1 324	−238	−912	−1 733	410	−251	304	−180	2 150	1 973	−1 007
Transactions in certificates of deposit (issued by UK monetary financial institutions)												
Issued by banks	HBRS	12 718	9 906	5 547	−16 985	11 500	34 653	19 911	4 080	−3 986	−1 359	−770
Issued by building societies	HBHH	–	23	157	−25	−6	301	−50	264	952	529	−1 067
Total certificates of deposit	HBQX	12 718	9 929	5 704	−17 010	11 494	34 954	19 861	4 344	−3 034	−830	−1 837
Transactions in commercial paper												
Issued by UK monetary financial Institutions												
Banks	HBHI	708	2 174	1 800	257	296	2 542	−599	14 950	−33	9 093	−3 567
Building societies	HBHL	2 768	−643	204	335	1 748	768	−182	−330	3 325	−259	556
Total monetary financial institutions	HBRU	3 476	1 531	2 004	592	2 044	3 310	−781	14 620	3 292	8 834	−3 011
Issued by other sectors[1]	HHZN	1 252	1 869	1 583	2 429	1 783	2 700	237	10 819	−4 323	259	−2 901
Total transactions in commercial paper	HBQW	4 728	3 400	3 587	3 021	3 827	6 010	−544	25 439	−1 031	9 093	−5 912
Total transactions in Money Market Instruments	HHZE	18 770	13 091	8 379	−15 722	15 731	40 713	19 621	29 603	−1 915	10 236	−8 756
Total transactions in debt securities	XBLX	32 204	36 923	21 993	−16 932	41 979	50 950	25 083	47 846	85 484	95 592	124 351
Total	HHZF	37 315	42 996	26 786	20 853	114 106	164 543	48 148	51 010	95 222	87 247	127 021

1 These series relate to non-governmental sectors other than monetary financial institutions.

7.5 Portfolio investment
continued

£ million

		1995	1996	1997	1998	1999	2000	2001	2002	2003	2004	2005
Net transactions (net credits less net debits)												
Equity securities (shares)	HBWV	–3 275	–4 288	8 995	35 072	57 672	93 072	–21 399	6 353	–9 946	–64 992	–62 096
Debt securities												
Bonds and notes	LTMY	–14 711	–27 169	–33 203	–29 303	33 581	–37 600	–24 949	9 954	83 101	–1 588	40 072
Money Market Instruments	LTMZ	16 027	14 693	–947	–16 989	1 463	43 508	7 945	33 692	–14 200	12 974	–11 665
Total debt securities	HBWX	1 316	–12 476	–34 150	–46 292	35 044	5 908	–17 004	43 646	68 901	11 386	28 407
Total	HHZD	–1 959	–16 764	–25 155	–11 220	92 716	98 980	–38 403	49 999	58 955	–53 606	–33 689

7.6 Portfolio investment
Sector analysis

£ million

		1995	1996	1997	1998	1999	2000	2001	2002	2003	2004	2005
Portfolio investment abroad (UK assets = net debits)												
Investment by:												
Monetary financial institutions												
Banks	-HBWF	26 066	20 057	21 599	51 751	20 840	40 239	43 017	–11 973	15 192	84 258	101 244
Building societies	HEPI	132	83	1 206	1 458	1 094	3 772	1 526	–339	–1 650	618	132
Total monetary financial institutions	-HBRJ	26 198	20 140	22 805	53 209	21 934	44 011	44 543	–12 312	13 542	84 876	101 376
Central government	LOFB	–	–	–	–	–	–	458	467	–925	–1	20
Insurance companies and pension funds	-HBRO	3 766	12 013	6 936	16 726	12 872	–5 963	27 548	23 479	4 460	18 711	36 043
Other financial intermediaries	-HBRP	10 070	27 309	24 076	–37 619	–13 270	16 744	10 889	–12 061	13 954	33 211	12 444
Private non-financial corporations	-HBRQ	–359	1 466	–2 153	–319	–336	11 336	2 922	1 358	5 107	432	850
Household sector[1]	-HBRR	–401	–1 168	277	76	190	–565	191	80	129	3 624	9 977
Total	-HHZC	39 274	59 760	51 941	32 073	21 390	65 563	86 551	1 011	36 267	140 853	160 710
Portfolio investment in the UK (UK liabilities = net credits)												
Investment in securities issued by:												
Monetary financial institutions (banks and building societies)	CGPH	19 686	21 070	15 826	–21 638	21 873	39 973	23 776	24 459	29 846	39 636	37 279
Central government	HBSO	296	5 995	–3 312	101	–4 877	–552	–369	–3 812	13 347	14 592	29 675
Local authorities	HBQT	–	–	–	–	–	–	–	–	–	–	–
Public corporations	HCEW	–7	–	–	–	–	–	–	–	–	–	–
Other sectors	CGPL	17 340	15 931	14 272	42 390	97 110	125 122	24 741	30 363	52 029	33 019	60 067
Total	HHZF	37 315	42 996	26 786	20 853	114 106	164 543	48 148	51 010	95 222	87 247	127 021
Net transactions net credits less net debits)												
In assets and liabilities of:												
Monetary financial institutions	LTNA	–6 512	930	–6 979	–74 847	–61	–4 038	–20 767	36 771	16 304	–45 240	–64 097
Central government	ZPOG	296	5 995	–3 312	101	–4 877	–552	–827	–4 279	14 272	14 593	29 655
Local authorities	HBQT	–	–	–	–	–	–	–	–	–	–	–
Public corporations	HCEW	–7	–	–	–	–	–	–	–	–	–	–
Other sectors	LTNB	4 264	–23 689	–14 864	63 526	97 654	103 570	–16 809	17 507	28 379	–22 959	753
Total	HHZD	–1 959	–16 764	–25 155	–11 220	92 716	98 980	–38 403	49 999	58 955	–53 606	–33 689

1 The household sector includes non-profit institutions serving households.

7.7 Other investment

£ million

		1995	1996	1997	1998	1999	2000	2001	2002	2003	2004	2005
Other investment abroad (UK assets = net debits)												
Trade credit												
Long-term												
Central government	-XBMC	400	400	–	–	–	–	–	–	–	–	–
Other sectors[1]	-HCQK	–407	–19	–	–	–	–	–	–	–	–	–
Total long-term trade credit	-HBRZ	–7	381	–	–	–	–	–	–	–	–	–
Short-term												
Other sectors[1]	-XBMF	49	1 698	–635	–1 119	102	–42	–315	292	573	–336	439
Total trade credit	-XBMB	42	2 079	–635	–1 119	102	–42	–315	292	573	–336	439
Loans												
Long-term												
Bank loans under ECGD guarantee	-HGBS	1 128	–626	643	–7	–355	–1 476	187	–1 017	113	231	224
Inter-government loans by the UK	-HEUC	–59	–44	–51	–176	–19	–27	–20	–19	–19	–15	–18
Loans by Commonwealth Development Corporation (public corporations)	-HETB	99	101	54	47	25	2	–	–	–	–	–
Loans by the Export Credit Guarantee Department	CY93	–	–	–31	–47	48	61	49	–186	–259	–229	–611
Loans by specialist leasing companies[1]	-HGKU	–	–	–	–	–	–	–	–	–	–	–
Total long-term loans	-HBSG	1 168	–569	615	–183	–301	–1 440	216	–1 222	–165	–13	–405
Short-term loans												
By monetary financial institutions												
By banks												
Sterling loans	NFBE	619	4 802	3 342	–613	2 621	1 869	4 863	4 768	360	6 871	20 209
Foreign currency loans	ZPON	11 183	34 157	27 803	1 581	14 299	55 631	43 228	12 416	70 447	105 144	115 008
Total banks	HEQO	11 802	38 959	31 145	968	16 920	57 500	48 091	17 184	70 807	112 015	135 217
By building societies	NFBG	–	–9	–	–	–	–	1	3	2	3	2
Total monetary financial institutions	ZPOL	11 802	38 950	31 145	968	16 920	57 500	48 092	17 187	70 809	112 018	135 219
By other sectors	-XBLN	34	125	8	–133	3	–	–	–	–	–	–
Total short-term loans	VTUL	11 836	39 075	31 153	835	16 923	57 500	48 092	17 187	70 809	112 018	135 219
Total loans	-XBMG	13 004	38 506	31 768	652	16 622	56 060	48 308	15 965	70 644	112 005	134 814
Currency and deposits												
Transactions in foreign notes and coin												
Monetary financial institutions (banks)	TAAG	–5	35	42	30	–63	–44	1	21	10	–2	–10
Other sectors[1]	-HETF	34	50	76	10	40	28	–4	33	20	48	11
Total foreign notes and coin	HEOV	29	85	118	40	–23	–16	–3	54	30	46	1
Deposits abroad by UK residents												
Deposits by monetary financial institutions												
Deposits by banks												
Sterling deposits	-HBQY	893	3 726	28 254	6 032	–12 470	20 713	7 296	–6 612	18 173	–2 946	34 282
Foreign currency deposits	-HBQZ	8 874	23 187	85 277	12 679	–16 617	110 531	28 883	60 761	69 519	108 415	125 213
Total deposits by UK banks	-XBMI	9 767	26 913	113 531	18 711	–29 087	131 244	36 179	54 149	87 692	105 469	159 495
Deposits by building societies	TAID	397	–443	1 351	–124	–145	610	–51	–567	–138	85	–96
Total deposits by monetary financial institutions	HCES	10 164	26 470	114 882	18 587	–29 232	131 854	36 128	53 582	87 554	105 554	159 399

1 These series relate to non-governmental sectors other than monetary financial institutions.

7.7 Other investment
continued

£ million

		1995	1996	1997	1998	1999	2000	2001	2002	2003	2004	2005
Other investment abroad - *continued*												
Currency and deposits - *continued*												
Deposits abroad by UK residents - *continued*												
Deposits by securities dealers	-HGTF	18 328	61 179	5 660	−6 117	45 920	47 567	58 756	−13 153	53 172	36 186	167 581
Deposits by other UK residents[1]	-HBSI	5 799	6 805	17 483	2 776	25 870	40 297	31 373	40 060	43 839	71 911	32 816
Total deposits abroad by UK residents	-HBXV	34 291	94 454	138 025	15 246	42 558	219 718	126 257	80 489	184 565	213 651	359 796
Total currency and deposits	-HBVN	34 320	94 539	138 143	15 286	42 535	219 702	126 254	80 543	184 595	213 697	359 797
Other assets												
Central government subscriptions to international organisations												
Regional development banks	-HEUD	65	56	60	65	50	50	53	69	75	61	42
European Investment Bank (EIB)	-HEUE	16	16	–	–	–	–	–	–	–	–	–
Other subscriptions	-HEUF	4	9	3	2	41	3	3	21	51	37	58
Total central government subscriptions	-HGLR	85	81	63	67	91	53	56	90	126	98	100
Short-term central government assets	-LOEL	43	1 460	105	28	256	310	−29	531	52	233	230
Total central government other assets	-LOES	128	1 541	168	95	347	363	27	621	178	331	330
Debt forgiveness (monetary financial institutions)[2]	-FJNF	–	–	–	–	–	–	–	–	–	–	–
Other sectors (excluding monetary financial institutions)												
Long-term assets	-HHZH	–	–	–	–	–	–	–	–	–	–	–
Short-term assets												
Public corporations assets abroad	-HBSR	–	–	–	–	–	–	–	–	–	–	–
Public corporations debt forgiveness	HMLW	–	–	−24	−27	−49	−55	−188	−236	−127	−109	−119
Other[1]	-HBSK	4	4	–	–	–	–	–	–	–	–	–
Total short-term assets of other sectors	-HHZI	4	4	−24	−27	−49	−55	−188	−236	−127	−109	−119
Total other sectors	-XBLP	4	4	−24	−27	−49	−55	−188	−236	−127	−109	−119
Total other assets	-XBMK	132	1 545	144	68	298	308	−161	385	51	222	211
Total	-XBMM	47 498	136 669	169 420	14 887	59 557	276 028	174 086	97 185	255 863	325 588	495 261

1 This series relates to non-governmental sectors other than monetary financial institutions.
2 This series also appears in the capital account (see Table 6.1).

7.7 Other investment
continued

£ million

		1995	1996	1997	1998	1999	2000	2001	2002	2003	2004	2005
Other investment in the UK (UK liabilities = net credits)												
Trade credit												
Long-term[1]	CGJF	265	18	–	–	–	–	–	–	–	–	–
Short-term[1]	XBLQ	–2	13	–7	–	–	–	–	–	–	–	–
Total trade credit	XBMO	263	31	–7	–	–	–	–	–	–	–	–
Loans												
Long-term												
Drawings by:												
Central government	HBSP	–	–	–	–	–	–	–	–	–	–	–
Local authorities	HBSQ	120	150	58	9	17	–	–	–	–	–	–
Public corporations	HHYT	–	–	–	–	–	–	–	–	–	–	–
Other[1]	HIBY	–	–	–	–	–	–	–	–	–	–	–
Total long-term drawings	HBST	120	150	58	9	17	–	–	–	–	–	–
Repayments from:												
Central government	HBSW	–103	–97	–254	–91	–105	–114	–45	–48	–45	–46	–65
Local authorities	HBSX	–174	–139	–109	–96	–123	–188	22	26	204	633	200
Public corporations	HHYU	–144	–14	–206	–5	–	–	–	–	–	–	–
Other[1]	HIBZ	–1	–	–	–	–	–	–	–	–	–	–
Total long-term repayments	HBSY	–422	–250	–569	–192	–228	–302	–23	–22	159	587	135
Total long-term loans	HBSZ	–302	–100	–511	–183	–211	–302	–23	–22	159	587	135
Short-term loans to:												
Central government	HBTA	–	–	–	–	–	–	–	–	–	–	–
Local authorities	HBTB	3	2	–2	–	–	–	–	–	–	–	–
Public corporations	HIAW	–	–	–	–	–	–	–	–	–	–	–
Securities dealers	HBTD	34 398	73 904	32 764	14 901	28 746	65 410	60 790	–38 813	34 054	44 239	223 816
Other[1]	HBSS	5 033	8 314	16 929	266	14 429	1 289	38 503	20 413	42 822	51 708	14 171
Total short-term loans	HBTC	39 434	82 220	49 691	15 167	43 175	66 699	99 293	–18 400	76 876	95 947	237 987
Total loans	XBMP	39 132	82 120	49 180	14 984	42 964	66 397	99 270	–18 422	77 035	96 534	238 122
Currency and deposits												
Sterling notes and coin												
Notes (issued by Bank of England)	HLYV	60	32	45	98	77	67	–51	78	74	120	33
Coins (issued by Royal Mint)	HMAT	6	3	5	11	8	8	–6	8	7	13	5
Total notes and coin	AASD	66	35	50	109	85	75	–57	86	81	133	38
Deposits from abroad with UK residents												
Deposits with monetary financial institutions												
Deposits with banks												
Sterling deposits	NWXP	10 248	–431	16 550	13 800	23 179	32 508	16 381	11 181	22 785	26 660	45 680
Foreign currency deposits	NFAS	15 766	70 488	131 530	37 421	–11 261	166 107	107 666	78 337	146 075	281 571	232 594
Total deposits with banks	HBWA	26 014	70 057	148 080	51 221	11 918	198 615	124 047	89 518	168 860	308 231	278 274
Deposits with building societies	NEWS	444	1 241	–165	883	542	567	523	308	487	305	1 296
Total deposits with UK monetary financial institutions	HDKE	26 458	71 298	147 915	52 104	12 460	199 182	124 570	89 826	169 347	308 536	279 570
Deposit liabilities of UK central government	HEUN	484	–608	–759	304	693	528	–178	–24	232	–877	–57
Total deposits from abroad with UK residents	HBXY	26 942	70 690	147 156	52 408	13 153	199 710	124 392	89 802	169 579	307 659	279 513
Total currency and deposits	HMAO	27 008	70 725	147 206	52 517	13 238	199 785	124 335	89 888	169 660	307 792	279 551

1 These series relate to non-governmental sectors other than monetary financial institutions.

7.7 Other investment continued

£ million

		1995	1996	1997	1998	1999	2000	2001	2002	2003	2004	2005
Other investment in the UK - *continued*												
Other liabilities												
Long-term												
Net equity of foreign households in life insurance reserves and in pension funds	QZEP	–2	–2	–2	–2	–2	–4	–5	–1	–12	–20	–67
Prepayments of premiums and reserves against outstanding claims	NQMC	973	6 793	264	3	–602	942	–157	335	–1 371	178	581
Total long-term liabilities	VTUG	971	6 791	262	1	–604	938	–162	334	–1 383	158	514
Short-term	HJYF	48	1 056	29	138	–129	–90	526	–613	58	–163	160
Total other liabilities	XBMX	1 019	7 847	291	139	–733	848	364	–279	–1 325	–5	674
Total	XBMN	**67 422**	**160 723**	**196 670**	**67 640**	**55 469**	**267 030**	**223 969**	**71 187**	**245 370**	**404 321**	**518 347**
Net transactions (net credits less net debits)												
Trade credit	LTNC	221	–2 048	628	1 119	–102	42	315	–292	–573	336	–439
Loans	LTND	26 128	43 614	17 412	14 332	26 342	10 337	50 962	–34 387	6 391	–15 471	103 308
Deposits	LTNE	–7 312	–23 814	9 063	37 231	–29 297	–19 917	–1 919	9 345	–14 935	94 095	–80 246
Other	LTNF	887	6 302	147	71	–1 031	540	525	–664	–1 376	–227	463
Total	HHYR	**19 924**	**24 054**	**27 250**	**52 753**	**–4 088**	**–8 998**	**49 883**	**–25 998**	**–10 493**	**78 733**	**23 086**

7.8 Other investment
Sector analysis

£ million

		1995	1996	1997	1998	1999	2000	2001	2002	2003	2004	2005
Other investment abroad (UK assets = net debits)												
Investment by:													
Monetary financial institutions													
Banks	-HBSL	22 692	65 281	145 361	19 702	−12 585	187 224	84 458	70 337	158 622	217 713	294 926
Building societies	HEQR	397	−452	1 351	−124	−145	610	−50	−564	−136	88	−94
Total monetary financial institutions	HCET	23 089	64 829	146 712	19 578	−12 730	187 834	84 408	69 773	158 486	217 801	294 832
Central government	-HBSM	469	1 897	117	−81	328	336	7	602	159	316	312
Public corporations	-HBSV	99	101	−1	−27	24	8	−139	−422	−386	−338	−730
Other sectors	-HBSN	23 841	69 842	22 592	−4 583	71 935	87 850	89 810	27 232	97 604	107 809	200 847
Total	-XBMM	**47 498**	**136 669**	**169 420**	**14 887**	**59 557**	**276 028**	**174 086**	**97 185**	**255 863**	**325 588**	**495 261**
Other investment in the UK (UK liabilities = net credits)												
Investment in:												
Monetary financial institutions												
Banks	CGOT	26 074	70 089	148 125	51 319	11 995	198 682	123 996	89 596	168 934	308 351	278 307
Building societies	NEWS	444	1 241	−165	883	542	567	523	308	487	305	1 296
Total monetary financial institutions	HBWG	26 518	71 330	147 960	52 202	12 537	199 249	124 519	89 904	169 421	308 656	279 603
Central government	HBWH	423	−694	−1 003	338	443	308	273	−722	310	−1 053	49
Local authorities	HBWJ	−51	13	−53	−87	−106	−188	22	26	204	633	200
Public corporations	HBWL	−144	−14	−206	−5	–	–	–	–	–	–	–
Other sectors	HBWM	40 676	90 088	49 972	15 192	42 595	67 661	99 155	−18 021	75 435	96 085	238 495
Total	XBMN	**67 422**	**160 723**	**196 670**	**67 640**	**55 469**	**267 030**	**223 969**	**71 187**	**245 370**	**404 321**	**518 347**
Net transactions (net credits less net debits)												
In assets and liabilities of:												
Monetary financial institutions												
Banks	LTNG	3 382	4 808	2 764	31 617	24 580	11 458	39 538	19 259	10 312	90 638	−16 619
Building societies	LTNH	47	1 693	−1 516	1 007	687	−43	573	872	623	217	1 390
Total monetary financial institutions	LTNI	3 429	6 501	1 248	32 624	25 267	11 415	40 111	20 131	10 935	90 855	−15 229
Central government	LTNJ	−46	−2 591	−1 120	419	115	−28	266	−1 324	151	−1 369	−263
Local authorities	HBWJ	−51	13	−53	−87	−106	−188	22	26	204	633	200
Public corporations	LTNK	−243	−115	−205	22	−24	−8	139	422	386	338	730
Other sectors	LTNL	16 835	20 246	27 380	19 775	−29 340	−20 189	9 345	−45 253	−22 169	−11 724	37 648
Total	HHYR	**19 924**	**24 054**	**27 250**	**52 753**	**−4 088**	**−8 998**	**49 883**	**−25 998**	**−10 493**	**78 733**	**23 086**

7.9 Reserve assets
Central government sector
Net debits

£ million

		1995	1996	1997	1998	1999	2000	2001	2002	2003	2004	2005
Monetary gold	-HBOX	−72	−23	1 115	931	−412	−883	−786	−266	–	−2	–
Special drawing rights	-HBOY	−48	−31	84	−16	38	−73	−22	26	−2	−35	−8
Reserve position in the Fund	-HBOZ	622	57	410	751	626	−478	633	469	−251	−558	−1 911
Foreign Exchange												
Currency and deposits												
With central banks	-HBPC	..	−418	−675	−1 822	239	−368	6	95	−79	33	28
With other banks	-HBPD	..	1 509	400	−733	2 312	6	−900	−863	−586	−882	367
Total currency and deposits	-HBPB	..	1 091	−275	−2 555	2 551	−363	−892	−767	−664	−849	395
Securities												
Bonds and notes	-HBPG	–	−1 108	−2 937	−214	−3 105	5 418	−1 838	2 280	−390	1 551	370
Money market instruments	-HBPH	–	−496	−777	939	−337	244	−185	−2 043	−62	107	1 465
Total securities	-HBPE	..	−1 604	−3 714	725	−3 442	5 662	−2 023	237	−452	1 658	1 835
Total foreign exchange	-HBPA	−701	−513	−3 989	−1 830	−891	5 299	−2 915	−530	−1 116	809	2 230
Other claims	-HBPI	–	–	–	–	–	50	5	−158	−190	−18	345
Total	-LTCV	−200	−510	−2 380	−164	−639	3 915	−3 085	−459	−1 559	196	656

Chapter 8
International Investment Position

Summary

The international investment position is the balance sheet of the stock of external assets and liabilities. Between 1966 and 1994 the UK's assets tended to exceed its liabilities, by up to a record £86.4 billion in 1986. From 1995 however, the UK has recorded a net liability position in every year, reaching a record £168.9 billion in 2005.

The value of UK assets and liabilities grew rapidly between 1996 and 2001, when they broadly doubled. This period corresponded with a surge in cross-border investment, much of it associated with merger and acquisition activity. In 2002 the level of assets and liabilities fell slightly as, although there was continued inward and outward investment, these flows were more than offset by revaluation changes resulting from the falls in the value of global equity markets. From 2003 the level of both UK external assets and liabilities increased strongly again, due to a rise in cross-border investment and upward revaluations in the value of equity prices.

Half of all UK assets and just over half of all UK liabilities at end-2005 were allocated to UK monetary financial institutions (MFIs – mostly banks). UK banks' liabilities have consistently exceeded their assets in the last ten years, the net liability position reaching a record £190.0 billion in 2004 before falling back to £153.6 billion in the latest year. In contrast, other sectors assets have historically exceeded liabilities (except for central government, due to non-residents holdings of British Government Stock).

UK assets include reserve assets held by central government. Reserves are mainly held in the form of foreign exchange – in particular bonds and notes. Reserve assets in 2005 account for 0.5 per cent of total UK assets, down from around 4 per cent in the late 1980s.

UK assets

The proportion of direct investment abroad remained fairly constant through much of the 1990s at around 12 to 14 per cent of total UK assets. Between 1998 and 2000 it increased to over 20 per cent, reflecting the high level of merger and acquisition activity by UK companies in those years. It has since declined to around 16 per cent in 2005. Portfolio investment assets remained at around a third of total UK assets from 1993 until 1999, since when they have declined to 28 per cent in 2005, mainly due to the falls in world stock markets in 2001 and 2002. From high proportions of total investment in the early 1980s (around 75 per cent), the proportion of other investment assets declined to 47 per cent of total assets in 1999, since when it has increased to 56 per cent in 2005. This may be due to the relative security of such assets during a period of flat or falling equity prices.

Figure 8.2
UK assets

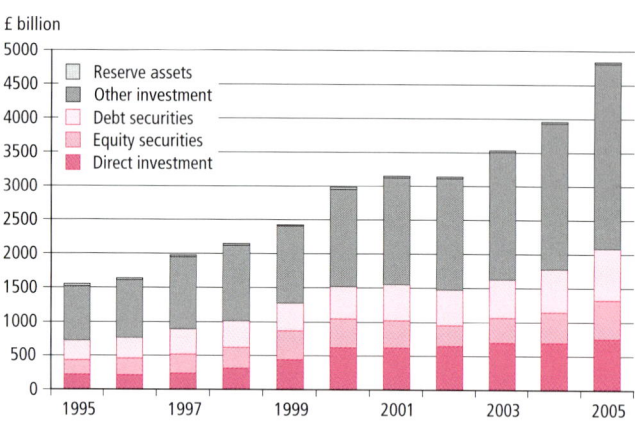

UK liabilities

Direct investment in the UK accounted for around 10 per cent of the total value of UK liabilities throughout the last decade. Portfolio investment increased from 26 per cent in 1995 to 33 per cent in 2000, before falling back to 28 per cent in 2005,

Figure 8.1
International investment position

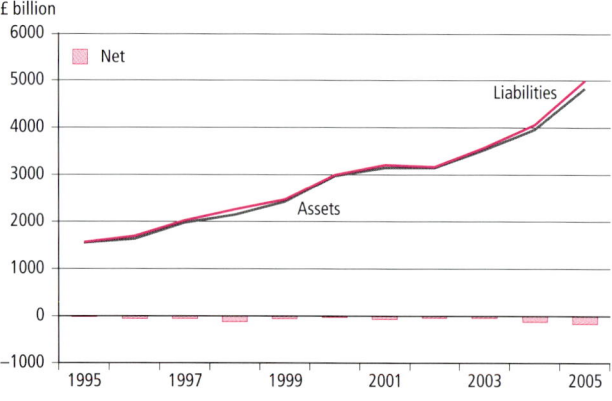

largely due to falls in the UK stock markets in 2001 and 2002 and the impact on the value of equity liabilities. Similarly to the asset position, the share of the value of other investment liabilities in the UK fell from around two-thirds in 1995 to around 56 per cent in 2000, since when it has increased to 62 per cent at end-2005.

Figure 8.3
UK liabilities

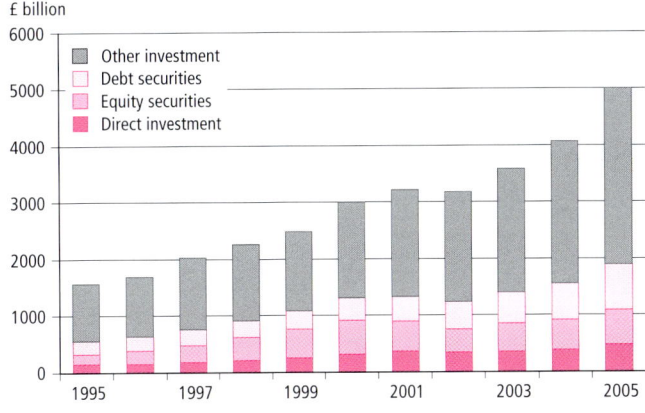

Direct investment

Direct investment assets have nearly quadrupled over the last decade, to reach £753.2 billion at end-2005. Investments by UK private non-financial corporations (PNFC's) accounted for 80 per cent of UK direct investment assets at end 2005, while banks accounted for 6 per cent and other financial corporations a further 8 per cent. The value of PNFC's assets almost trebled between 1997 and 2000, reflecting the substantial foreign acquisitions by UK oil and telecom companies in this period. Since 2000, the value of PNFC assets has generally continued to rise, but at a less dramatic rate.

Figure 8.4
Direct investment

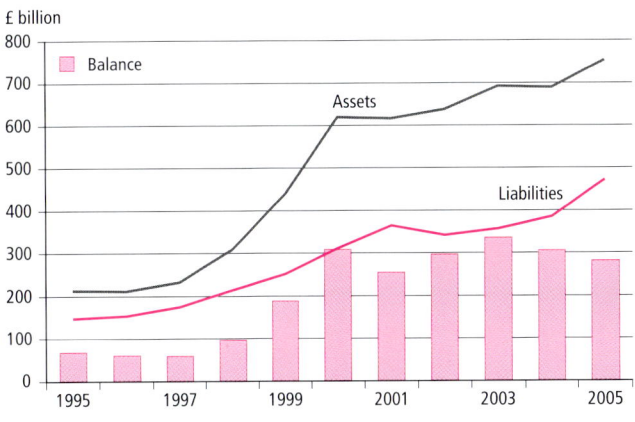

Inward direct investment grew sharply in the late 1990s, with the total value of UK liabilities doubling between 1997 and 2001. PNFC's share of the value of total foreign direct investment liabilities fell from 84 per cent in 1992 to 75 per cent in 1997 before rising again in recent years to reach 80 per cent by the end of 2005. Direct investment in UK banks increased to 9 per cent of total inward direct investment in 2004, following the merger and acquisition activity in that sector during the year, but has fallen back to 8 per cent in 2005.

Portfolio investment

Between 1995 and 2005 UK portfolio investment assets more than doubled to £1,322.1 billion. The pattern of growth in equities has been much more erratic than the growth in debt, as the value of equity assets is heavily influenced by changes in global equity prices. Between the end of 2001 and the end of 2002 for instance, the value of portfolio investment equity assets fell 24 per cent to £305.9 billion, mirroring the fall in world equity prices over the same period. The value of foreign debt securities held by UK investors more than doubled between 1995 and 2005, with a £118.3 billion increase in 2005 alone. UK banks hold nearly 60 per cent of total UK debt securities assets. They have also increased their holdings of foreign equities from 1 per cent of total UK equity holdings in 2002 to 15 per cent at the end of 2005. UK insurance companies, pension funds and other financial intermediaries hold the vast majority of UK equity assets and around a third of UK debt security assets.

Figure 8.5
Portfolio investment

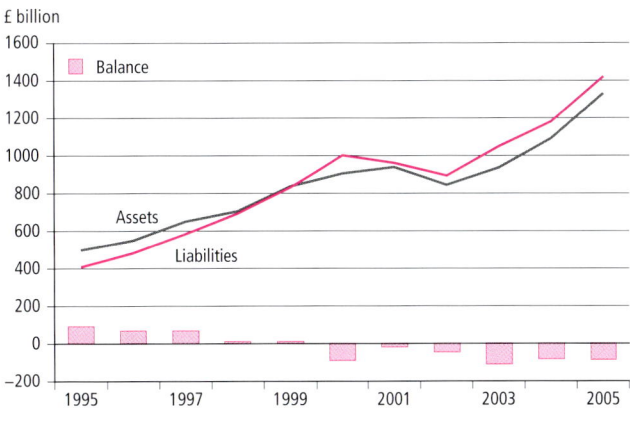

The total value of UK portfolio investment liabilities increased strongly in 2005, to £1,418.3 billion, due to continued foreign acquisitions of UK debt securities and an 18 per cent rise in the value of equity on the London Stock Exchange. The fall in the value of portfolio investment liabilities in 2001 and 2002 mirror the fall in the price of UK equity on the London stock exchange – which fell around 20 per cent in both 2001 and 2002.

Other investment

Other investment accounts for over half of total UK external assets, with the level of investment more than doubling since 1999. UK banks deposits and short term loans to non-residents accounted for over two-thirds of total other investment abroad in 2005; this proportion has declined from around 90 per cent of total other investment in the late 1980s. The bulk of UK bank deposits abroad were in foreign currencies: only 12 per cent was held in sterling at end-2005.

Figure 8.6
Other investment

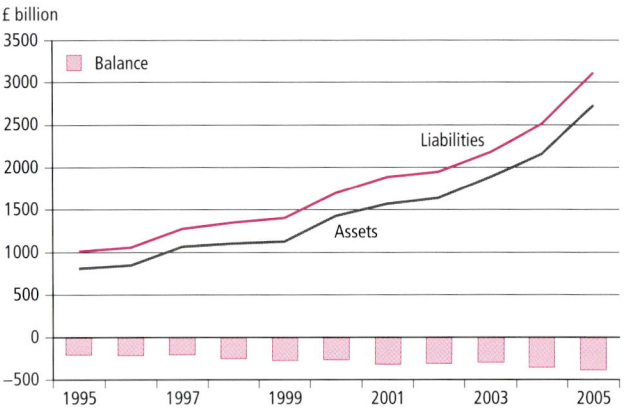

Deposits from abroad held with UK banks represent the largest item in other investment liabilities, although these have declined from over 80 per cent in the early 1990s to 70 per cent at end-2005. Of total deposits with UK banks of £2192.8 billion in 2005, only 15 per cent were held in sterling. The fall in the proportionate value of deposits with banks is largely the result of the increase in short term loans to UK securities dealers and other non-bank sectors – increasing from £150.0 billion in 1994 to £897.5 billion by end 2005.

8.1 International investment position
Summary table
Balance sheets valued at end of year

£ billion

		1995	1996	1997	1998	1999	2000	2001	2002	2003	2004	2005
UK Assets												
Direct investment abroad												
Equity capital and reinvested earnings	CGMO	197.6	201.0	223.3	283.9	412.3	586.3	582.2	619.3	670.7	664.2	727.3
Other capital assets	HBUW	15.7	10.7	9.0	25.9	26.0	32.5	34.8	17.9	20.5	24.7	25.8
Total direct investment abroad	HBWD	213.3	211.7	232.4	309.8	438.3	618.8	616.9	637.2	691.1	689.0	753.2
Portfolio investment abroad												
Equity securities	HEPX	217.0	238.3	282.3	303.7	419.9	429.3	404.6	305.9	372.0	455.3	576.7
Debt securities	HHZX	282.3	310.0	368.6	400.2	418.4	476.8	532.8	538.1	563.9	637.0	755.3
Total portfolio investment abroad	HHZZ	499.3	548.3	651.0	703.8	838.3	906.1	937.4	844.0	935.8	1 092.3	1 332.1
Other investment abroad	HLXV	808.1	851.7	1 070.4	1 107.7	1 129.7	1 427.5	1 573.1	1 635.8	1 885.1	2 156.2	2 727.1
Reserve assets	LTEB	31.8	27.3	22.8	23.3	22.2	28.8	25.6	25.5	23.8	23.3	24.7
Total	HBQA	**1 552.5**	**1 638.9**	**1 976.5**	**2 144.7**	**2 428.5**	**2 981.2**	**3 153.1**	**3 142.4**	**3 535.8**	**3 960.7**	**4 837.1**
UK Liabilities												
Direct investment in the UK												
Equity capital and reinvested earnings	HBUY	118.9	120.6	134.3	159.8	192.3	240.6	259.7	229.2	245.7	278.2	365.7
Other capital liabilities	HBVC	27.3	32.1	39.4	53.8	57.9	69.8	103.7	111.4	109.9	106.2	106.3
Total direct investment in the UK	HBWI	146.2	152.6	173.7	213.6	250.2	310.4	363.5	340.6	355.5	384.4	472.0
Portfolio investment in the UK												
Equity securities	HLXX	172.7	226.1	301.9	402.1	509.8	604.4	529.4	409.8	491.5	524.4	613.3
Debt securities	HLXY	233.6	253.9	281.4	290.6	319.0	393.8	429.1	482.6	555.8	653.4	805.0
Total portfolio investment in the UK	HLXW	406.3	480.0	583.3	692.7	828.8	998.2	958.5	892.3	1 047.3	1 177.8	1 418.3
Other investment in the UK	HLYD	1 013.0	1 061.7	1 274.3	1 355.0	1 403.9	1 696.4	1 889.6	1 945.8	2 177.1	2 509.4	3 115.6
Total	HBQB	**1 565.5**	**1 694.4**	**2 031.3**	**2 261.4**	**2 482.9**	**3 005.0**	**3 211.5**	**3 178.7**	**3 579.9**	**4 071.6**	**5 006.0**
Net International Investment Position												
Direct investment												
Equity capital and reinvested earnings	HBSH	78.7	80.4	89.0	124.1	220.0	345.7	322.4	390.1	425.0	386.0	361.6
Other capital	CGKF	−11.5	−21.4	−30.4	−27.9	−31.9	−37.3	−69.0	−93.5	−89.4	−81.4	−80.4
Total net direct investment	HBWQ	67.1	59.0	58.6	96.2	188.1	308.4	253.5	296.6	335.6	304.6	281.2
Portfolio investment												
Equity securities	CGNE	44.3	12.2	−19.5	−98.4	−90.0	−175.1	−124.8	−103.9	−119.5	−69.1	−36.6
Debt securities	CGNF	48.7	56.0	87.2	109.6	99.5	82.9	103.7	55.6	8.1	−16.4	−49.6
Total net portfolio investment	CGNH	93.0	68.3	67.7	11.1	9.5	−92.2	−21.1	−48.3	−111.5	−85.5	−86.3
Other investment	CGNG	−204.9	−210.1	−204.0	−247.3	−274.2	−268.9	−316.5	−310.0	−292.0	−353.2	−388.5
Reserve assets	LTEB	31.8	27.3	22.8	23.3	22.2	28.8	25.6	25.5	23.8	23.3	24.7
Total	HBQC	**−13.0**	**−55.5**	**−54.8**	**−116.7**	**−54.4**	**−23.9**	**−58.4**	**−36.3**	**−44.1**	**−110.9**	**−168.9**
Allocations of Special Drawing Rights to the UK by the IMF	HEVP	1.8	1.6	1.6	1.6	1.6	1.7	1.7	1.6	1.6	1.5	1.6

8.2 International investment position
Sector analysis
Balance sheets valued at end of year

£ billion

		1995	1996	1997	1998	1999	2000	2001	2002	2003	2004	2005
UK Assets												
Monetary financial institutions												
Banks	CGNI	805.5	798.9	1 030.1	1 140.0	1 132.2	1 411.5	1 523.1	1 561.8	1 728.8	1 981.8	2 420.1
Building societies	VTXF	3.4	3.1	2.9	4.3	5.0	8.5	10.0	9.1	7.3	8.0	8.0
Total monetary financial institutions	CGNJ	808.9	801.9	1 033.0	1 144.3	1 137.2	1 419.9	1 533.0	1 570.9	1 736.0	1 989.7	2 428.2
Central government	CGNK	41.2	38.7	34.4	34.8	25.7	32.7	29.6	30.0	28.6	28.4	30.1
Public corporations	CGNL	1.9	1.9	4.4	4.2	4.5	4.6	3.8	4.2	4.2	4.5	4.3
Other sectors	CGNM	700.4	796.4	904.8	961.4	1 261.1	1 524.0	1 586.7	1 537.3	1 767.0	1 938.1	2 374.6
Total	HBQA	**1 552.5**	**1 638.9**	**1 976.5**	**2 144.7**	**2 428.5**	**2 981.2**	**3 153.1**	**3 142.4**	**3 535.8**	**3 960.7**	**4 837.1**
UK Liabilities												
UK Monetary financial institutions (banks and building societies)	HBYJ	919.9	916.5	1 113.0	1 182.8	1 197.0	1 485.9	1 623.4	1 707.0	1 892.8	2 179.8	2 581.7
Central government	CGOG	65.7	69.2	67.9	76.3	63.6	65.6	63.3	59.2	70.9	89.5	116.0
Local authorities	CGOH	1.2	1.2	1.1	1.2	1.1	0.8	0.8	0.9	1.1	1.7	1.9
Public corporations	CGOI	0.2	0.2	–	–	–	–	–	–	–	–	–
Other sectors	HCON	578.5	707.3	849.3	1 001.1	1 221.2	1 452.8	1 523.9	1 411.7	1 615.1	1 800.6	2 306.4
Total	HBQB	**1 565.5**	**1 694.4**	**2 031.3**	**2 261.4**	**2 482.9**	**3 005.0**	**3 211.5**	**3 178.7**	**3 579.9**	**4 071.6**	**5 006.0**
Net International Investment Position												
Monetary financial institutions (banks and building societies)	HDIJ	−111.0	−114.6	−80.0	−38.5	−59.9	−66.0	−90.4	−136.1	−156.8	−190.0	−153.6
Central government	CGOK	−24.4	−30.5	−33.5	−41.6	−37.9	−32.9	−33.8	−29.2	−42.3	−61.1	−85.9
Local authorities	-CGOH	−1.2	−1.2	−1.1	−1.2	−1.1	−0.8	−0.8	−0.9	−1.1	−1.7	−1.9
Public corporations	CGOL	1.7	1.7	4.3	4.2	4.5	4.6	3.8	4.2	4.2	4.4	4.3
Other sectors	HDKB	121.9	89.1	55.5	−39.6	39.9	71.2	62.8	125.6	151.8	137.6	68.2
Total	HBQC	**−13.0**	**−55.5**	**−54.8**	**−116.7**	**−54.4**	**−23.9**	**−58.4**	**−36.3**	**−44.1**	**−110.9**	**−168.9**

8.3 Direct investment
Balance sheets valued at end of year

£ billion

		1995	1996	1997	1998	1999	2000	2001	2002	2003	2004	2005
Direct investment abroad (UK assets)												
Equity capital and reinvested earnings												
Ordinary share capital and reinvested earnings	CVWF	185.4	188.5	212.3	271.9	399.8	570.6	565.7	600.4	646.1	633.6	689.3
Holdings of property	HCHP	12.2	12.4	11.0	12.0	12.5	15.7	16.5	18.9	24.6	30.6	38.1
Total equity capital and reinvested earnings	CGMO	197.6	201.0	223.3	283.9	412.3	586.3	582.2	619.3	670.7	664.2	727.3
Other capital												
Claims on affiliated enterprises												
Debt securities issued by affiliated enterprises	CVWG	3.0	1.5	2.9	6.3	8.8	28.1	31.0	31.8	31.9	30.8	31.7
Other claims on affiliated enterprises												
Inter-company balance	CVOK	45.9	44.8	47.7	66.1	71.4	80.7	88.8	103.1	105.2	132.7	141.8
Branch indebtedness balance	CVOP	5.9	5.7	6.3	10.4	10.2	7.9	12.4	11.4	11.1	12.4	8.2
Total claims on affiliated enterprises	CGLS	54.8	52.0	57.0	82.8	90.3	116.8	132.2	146.3	148.2	176.0	181.6
Liabilities to affiliated enterprises												
Inter-company balance	-CVOL	−37.0	−40.3	−47.1	−53.2	−61.8	−79.1	−93.2	−124.6	−122.4	−147.1	−151.5
Branch indebtedness balance	-CVOQ	−2.0	−1.0	−0.8	−3.7	−2.5	−5.2	−4.2	−3.9	−5.3	−4.1	−4.3
Total liabilities to affiliated enterprises	-HHDJ	−39.1	−41.3	−48.0	−56.9	−64.4	−84.3	−97.4	−128.4	−127.7	−151.2	−155.8
Total other capital assets	HBUW	15.7	10.7	9.0	25.9	26.0	32.5	34.8	17.9	20.5	24.7	25.8
Total	HBWD	**213.3**	**211.7**	**232.4**	**309.8**	**438.3**	**618.8**	**616.9**	**637.2**	**691.1**	**689.0**	**753.2**
Direct investment in the UK (UK liabilities)												
Equity capital and reinvested earnings												
Share capital and reinvested earnings												
Quoted share capital and reinvested earnings[1]	CVVB	..	..	..	..	..	..	..	20.7	25.8	44.4	54.2
Unquoted share capital and reinvested earnings	CVVC	110.2	111.3	124.1	149.0	180.7	227.1	245.6	192.6	204.0	216.0	293.2
Total share capital and reinvested earnings	HBUX	110.2	111.3	124.1	149.0	180.7	227.1	245.6	213.3	229.8	260.4	347.4
Holdings of UK property	HCQM	8.7	9.3	10.2	10.9	11.7	13.5	14.1	15.9	15.9	17.8	18.3
Total equity capital and reinvested earnings	HBUY	118.9	120.6	134.3	159.8	192.3	240.6	259.7	229.2	245.7	278.2	365.7
Other capital												
Liabilities to direct investors												
Debt securities issued by affiliated enterprises	CVVD	4.2	6.3	6.7	6.6	7.0	11.0	17.5	17.3	16.0	18.9	20.6
Other liabilities to direct investors												
Inter-company balance	CVVJ	38.9	43.4	53.4	78.4	96.0	103.8	133.1	147.0	142.2	144.1	142.6
Branch indebtedness balance	CVVM	4.4	6.2	7.4	8.1	6.8	8.5	9.4	8.7	10.7	8.4	9.2
Total liabilities to direct investors	HBVB	47.5	55.9	67.5	93.2	109.8	123.2	160.0	173.0	168.9	171.4	172.4
Claims on direct investors												
Inter-company balance	-CVVI	−20.1	−23.7	−27.7	−39.2	−51.8	−51.3	−53.4	−60.5	−55.6	−62.9	−63.5
Branch indebtedness balance	-CVVL	−0.1	−0.2	−0.3	−0.2	−0.1	−2.1	−2.9	−1.1	−3.5	−2.4	−2.6
Total claims on direct investors	-HBVA	−20.2	−23.9	−28.0	−39.4	−51.9	−53.4	−56.3	−61.6	−59.0	−65.3	−66.1
Total other capital liabilities	HBVC	27.3	32.1	39.4	53.8	57.9	69.8	103.7	111.4	109.9	106.2	106.3
Total	HBWI	**146.2**	**152.6**	**173.7**	**213.6**	**250.2**	**310.4**	**363.5**	**340.6**	**355.5**	**384.4**	**472.0**
Net international investment position (UK assets less UK liabilities)												
Equity capital												
Ordinary share capital and reinvested earnings	LTNM	75.1	77.3	88.2	123.0	219.1	343.5	320.0	387.1	416.3	373.3	341.9
Holdings of property	LTNN	3.5	3.1	0.9	1.1	0.9	2.2	2.4	3.0	8.7	12.8	19.7
Total equity capital and reinvested earnings	HBSH	78.7	80.4	89.0	124.1	220.0	345.7	322.4	390.1	425.0	386.0	361.6
Total other capital	CGKF	−11.5	−21.4	−30.4	−27.9	−31.9	−37.3	−69.0	−93.5	−89.4	−81.4	−80.4
Total	HBWQ	**67.1**	**59.0**	**58.6**	**96.2**	**188.1**	**308.4**	**253.5**	**296.6**	**335.6**	**304.6**	**281.2**

1 Prior to 2003 holdings of quoted share capital were included in series CVVC

8.4 Direct investment
Sector analysis
Balance sheets valued at end of year

£ billion

		1995	1996	1997	1998	1999	2000	2001	2002	2003	2004	2005
Direct investment abroad (UK assets)												
By:												
Monetary financial institutions (banks)	CVKH	5.8	5.3	3.2	9.9	11.7	18.1	25.5	27.7	27.6	39.1	48.7
Insurance companies	DPYH	17.0	17.5	22.5	22.0	21.2	24.3	22.8	22.0	24.9	27.2	28.5
Other financial intermediaries	CVWH	9.3	11.5	15.8	26.9	26.8	34.9	37.8	42.0	44.2	31.4	34.9
Private non-financial corporations	CVLX	170.1	165.9	180.6	239.5	366.7	527.9	515.8	527.5	569.9	560.9	603.5
Public corporations	CVOF	0.8	0.7	0.8	0.8	1.1	1.7	0.8	1.5	1.4	1.6	1.7
Household sector[1]	AQHH	10.3	10.7	9.4	10.7	10.8	12.0	14.3	16.5	23.2	28.7	35.7
Total	HBWD	**213.3**	**211.7**	**232.4**	**309.8**	**438.3**	**618.8**	**616.9**	**637.2**	**691.1**	**689.0**	**753.2**
Direct investment in the UK (UK liabilities)												
In:												
Monetary financial institutions (banks)	CVJW	15.4	17.8	21.3	20.3	19.8	26.0	27.2	28.1	30.1	34.2	38.1
Insurance companies	CVSM	3.1	6.8	9.0	9.4	13.7	11.7	13.0	14.1	19.4	18.8	19.3
Other financial intermediaries												
Securities dealers	CVTC	7.1	8.0	8.4	7.2	8.2	9.5	11.0	11.9	12.4	13.0	15.1
Other	CVTS	2.2	2.8	4.9	8.1	7.4	15.8	27.3	29.1	30.3	26.9	23.3
Total other financial intermediaries	CVUI	9.4	10.8	13.3	15.2	15.6	25.2	38.3	41.0	42.7	39.9	38.4
Private non-financial corporations	CVKW	118.3	117.2	130.2	168.7	201.2	247.4	284.9	257.3	263.3	291.5	376.2
Total	HBWI	**146.2**	**152.6**	**173.7**	**213.6**	**250.2**	**310.4**	**363.5**	**340.6**	**355.5**	**384.4**	**472.0**
Net international investment position (UK assets less UK liabilities)												
Monetary financial institutions	LTNO	−9.5	−12.5	−18.1	−10.5	−8.1	−8.0	−1.7	−0.5	−2.5	4.9	10.7
Insurance companies	LTNP	13.9	10.6	13.4	12.6	7.6	12.6	9.8	7.9	5.4	8.4	9.2
Other financial intermediares	LTNQ	–	0.7	2.6	11.7	11.2	9.7	−0.5	1.0	1.4	−8.4	−3.4
Private non-financial corporations	LTNR	51.7	48.7	50.4	70.8	165.5	280.4	230.8	270.2	306.6	269.4	227.3
Public corporations	CVOF	0.8	0.7	0.8	0.8	1.1	1.7	0.8	1.5	1.4	1.6	1.7
Household sector[1]	AQHH	10.3	10.7	9.4	10.7	10.8	12.0	14.3	16.5	23.2	28.7	35.7
Total	HBWQ	**67.1**	**59.0**	**58.6**	**96.2**	**188.1**	**308.4**	**253.5**	**296.6**	**335.6**	**304.6**	**281.2**

1 The household sector includes non-profit institutions serving households.

8.5 Portfolio investment
Balance sheets valued at end of year

£ billion

Portfolio investment abroad (UK assets)		1995	1996	1997	1998	1999	2000	2001	2002	2003	2004	2005
Investment in equity securities (shares) by:												
Monetary financial Institutions (banks)	VTWF	4.8	5.0	2.7	8.8	6.8	19.7	14.3	2.7	20.8	53.0	86.2
Central Government	LOER	–	–	–	–	–	–	–	–	–	–	0.2
Insurance companies and pension funds												
Insurance companies	CGPB	59.9	62.0	72.3	77.3	115.7	100.7	106.2	82.1	79.1	91.4	125.6
Pension funds[1]	ZPOR	82.2	84.2	104.2	108.9	148.3	135.5	127.9	104.4	125.7	140.3	162.8
Total insurance companies and pension funds	CGPV	142.0	146.2	176.5	186.2	264.0	236.2	234.1	186.5	204.9	231.7	288.4
Other financial intermediaries												
Securities dealers	HCEA	4.3	8.5	31.4	27.0	38.3	49.3	46.8	22.9	32.4	46.3	37.3
Unit and Investment Trusts	CGSN	55.5	68.4	60.7	69.0	93.6	99.1	88.0	77.3	94.1	100.0	123.0
Other	CGTV	–	–	–	–	–	–	–	–	–	–	–
Total other financial intermediaries	HDIG	59.8	76.9	92.1	96.0	131.8	148.4	134.8	100.2	126.5	146.2	160.3
Private non-financial corporations	XBNN	0.3	0.5	0.5	0.7	1.1	10.0	8.9	6.5	7.8	7.9	9.2
Household sector[2]	HFLX	10.1	9.8	10.6	11.9	16.1	15.0	12.5	10.0	12.0	16.5	32.6
Total investment in equity securities	HEPX	217.0	238.3	282.3	303.7	419.9	429.3	404.6	305.9	372.0	455.3	576.7
Investment in debt securities												
Investment in bonds and notes by:												
Monetary financial institutions												
Banks	VTWJ	144.9	154.7	181.9	224.8	239.0	282.8	312.9	326.0	318.2	349.5	403.2
Building societies	HPEG	1.6	1.6	1.6	3.0	4.1	5.8	6.7	6.3	4.8	5.6	5.9
Total monetary financial institutions	HPCO	146.5	156.4	183.5	227.8	243.1	288.7	319.6	332.3	323.1	355.1	409.1
Central Government	MDZI	–	–	–	–	–	–	–	–	–	–	–
Insurance companies and pension funds												
Insurance companies	CGTU	19.8	22.5	24.4	41.4	37.8	39.8	55.9	62.9	64.5	77.0	88.2
Pension funds[1]	JIRX	16.7	22.3	21.6	23.9	36.4	44.2	49.9	45.7	53.9	64.6	83.9
Total insurance companies and pension funds	HBUM	36.6	44.9	46.0	65.3	74.1	84.0	105.8	108.6	118.3	141.7	172.2
Other financial intermediaries												
Securities dealers	HCDZ	61.4	80.4	103.4	68.1	45.6	45.2	34.9	31.1	38.7	57.3	82.7
Unit and investment trusts	HBXZ	4.3	3.0	3.6	4.7	5.8	6.8	8.4	7.7	10.7	13.4	18.9
Other	HCNA	–	–	0.1	–	–	–	–	–	–	–	–
Total other financial intermediaries	HCOR	65.6	83.5	107.1	72.8	51.4	52.1	43.3	38.8	49.4	70.7	101.6
Private non-financial corporations	XBNK	2.5	3.1	0.8	1.4	0.4	1.6	2.0	2.2	3.5	3.7	3.0
Household sector[2]	HCJC	11.9	6.4	6.7	7.1	6.9	7.5	7.6	7.8	7.7	7.7	7.7
Total investment in bonds and notes	HEPW	263.1	294.2	344.2	374.4	376.0	433.8	478.3	489.6	502.0	578.8	693.6
Investment in Money Market Instruments												
Investment in commercial paper by:												
Monetary financial institutions												
Banks	HBMW	13.2	8.6	16.6	21.3	31.3	33.2	39.7	32.3	40.6	35.2	37.9
Building societies	TAIG	0.1	0.1	0.3	0.2	0.2	1.1	1.8	1.2	1.0	1.1	1.1
Central government	LSPI	–	–	–	–	–	0.5	0.9	–	–	–	–
Insurance companies and pension funds	HBXX	1.1	1.4	2.0	1.1	1.4	1.3	1.1	1.4	1.5	2.1	2.1
Other financial intermediaries	HGRJ	2.6	3.4	3.0	1.1	4.1	2.2	4.7	4.2	7.2	7.9	7.1
Private non-financial corporations	HFBN	1.7	1.8	2.1	1.2	1.9	3.0	4.9	6.0	9.8	10.4	12.0
Total investment in commercial paper	HGRK	18.7	15.3	24.0	24.8	38.9	40.8	52.6	46.0	60.0	56.6	60.3
Investment in certificates of deposit by:												
Monetary financial institutions												
(Building societies)	TAIE	–	–	–	0.2	0.1	0.6	0.6	1.2	1.2	0.9	0.8
Other financial intermediaries	CDHB	0.5	0.5	0.5	0.8	3.4	1.6	1.2	1.4	0.7	0.7	0.7
Total transactions in certificates of deposit	VTWN	0.5	0.5	0.5	1.0	3.6	2.2	1.8	2.5	1.9	1.6	1.5
Total investment in Money Market Instruments	HLYR	19.2	15.8	24.5	25.8	42.5	43.0	54.5	48.5	61.9	58.2	61.8
Total investment in debt securities	HHZX	282.3	310.0	368.6	400.2	418.4	476.8	532.8	538.1	563.9	637.0	755.3
Total	HHZZ	**499.3**	**548.3**	**651.0**	**703.8**	**838.3**	**906.1**	**937.4**	**844.0**	**935.8**	**1 092.3**	**1 332.1**

1 The pension funds data only covers self-administered funds, see glossary.   2 The household sector includes non-profit institutions serving households.

8.5 Portfolio investment
Balance sheets valued at end of year
continued

£ billion

		1995	1996	1997	1998	1999	2000	2001	2002	2003	2004	2005
Portfolio investment in the UK (UK liabilities)												
Investment in equity securities (shares) issued by:													
Monetary financial Institutions (banks and building societies)	HBQD	6.3	9.7	14.8	11.0	9.1	5.9	5.1	3.0	4.2	4.2	4.9
Other sectors[1]	HBQE	166.4	216.3	287.1	391.1	500.7	598.5	524.4	406.8	487.3	520.2	608.4
Total investment in equity securities	HLXX	172.7	226.1	301.9	402.1	509.8	604.4	529.4	409.8	491.5	524.4	613.3
Investment in debt securities													
Investment in bonds and notes													
Issues by central government													
UK foreign currency bonds and notes	HEWE	13.1	10.0	6.4	5.1	4.7	6.5	3.3	0.9	1.6	1.5	1.7
Investment in British government stocks by:													
Foreign central banks (exchange reserves)	HCCH	14.7	14.7	14.1	18.0	16.7	18.1	18.7	17.3	15.9	21.0	21.0
Other foreign residents	HEQF	31.2	38.8	43.8	50.9	39.6	37.8	37.8	38.2	48.6	61.3	88.8
Total investment in British government stocks	HEWD	45.9	53.5	58.0	68.8	56.2	55.9	56.5	55.5	64.5	82.3	109.8
Total issues by central government	HHGF	58.9	63.5	64.4	73.9	60.9	62.4	59.9	56.4	66.1	83.8	111.4
Local authorities' bonds	HHGG	–	–	–	–	–	–	–	–	–	–	–
Public corporations' bonds	HEWM	–	–	–	–	–	–	–	–	–	–	–
Issues by monetary financial Institutions (banks and building societies)													
Bonds	HMBL	20.4	22.1	25.4	28.6	33.6	39.0	41.6	51.4	68.0	83.0	106.4
European medium term notes and other medium-term paper:													
Issued by UK banks	HCFA	15.5	19.6	24.9	27.7	33.5	35.8	39.2	40.4	49.5	64.5	85.6
Issued by UK building societies	HCFD	5.1	4.4	1.3	1.1	1.2	2.6	3.3	3.2	4.2	6.4	9.9
Total	HHGI	20.7	24.0	26.3	28.9	34.7	38.4	42.5	43.6	53.7	70.9	95.5
Total monetary financial institutions	HMBF	41.1	46.1	51.7	57.4	68.3	77.4	84.1	95.0	121.7	154.0	201.9
Issues by other sectors[1]	HHGJ	63.6	68.8	79.1	89.0	104.5	121.3	129.4	160.1	211.7	258.3	331.1
Total investment in bonds and notes	HLXZ	163.7	178.4	195.2	220.3	233.8	261.1	273.4	311.5	399.5	496.1	644.5
Investment in Money Market Instruments													
Investment in treasury bills (issued by central government)													
Sterling treasury bills	ACQJ	1.4	0.9	0.6	0.1	0.1	–	0.1	0.2	1.9	4.0	2.8
Euro treasury bills	HHNX	1.8	2.0	1.1	0.2	–	–	–	–	–	–	–
Total treasury bills	HLYU	3.2	3.0	1.7	0.3	0.1	–	0.1	0.2	1.9	4.0	2.8
Investment in certificates of deposit (issued by monetary financial institutions)													
Issued by UK banks	HHGK	46.3	51.1	59.1	41.6	53.9	92.8	115.0	108.4	96.2	87.9	94.8
Issued by UK building societies	HHGL	0.6	0.6	0.2	0.3	0.5	0.5	0.4	0.6	1.7	2.2	1.1
Total certificates of deposit	HHGM	46.9	51.7	59.3	42.0	54.4	93.3	115.4	108.9	97.8	90.1	95.9
Investment in commercial paper													
Issued by UK monetary financial Institutions													
UK banks	HHGN	4.7	6.3	11.1	11.4	10.1	14.7	14.9	28.9	27.0	35.1	33.9
Building societies	HHGO	3.9	3.2	0.7	1.0	2.7	2.9	2.8	2.4	5.7	5.5	6.0
Total monetary financial institutions	HHGP	8.6	9.5	11.8	12.4	12.8	17.7	17.7	31.4	32.8	40.6	39.9
Issued by other sectors[1]	HLYQ	11.2	11.3	13.4	15.6	17.8	21.7	22.5	30.6	23.7	22.6	22.0
Total investment in commercial paper	HHGR	19.8	20.9	25.2	28.0	30.6	39.4	40.2	62.0	56.5	63.3	61.9
Total investment in Money Market Instruments	HLYB	69.9	75.6	86.2	70.3	85.2	132.7	155.7	171.1	156.3	157.3	160.5
Total investment in debt securities	HLXY	233.6	253.9	281.4	290.6	319.0	393.8	429.1	482.6	555.8	653.4	805.0
Total	HLXW	**406.3**	**480.0**	**583.3**	**692.7**	**828.8**	**998.2**	**958.5**	**892.3**	**1 047.3**	**1 177.8**	**1 418.3**

1 These series relate to non-governmental sectors other than monetary financial institutions.

Chapter 8: International Investment Position — The Pink Book: 2006 edition

8.5 Portfolio investment
Balance sheets valued at end of year
continued

£ billion

		1995	1996	1997	1998	1999	2000	2001	2002	2003	2004	2005
Net international investment position (UK assets less UK liabilities)												
Equity securities	CGNE	44.3	12.2	−19.5	−98.4	−90.0	−175.1	−124.8	−103.9	−119.5	−69.1	−36.6
Debt securities												
Bonds and notes	LTNS	99.4	115.8	149.0	154.0	142.2	172.7	204.9	178.1	102.5	82.7	49.1
Money market instruments	LTNT	−50.7	−59.8	−61.8	−44.5	−42.7	−89.8	−101.2	−122.6	−94.4	−99.1	−98.7
Total debt securities	CGNF	48.7	56.0	87.2	109.6	99.5	82.9	103.7	55.6	8.1	−16.4	−49.6
Total	CGNH	**93.0**	**68.3**	**67.7**	**11.1**	**9.5**	**−92.2**	**−21.1**	**−48.3**	**−111.5**	**−85.5**	**−86.3**

8.6 Porfolio investment
Sector analysis
Balance sheets valued at end of year

£ billion

		1995	1996	1997	1998	1999	2000	2001	2002	2003	2004	2005
Portfolio investment abroad (UK assets)												
Investment by:												
Monetary financial institutions												
Banks	HBRW	163.0	168.3	201.1	254.8	277.1	335.8	367.0	361.0	379.6	437.6	527.3
Building societies	VTWM	1.7	1.7	1.9	3.4	4.5	7.5	9.0	8.7	7.0	7.7	7.8
Total monetary financial institutions	HHGQ	164.6	170.1	203.1	258.2	281.6	343.2	376.0	369.7	386.7	445.3	535.1
Central government	LOFC	–	–	–	–	–	–	0.5	0.9	–	–	0.2
Insurance companies and pension funds	HHHH	179.7	192.4	224.4	252.7	339.5	321.4	341.0	296.5	324.7	375.4	462.7
Other financial intermediaries	HHNH	128.6	164.2	202.7	170.7	190.8	204.3	184.0	144.5	183.7	225.5	269.6
Private non-financial corporations	AIMH	4.5	5.4	3.4	3.2	3.4	14.7	15.8	14.6	21.0	21.9	24.2
Household sector[1]	AINA	22.0	16.2	17.3	19.0	23.0	22.4	20.2	17.7	19.7	24.2	40.3
Total	HHZZ	**499.3**	**548.3**	**651.0**	**703.8**	**838.3**	**906.1**	**937.4**	**844.0**	**935.8**	**1 092.3**	**1 332.1**
Portfolio investment in the UK (UK liabilities)												
Investment in securities issued by:												
Monetary financial institutions (banks and building societies)	CGPC	103.0	117.1	137.6	122.7	144.6	194.3	222.2	238.3	256.5	288.9	342.6
Central government	HHGS	62.2	66.5	66.1	74.3	61.1	62.4	60.0	56.5	68.0	87.8	114.2
Local authorities	HHGG	–	–	–	–	–	–	–	–	–	–	–
Public corporations	HEWM	–	–	–	–	–	–	–	–	–	–	–
Other sectors	CGPG	241.2	296.5	379.6	495.7	623.1	741.5	676.3	597.5	722.7	801.1	961.5
Total	HLXW	**406.3**	**480.0**	**583.3**	**692.7**	**828.8**	**998.2**	**958.5**	**892.3**	**1 047.3**	**1 177.8**	**1 418.3**
Net international investment position (UK assets less UK liabilities)												
Monetary financial institutions	LTNU	61.7	53.0	65.5	135.5	137.0	149.0	153.8	131.3	130.1	156.3	192.5
Central government	ZPOH	−62.2	−66.5	−66.1	−74.3	−61.1	−62.4	−59.5	−55.6	−68.0	−87.8	−114.1
Local authorities	HHGG	–	–	–	–	–	–	–	–	–	–	–
Public corporations	-HEWM	–	–	–	–	–	–	–	–	–	–	–
Other sectors	LTNV	93.5	81.7	68.3	−50.1	−66.5	−178.7	−115.4	−124.0	−173.6	−154.1	−164.8
Total	CGNH	**93.0**	**68.3**	**67.7**	**11.1**	**9.5**	**−92.2**	**−21.1**	**−48.3**	**−111.5**	**−85.5**	**−86.3**

1 The household sector includes non-profit institutions serving households.

8.7 Other investment
Balance sheets valued at end of year

£ billion

		1995	1996	1997	1998	1999	2000	2001	2002	2003	2004	2005
Other investment abroad (UK assets)												
Trade credit												
Long-term												
Central government	ZPOC	7.8	8.2	8.2	8.2	–	–	–	–	–	–	–
Other sectors[1]	HCLK	0.5	0.5	0.5	0.5	–	–	–	–	–	–	–
Total long-term trade credit	HHGU	8.3	8.7	8.7	8.7	–	–	–	–	–	–	–
Short-term												
Other sectors[1]	HLXU	1.6	3.2	2.6	1.4	0.5	0.4	0.1	0.4	1.0	0.6	1.1
Total trade credit	HLXP	9.9	11.9	11.3	10.1	0.5	0.4	0.1	0.4	1.0	0.6	1.1
Loans												
Long-term												
Bank loans under ECGD guarantee	HCFQ	6.3	5.2	5.8	6.0	6.0	4.8	5.1	3.8	3.7	3.7	4.1
Inter-government loans by the UK and other central government assets	HCFN	0.6	0.5	0.5	0.3	0.3	0.3	0.2	0.2	0.2	0.2	0.2
Loans by Commonwealth Development Corporation (public corporations)	HEWZ	1.1	1.2	1.2	1.1	1.1	0.5	0.4	0.4	0.4	0.3	0.3
Loans by the Export Credit Guarantee Department	CY94	..	..	2.4	2.3	2.4	2.4	2.6	2.4	2.4	2.5	2.2
Loans by specialist leasing companies[1]	HGIH	–	–	–	–	–	–	–	–	–	–	–
Total long-term loans	HFAX	8.0	6.9	9.8	9.7	9.7	8.0	8.3	6.8	6.6	6.8	6.8
Short-term												
By monetary financial institutions												
By banks												
Sterling loans	NLHN	13.9	18.7	24.0	23.4	26.1	27.6	32.3	37.6	40.2	47.4	66.9
Foreign currency loans	ZPOO	110.9	127.8	168.6	180.1	189.1	252.5	290.9	291.0	358.3	448.9	575.6
Total banks	HEQS	124.8	146.5	192.6	203.5	215.3	280.1	323.2	328.6	398.4	496.3	642.4
By building societies	NLHP	–	–	–	–	–	–	–	–	–	–	–
Total monetary financial institutions	ZPOM	124.8	146.5	192.6	203.5	215.3	280.1	323.2	328.6	398.5	496.3	642.5
By other sectors	HLXI	0.6	0.7	0.7	0.6	0.5	0.5	0.5	0.5	0.5	0.5	0.5
Total short-term loans	VTUM	125.4	147.2	193.3	204.0	215.8	280.6	323.8	329.1	399.0	496.9	643.0
Total loans	HLXQ	133.4	154.2	203.1	213.7	225.5	288.6	332.1	335.9	405.6	503.6	649.8
Currency and deposits												
Foreign notes and coin												
Monetary financial institutions (banks)	TAAF	0.1	0.1	0.1	0.2	0.1	0.1	0.1	0.1	0.1	0.1	0.1
Other sectors[1]	CGML	0.3	0.3	0.3	0.3	0.4	0.4	0.4	0.4	0.5	0.5	0.5
Total foreign notes and coin	HEOX	0.4	0.4	0.5	0.5	0.5	0.5	0.5	0.5	0.6	0.6	0.6
Deposits abroad by UK residents												
Deposits by monetary financial institutions												
Deposits by banks												
Sterling deposits	HFBB	42.9	47.0	83.2	89.2	75.6	94.8	102.1	94.6	113.0	110.1	144.3
Foreign currency deposits	HFBG	462.6	426.3	544.1	576.5	546.4	677.8	700.1	746.1	806.3	894.8	1 053.1
Total deposits by UK banks	HLXL	505.5	473.4	627.3	665.7	621.9	772.7	802.2	840.7	919.3	1 004.9	1 197.4
Deposits by building societies	TAIC	1.8	1.3	1.0	0.9	0.5	1.0	0.9	0.4	0.2	0.3	0.2
Total deposits by monetary financial institutions	VTWL	507.3	474.7	628.3	666.5	622.4	773.7	803.2	841.1	919.5	1 005.2	1 197.6
Deposits by securities dealers	HGUX	79.2	129.8	129.1	111.5	152.2	206.1	261.9	242.0	289.7	315.7	497.4
Deposits by other UK residents[1]	HHGW	76.5	77.6	95.0	102.0	124.6	153.5	170.6	210.2	262.8	324.3	374.3
Total deposits abroad	HBXS	663.0	682.1	852.4	880.0	899.2	1 133.3	1 235.7	1 293.3	1 472.0	1 645.3	2 069.3
Total currency and deposits	HBVS	663.3	682.5	852.8	880.5	899.7	1 133.7	1 236.1	1 293.8	1 472.6	1 645.9	2 069.9

1 These series relate to non-governmental sectors other than monetary financial institutions.

8.7 Other investment
Balance sheets valued at end of year
continued

£ billion

		1995	1996	1997	1998	1999	2000	2001	2002	2003	2004	2005
Other investment abroad - *continued* (UK assets)												
Other assets													
Central government assets												
Central government subscriptions to international organisations												
Regional development banks	HEXW	0.8	0.8	0.9	1.0	1.0	1.1	1.1	1.2	1.3	1.3	1.4
European Investment Bank (EIB)	HEXX	0.4	0.4	0.4	0.4	0.4	0.4	0.4	0.4	0.4	0.4	0.4
Other subscriptions	HEXZ	0.3	0.3	0.3	0.3	0.4	0.4	0.4	0.4	0.4	0.5	0.5
Total central government subscriptions	HLXO	1.5	1.6	1.6	1.7	1.8	1.8	1.9	2.0	2.1	2.2	2.3
Other long-term central government assets	XBJL	–	–	–	–	–	–	–	–	–	–	–
Other short-term central government assets	LOEM	–0.4	1.1	1.2	1.2	1.5	1.8	1.7	2.3	2.5	2.7	2.9
Total central government	LOET	1.1	2.7	2.8	2.9	3.3	3.6	3.7	4.3	4.6	5.0	5.2
Other sectors assets												
Long-term assets[1]	HLXM	–	–	–	–	–	–	–	–	–	–	–
Short-term assets												
Public corporations assets abroad	HGJM	–	–	–	–	–	–	–	–	–	–	–
Other[1]	HHGY	0.4	0.5	0.4	0.5	0.8	1.1	1.1	1.4	1.3	1.1	1.1
Total short-term assets	HLXJ	0.4	0.5	0.4	0.5	0.8	1.1	1.1	1.4	1.3	1.1	1.1
Total other sectors	HLXN	0.4	0.5	0.4	0.5	0.8	1.1	1.1	1.4	1.3	1.1	1.1
Total other assets	HLXS	1.5	3.2	3.2	3.4	4.0	4.7	4.8	5.7	5.9	6.0	6.3
Total	HLXV	808.1	851.7	1 070.4	1 107.7	1 129.7	1 427.5	1 573.1	1 635.8	1 885.1	2 156.2	2 727.1

1 These series relate to non-governmental sectors other than monetary financial institutions.

8.7 Other investment
Balance sheets valued at end of year
continued

£ billion

		1995	1996	1997	1998	1999	2000	2001	2002	2003	2004	2005
Other investment in the UK (UK liabilities)												
Trade credit												
Long-term [1]	HBWC	1.5	1.5	1.5	1.5	–	–	–	–	–	–	–
Short-term [1]	HCGB	1.3	1.2	1.2	1.2	1.0	1.1	1.1	1.0	0.9	0.9	1.0
Total trade credit	HLYL	2.7	2.7	2.7	2.7	1.0	1.1	1.1	1.0	0.9	0.9	1.0
Loans												
Long-term loans to:												
Central government	HHGZ	1.0	0.8	0.6	0.4	0.4	0.6	0.5	0.4	0.2	0.1	0.1
Local authorities	HHHA	1.2	1.2	1.1	1.2	1.1	0.8	0.8	0.9	1.1	1.7	1.9
Public corporations	HHHB	0.2	0.2	–	–	–	–	–	–	–	–	–
Other [1]	AQBX	–	–	–	–	–	–	–	–	–	–	–
Total long-term loans	HHHC	2.3	2.2	1.7	1.6	1.4	1.4	1.3	1.2	1.3	1.9	1.9
Short-term loans to:												
Central government	HHHD	–	–	–	–	–	–	–	–	–	–	–
Local authorities	HHHE	–	–	–	–	–	–	–	–	–	–	–
Securities dealers	HHHF	113.4	178.0	204.3	198.3	235.7	286.6	344.6	307.6	337.0	372.0	610.6
Other [1]	HHHG	81.5	80.1	94.2	95.5	116.3	127.8	154.3	179.9	218.2	265.5	286.9
Total short-term loans	HHHJ	194.9	258.1	298.5	293.8	352.0	414.4	498.9	487.4	555.2	637.5	897.5
Total loans	HLYI	197.2	260.2	300.2	295.4	353.4	415.8	500.2	488.6	556.5	639.3	899.4
Currency and deposits												
Sterling notes and coin												
Notes (issued by Bank of England)	HLVG	0.7	0.7	0.8	0.9	1.0	1.0	1.0	1.1	1.1	1.2	1.3
Coins (issued by Royal Mint)	HLVH	0.1	0.1	0.1	0.1	0.1	0.1	0.1	0.1	0.1	0.1	0.1
Total notes and coin	APME	0.8	0.8	0.9	1.0	1.1	1.1	1.1	1.2	1.3	1.4	1.4
Deposits from abroad with UK residents												
Deposits with monetary financial institutions												
Deposits with banks												
Sterling deposits	NLCZ	103.7	106.5	134.4	147.2	167.5	200.4	215.9	228.0	251.7	279.6	331.3
Foreign currency deposits	NLDA	688.2	664.6	814.9	886.7	859.0	1 060.0	1 152.5	1 206.6	1 347.9	1 570.0	1 861.5
Total deposits with banks	CGEH	791.9	771.1	949.3	1 033.9	1 026.5	1 260.4	1 368.4	1 434.6	1 599.6	1 849.6	2 192.8
Deposits with building societies	NLDB	8.9	9.9	4.0	4.9	5.2	4.1	4.6	4.9	5.4	5.7	6.9
Total deposits with UK monetary financial institutions	HDKG	800.8	781.0	953.4	1 038.8	1 031.7	1 264.6	1 373.0	1 439.5	1 605.1	1 855.4	2 199.8
Deposit liabilities of UK central government	HEYH	1.7	1.1	0.3	0.6	1.3	1.8	1.7	1.6	1.9	1.0	0.9
Total deposits from abroad with UK residents	HBYA	802.5	782.1	953.7	1 039.5	1 033.0	1 266.4	1 374.7	1 441.1	1 606.9	1 856.4	2 200.7
Total currency and deposits	HLVI	803.3	782.9	954.6	1 040.4	1 034.1	1 267.5	1 375.8	1 442.3	1 608.2	1 857.8	2 202.1
Other liabilities												
Long-term												
Net equity of foreign households in life insurance reserves and in pension funds	VTUE	0.2	0.2	0.2	0.2	0.2	0.2	0.2	0.2	0.2	0.2	0.2
Prepayments of premiums and reserves against oustanding claims	NQLR	9.5	14.7	15.5	15.0	14.1	10.8	10.7	12.6	10.2	10.4	11.9
Total long-term liabilities[1]	VTUF	9.6	14.8	15.7	15.2	14.3	11.0	10.9	12.9	10.4	10.6	12.1
Short-term[1]	HBMV	–	1.1	1.1	1.3	1.1	1.1	1.6	1.0	1.0	0.9	1.0
Total other liabilities	HLYM	9.7	16.0	16.8	16.5	15.4	12.0	12.5	13.8	11.4	11.5	13.2
Total	HLYD	1 013.0	1 061.7	1 274.3	1 355.0	1 403.9	1 696.4	1 889.6	1 945.8	2 177.1	2 509.4	3 115.6

1 These series relate to non-governmental sectors other than monetary financial institutions.

8.7 Other investment
Balance sheets valued at end of year
continued

£ billion

		1995	1996	1997	1998	1999	2000	2001	2002	2003	2004	2005
Net international investment position (UK assets less UK liabilities)												
Trade credit	LTNW	7.2	9.2	8.6	7.4	−0.5	−0.7	−1.0	−0.6	0.1	−0.2	0.1
Loans	LTNX	−63.9	−106.1	−97.2	−81.7	−127.8	−127.1	−168.1	−152.8	−150.9	−135.7	−249.6
Currency and deposits	LTNY	−140.0	−100.4	−101.7	−160.0	−134.4	−133.8	−139.7	−148.5	−135.6	−211.9	−132.2
Other	LTNZ	−8.2	−12.8	−13.7	−13.1	−11.4	−7.3	−7.7	−8.2	−5.5	−5.4	−6.8
Total	**CGNG**	**−204.9**	**−210.1**	**−204.0**	**−247.3**	**−274.2**	**−268.9**	**−316.5**	**−310.0**	**−292.0**	**−353.2**	**−388.5**

8.8 Other investment
Sector analysis
Balance sheets valued at end of year

£ billion

		1995	1996	1997	1998	1999	2000	2001	2002	2003	2004	2005
Other investment abroad (UK assets)												
Investment by:												
Monetary financial institutions												
Banks	CGEI	636.7	625.2	825.8	875.3	843.4	1 057.6	1 130.6	1 173.2	1 321.5	1 505.1	1 844.1
Building societies	HEQT	1.8	1.3	1.0	0.9	0.5	1.0	0.9	0.4	0.2	0.3	0.2
Total monetary financial institutions	VTXD	638.4	626.6	826.8	876.2	843.9	1 058.6	1 131.6	1 173.5	1 321.8	1 505.4	1 844.3
Central government	CGEN	9.5	11.4	11.5	11.4	3.5	3.9	3.9	4.5	4.8	5.1	5.4
Public corporations	CGEO	1.1	1.2	3.6	3.4	3.4	2.9	3.0	2.8	2.8	2.9	2.5
Other sectors	CGGH	159.1	212.5	228.5	216.7	278.9	362.1	434.6	454.9	555.7	642.8	874.9
Total	**HLXV**	**808.1**	**851.7**	**1 070.4**	**1 107.7**	**1 129.7**	**1 427.5**	**1 573.1**	**1 635.8**	**1 885.1**	**2 156.2**	**2 727.1**
Other investment in the UK (UK liabilities)												
Investment in:												
Monetary financial institutions												
Banks	CGOV	792.7	771.8	950.1	1 034.8	1 027.4	1 261.5	1 369.4	1 435.6	1 600.8	1 850.9	2 194.1
Building societies	NLDB	8.9	9.9	4.0	4.9	5.2	4.1	4.6	4.9	5.4	5.7	6.9
Total monetary financial institutions	CGHB	801.6	781.7	954.2	1 039.7	1 032.7	1 265.6	1 374.0	1 440.6	1 606.2	1 856.6	2 201.0
Central government	CGHG	3.5	2.7	1.8	2.1	2.5	3.2	3.4	2.6	2.8	1.7	1.8
Local authorities	CGHX	1.2	1.2	1.1	1.2	1.1	0.8	0.8	0.9	1.1	1.7	1.9
Public corporations	ZPOX	0.2	0.2	–	–	–	–	–	–	–	–	–
Other sectors	CGNC	206.5	275.9	317.3	312.0	367.7	426.9	511.3	501.8	567.0	649.3	911.0
Total	**HLYD**	**1 013.0**	**1 061.7**	**1 274.3**	**1 355.0**	**1 403.9**	**1 696.4**	**1 889.6**	**1 945.8**	**2 177.1**	**2 509.4**	**3 115.6**
Net international investment position (UK assets less UK liabilities)												
Monetary financial institutions												
Banks	LTOA	−156.0	−146.6	−124.3	−159.5	−184.1	−203.8	−238.8	−262.4	−279.2	−345.8	−350.0
Building societies	LTOB	−7.1	−8.6	−3.0	−4.1	−4.7	−3.1	−3.7	−4.6	−5.2	−5.4	−6.7
Total monetary financial institutions	LTOC	−163.1	−155.2	−127.4	−163.6	−188.8	−207.0	−242.5	−267.0	−284.4	−351.2	−356.7
Central government	LTOD	6.0	8.7	9.8	9.4	1.0	0.7	0.5	1.9	2.0	3.4	3.6
Local authorities	-CGHX	−1.2	−1.2	−1.1	−1.2	−1.1	−0.8	−0.8	−0.9	−1.1	−1.7	−1.9
Public corporations	LTOE	0.9	1.0	3.5	3.4	3.4	2.9	3.0	2.8	2.8	2.9	2.5
Other sectors	LTOF	−47.5	−63.3	−88.7	−95.3	−88.7	−64.8	−76.7	−46.8	−11.3	−6.5	−36.0
Total	**CGNG**	**−204.9**	**−210.1**	**−204.0**	**−247.3**	**−274.2**	**−268.9**	**−316.5**	**−310.0**	**−292.0**	**−353.2**	**−388.5**

8.9 Reserve assets
Central government sector
Balance sheets valued at end of year

£ billion

		1995	1996	1997	1998	1999	2000	2001	2002	2003	2004	2005
Monetary gold	HCGD	4.6	4.0	3.2	4.0	3.7	2.9	2.2	2.1	2.3	2.3	3.0
Special drawing rights	HCGE	0.3	0.2	0.3	0.3	0.3	0.2	0.2	0.2	0.2	0.2	0.2
Reserve position in the Fund	HCGF	1.6	1.4	1.8	2.6	3.3	2.9	3.5	3.8	3.5	2.9	1.0
Foreign exchange												
Currency and deposits												
With central banks	CGDE	3.5	2.9	3.0	0.8	0.4	0.1	0.1	0.2	0.1	0.1	0.1
With other banks	CGDF	2.2	3.1	2.9	2.6	5.0	3.7	2.8	1.9	1.3	0.3	0.8
Total currency and deposits	CGDD	5.7	5.9	5.9	3.4	5.5	3.7	2.9	2.1	1.4	0.4	1.0
Securities												
Bonds and notes	CGDH	17.0	14.1	10.6	10.9	7.6	16.7	14.4	16.8	16.2	17.1	17.5
Money market instruments	CGDL	2.6	1.7	1.0	2.1	1.8	2.3	2.2	0.2	0.2	0.3	1.7
Total securities	CGDG	19.6	15.8	11.6	13.0	9.5	19.0	16.6	17.0	16.4	17.4	19.2
Total foreign exchange	HCGG	25.3	21.7	17.6	16.4	14.9	22.7	19.4	19.1	17.7	17.8	20.2
Other claims	CGDM	–	–	–	–	–	0.1	0.4	0.2	–	0.1	0.4
Total	LTEB	31.8	27.3	22.8	23.3	22.2	28.8	25.6	25.5	23.8	23.3	24.7

8.10 External debt statement
Balance sheets valued at end of year

£ billion

		2000	2001	2002	2003	2004	2005
General Government							
Short-term							
Money market instruments	HLYU	–	0.1	0.2	1.9	4.0	2.8
Currency and deposits	HLVH	0.1	0.1	0.1	0.1	0.1	0.1
Other liabilities	VTZZ	1.8	1.7	1.6	1.9	1.0	0.9
Total short-term	ZAVF	2.0	1.9	1.9	3.9	5.1	3.8
Long-term							
Bonds and notes issued by central government	HHGF	62.4	59.9	56.4	66.1	83.8	111.4
Loans							
to central government	HHGZ	0.6	0.5	0.4	0.2	0.1	0.1
to local authorities	HHHA	0.8	0.8	0.9	1.1	1.7	1.9
Total long-term	ZAVG	63.7	61.2	57.6	67.4	85.7	113.4
Total General Government liabilities	ZAVH	65.7	63.0	59.5	71.4	90.8	117.2
Monetary Authorities							
Short-term							
Money market instruments	VTZS	0.3	1.4	2.8	3.5	3.4	3.6
Currency and deposits	VTZT	5.2	3.8	5.5	6.8	9.9	12.9
Total short-term	VTZY	5.5	5.2	8.3	10.3	13.3	16.4
Long-term							
Bonds and notes	VTZU	–	–	–	–	–	–
Total long-term	VTZV	–	–	–	–	–	–
Total Monetary Authorities liabilities	VTZW	5.5	5.2	8.3	10.3	13.3	16.4
Banks							
Short-term							
Money market instruments							
Banks	ZAVC	107.3	128.5	134.5	119.7	119.7	125.0
Building societies	ZAVD	3.4	3.2	3.0	7.4	7.7	7.2
Total money market instruments	ZAUX	110.7	131.7	137.5	127.1	127.4	132.2
Currency and deposits							
Banks	VTZX	1 255.2	1 364.6	1 429.1	1 592.8	1 839.7	2 179.9
Building societies	NLDB	4.1	4.6	4.9	5.4	5.7	6.9
Total short-term	ZAVI	1 370.0	1 500.9	1 571.5	1 725.4	1 972.9	2 319.1
Long-term							
Bonds and notes	HMBF	77.4	84.1	95.0	121.7	154.0	201.9
Total long-term	ZPOK	77.4	84.1	95.0	121.7	154.0	201.9
Total Banks liabilities	ZAVA	1 447.4	1 585.0	1 666.6	1 847.1	2 126.8	2 521.0
Other sectors							
Short-term							
Money market instruments	HLYQ	21.7	22.5	30.6	23.7	22.6	22.0
Loans	ZLBY	414.4	498.9	487.4	555.2	637.5	897.5
Trade credits	HCGB	1.1	1.1	1.0	0.9	0.9	1.0
Other liabilities	LSYR	1.1	1.6	1.0	1.0	0.9	1.0
Total short-term liabilities	ZAVB	438.3	524.1	520.0	580.9	661.8	921.4
Long-term							
Bond and notes	HHGJ	121.3	129.4	160.1	211.7	258.3	331.1
Loans	ZLBZ	–	–	–	–	–	–
Trade credits	HBWC	–	–	–	–	–	–
Other liabilities	VTUF	11.0	10.9	12.9	10.4	10.6	12.1
Total long-term liabilities	ZAUQ	132.3	140.3	172.9	222.1	268.9	343.3
Total other sectors liabilities	ZAUR	570.6	664.4	693.0	803.0	930.7	1 264.7
Direct investment							
Debt liabilities to affiliated enterprises	HHDJ	84.3	97.4	128.4	127.7	151.2	155.8
Debt liabilities to direct investors	HBVB	123.2	160.0	173.0	168.9	171.4	172.4
Total liabilities to direct investors	ZAUY	207.5	257.4	301.4	296.6	322.7	328.2
GROSS EXTERNAL DEBT	ZAUS	2 296.7	2 575.1	2 728.7	3 028.4	3 484.3	4 247.5

FD Financial derivatives[1]
Balance sheets valued at end of year

£ billion

		1999	2000	2001	2002	2003	2004	2005
Financial derivatives assets								
UK banks								
Sterling	ZPNP	29.4	49.8	43.5	56.7	44.1	46.1	51.3
Foreign currency	ZPNQ	360.8	340.6	481.0	626.2	579.3	663.3	768.6
Total UK banks	ZPNA	390.2	390.5	524.5	682.8	623.4	709.4	819.8
Other Financial Intermediaries								
UK securities dealers								
Sterling	RUVI	4.6	3.2	13.2	16.2	10.6	11.5	15.0
Foreign currency	RUVJ	58.0	52.6	51.9	70.9	144.0	104.7	186.5
Total UK securities dealers	RVAP	62.6	55.7	65.1	87.1	154.7	116.2	201.5
Other[2]	D4AG	0.2	0.5	0.6	0.7	1.1	0.4	..
Total Other Financial Intermediaries	D4AH	62.8	56.2	65.7	87.8	155.8	116.6	..
Insurance companies and pension funds								
Insurance companies[3]	D4AE	0.7	0.7	1.0	0.8	0.2	–	..
Pension funds[4]	GOJU	0.9	0.8	0.8	0.7	0.6	3.0	..
Total insurance companies and pension funds	D4AF	1.6	1.5	1.7	1.5	0.8	3.0	..
Total UK assets	ZPNC	454.6	448.2	592.0	772.1	780.0	828.9	..
Financial derivative liabilities								
UK banks								
Sterling	ZPNR	36.1	48.2	43.8	57.1	32.4	36.3	66.2
Foreign currency	ZPNS	351.9	351.8	485.8	631.5	600.0	678.8	764.6
Total UK banks	ZPNB	388.0	400.0	529.6	688.7	632.4	715.0	830.9
Other Financial Intermediaries								
UK securities dealers								
Sterling	RUXE	5.3	4.3	13.6	17.2	14.0	14.0	18.2
Foreign currency	RUXF	51.9	46.5	50.2	73.7	150.0	112.2	183.1
Total UK securities dealers	RVAV	57.1	50.9	63.8	90.9	163.9	126.2	201.3
Other[2]	D4AK	0.1	0.1	–	0.1	0.7	0.1	..
Total Other Financial Intermediaries	D4AL	57.2	51.0	63.8	91.0	164.6	126.3	..
Insurance companies and pension funds								
Insurance companies[3]	D4AI	0.3	0.3	0.2	0.5	0.2	0.1	..
Pension funds[4]	GKGR	1.0	0.6	0.7	0.4	0.4	3.1	..
Total insurance companies and pension funds	D4AJ	1.3	0.9	0.9	0.9	0.7	3.2	..
Total UK liabilities	ZPND	446.6	451.9	594.3	780.6	797.7	844.6	..
Net international investment position								
Banks	ZPNE	2.2	–9.5	–5.1	–5.8	–9.0	–5.6	–11.0
Other Financial Intermediaries								
Securities dealers	ZPNF	5.5	4.9	1.3	–3.8	–9.3	–10.0	0.2
Other[2]	D4AP	0.1	0.3	0.6	0.6	0.4	0.3	..
Total Other Financial Intermediaries	D4AQ	5.6	5.2	1.9	–3.2	–8.8	–9.8	..
Insurance companies and pension funds								
Insurance companies[3]	D4AM	0.4	0.4	0.8	0.3	–	–0.1	..
Pension funds[4]	D4AN	–0.2	0.2	–	0.3	0.1	–0.1	..
Total insurance companies and pension funds	D4AO	0.3	0.6	0.8	0.6	0.1	–0.3	..
Total	ZPNG	8.1	–3.7	–2.4	–8.5	–17.7	–15.7	..

1 The data in this table are not included in the main aggregates of the UK's international investment position as the data are developmental. Work is continuing to validate and improve the estimates and to obtain more information on the type of derivatives traded. An article assessing the current position can be found at http://www.statistics.gov.uk/articles/economic_trends/ET618Sem.pdf .
2 Includes unit and investment trusts and open-ended investment companies, finance leasing companies, credit grantors, factoring companies and building societies.
3 Includes both general and long-term insurance.
4 Includes self-administered pension funds only.

Part 3

Geographical breakdown

Chapter 9
Geographical breakdown of current account

The tables appearing in this chapter show a geographical breakdown of the current account. The data cover 66 individual countries as well as international organisations. These estimates are generally less firmly based than the world totals, and data for earlier years are less reliable than recent figures. In some cases estimates are unavailable for the first few years.

Changes to the pattern of trading associated with MTIC fraud can make it difficult to analyse trade by country, in particular the breakdown between EU and non-EU trade. Originally, most of the fraud only involved carousel chains (import/export chains) through EU member states. More recently, some chains include exports to non-EU countries, for example, Dubai and Switzerland. However, the MTIC trade adjustments are added to the EU import estimates as it is this part of the trading chain that is not recorded. For more information, see the methodological notes relating to chapter 2.

Data are presented as if the EU expanded to 25 countries on 01 January 1999.

Current account by region

Current account surpluses were recorded with the Americas and Australasia & Oceania in all years since 1992. In contrast, the UK has recorded a rising current account deficit with Europe, particularly since 2002, reaching £46.1 billion in 2005. The current account surplus with the Americas rose to £25.7 billion in 2005. There was a surplus with Asia for the years 1995 to 1997 but an overall deficit in all years since then.

Figure 9.1
Current account
Credit less debits

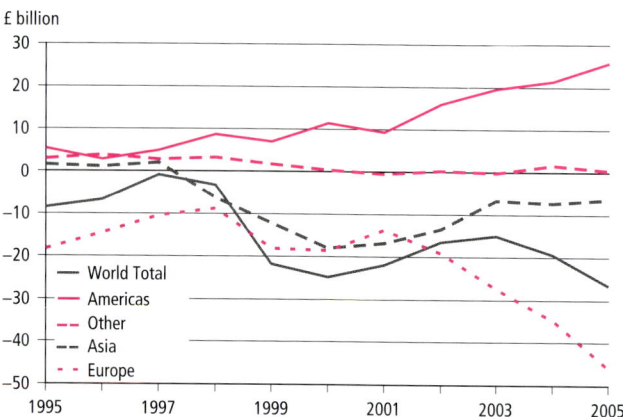

In 2005, around half the value of current account credit and debit transactions were with the 25 European Union (EU) member states. EU countries accounted for about 85 per cent of current account credits and debits with Europe. By component, trade in goods accounted for about half the value of current account credits and debits with Europe, with income accounting for around a further 32 per cent of credits and 26 per cent of debits.

The continent of America accounted for around a quarter of total credits and a fifth of total debits in 2005. Income and goods transactions together accounted for around 75 per cent of total credit and debit flows with the Americas. The United States of America (USA) was the most significant country, representing almost 80 per cent of total current account credits and debits in the region.

Figure 9.2
Current account by continent, 2005

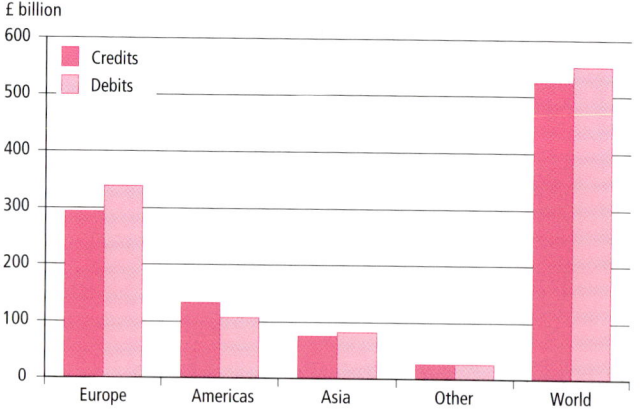

Asia accounted for 14 per cent of UK current account credits in 2005, down from 17 per cent in 1995, largely due to the region accounting for a lower proportion of total income receipts. Similarly, Asia accounts for a lower proportion of total UK debits in recent years, also due to income. The largest component of current account debits was imports of goods, which at £52.5 billion represent 65 percent of the total debits with the region in 2005. Whilst Japan remains the UK's largest current account partner country in Asia, transactions with China have grown the fastest in recent years, largely due to increased exports and – particularly – imports of goods.

The current account with Africa was in surplus until 1999, with the first deficit being recorded in 2000. In 2005 the deficit increased to £2.1 billion. These deficits have mainly been driven by higher imports of goods to the UK.

Current account with EU25, USA and Japan

A current account deficit has been recorded with the EU25 in every year for which data are available (1999 onwards). Broadly speaking, surpluses on the income account are offset by deficits on all the remaining components of the current account. In 2005 the balance with the EU fell to a deficit of £32.2 billion, largely due to a rise in imports of goods from EU countries.

Figure 9.3
Current account with the European Union
Credits less debits

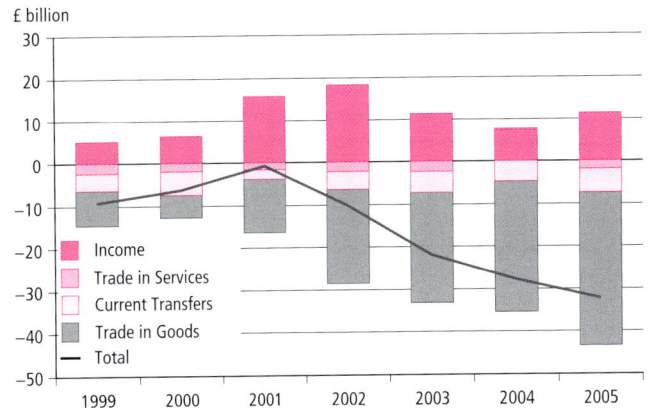

The current account deficit with the EU narrowed from £9.2 billion in 1999 to only £0.7 billion in 2001 before increasing rapidly to £32.2 billion in 2005.

The trade in goods and services deficit with the EU increased to £37.9 billion in 2005, largely due to higher imports of goods from Germany and the Netherlands. Net income received from the EU grew from £5.2 billion in 1999 to £18.3 billion in 2002, before falling back to £11.3 billion in 2005. The deficit on current transfers has remained relatively stable at between £2 billion and £6 billion since 1999. The main components of current transfers are payments to, and receipts from, EU institutions.

The USA is consistently the single largest counterpart country within the UK's balance of payments, representing 19 per cent of current account credits and 15 per cent of debits in 2005. There has been a current account surplus with the USA in all years for which data are available. Prior to 2000 these were typically between £2 billion and £5 billion, whereas the most recent periods have seen significantly higher surpluses. The surplus in 2005 is £19.5 billion, with

Figure 9.4
Current account with the USA
Credits less debits

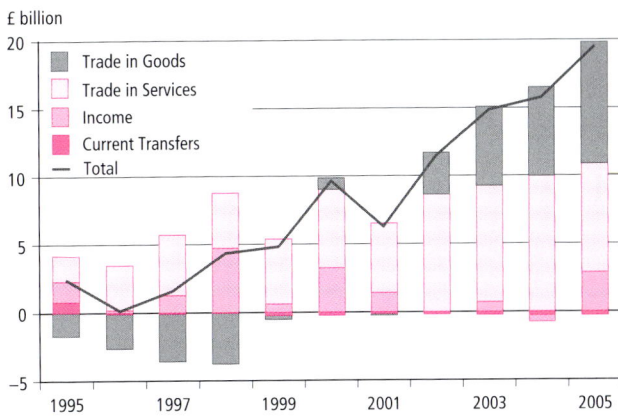

a £9.0 billion surplus on trade in goods, £8.0 billion surplus on services and £2.9 billion surplus on income.

The UK has recorded a current account deficit with Japan in every year since 1992, peaking at £6.6 billion in 2000. Since then the deficit has narrowed to £3.0 billion in 2005.

When ranking individual countries by the size of the current account balance in 2005, the largest surpluses were recorded with: the United States of America (£19.5 billion), Ireland (£6.9 billion), Australia (£3.5 billion), Netherlands (£2.6 billion) and Mexico (£1.0 billion).

The current account has been in surplus with the USA since the geographic split of the data began in 1992. The surpluses with the Netherlands and Australia are driven by a positive balance on investment income; while the surplus with Ireland is largely due to exports exceeding imports of goods and services.

Figure 9.5
Current account: largest five surpluses in 2005
£ billion

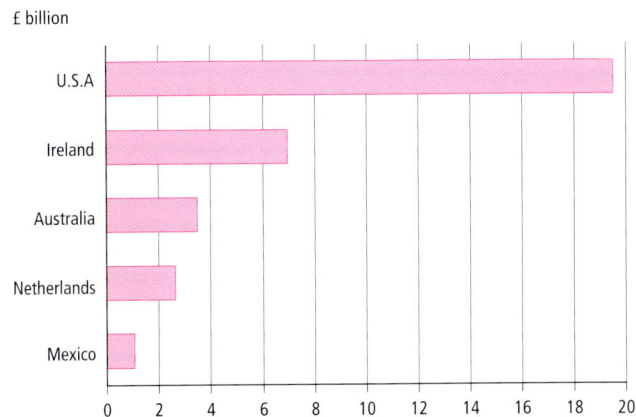

Figure 9.6
Current account: largest 5 deficits in 2005

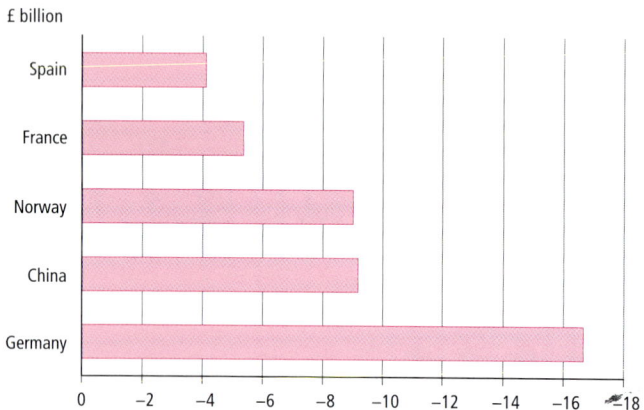

When ranking individual countries by the size of the current account balance, the largest deficits were recorded with: Germany (£16.6 billion), China (£9.2 billion), Norway (£9.0 billion), France (£5.3 billion) and Spain (£4.1 billion).

The largest current account deficit was with Germany, with imports of goods exceeding exports of goods (by £16.0 billion), combined with a deficit on income (£1.6 billion), being partly offset by a surplus on trade in services (£0.9 billion). The deficits with China, Norway and France are all a result of trade in goods deficits, with France also having a deficit on trade in services. A trade in services deficit is the main factor in the balance with Spain. The remaining deficits are largely due to high levels of UK imports of goods from these nations.

2005 Balances/£m
4,000 and above
2,000 to 3,999
0 to 1,999
-1 to -1,999
-2,000 and below

EU current account
£ billion
Balance
Debits
Credits
320
280
240
200
160
120
80
40
0
−40
1999
2000
2001
2002
2003
2004
2005

UK balance with EU
£ billion
20
10
0
−10
−20
−30
−40
−50
Income
Trade in Services
Current Transfers
Trade in Goods
Total
1999
2000
2001
2002
2003
2004
2005

Largest 5 surpluses in 2005
£ billion
Ireland
Netherlands
Sweden
Luxembourg
Slovenia
0
1
2
3
4
5
6
7

Largest 5 deficits in 2005
£ billion
Germany
France
Spain
Italy
Belgium
−18
−16
−14
−12
−10
−8
−6
−4
−2
0

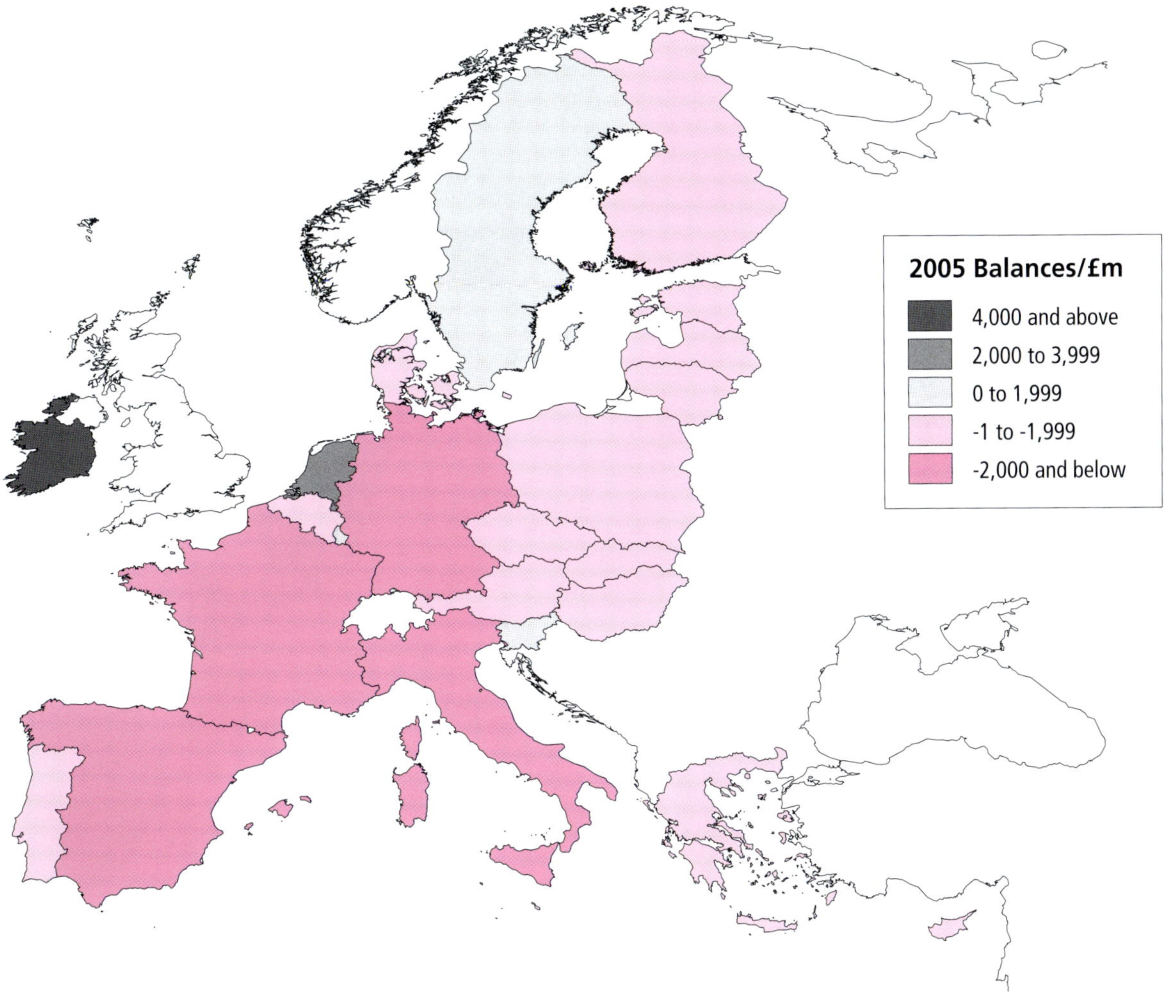

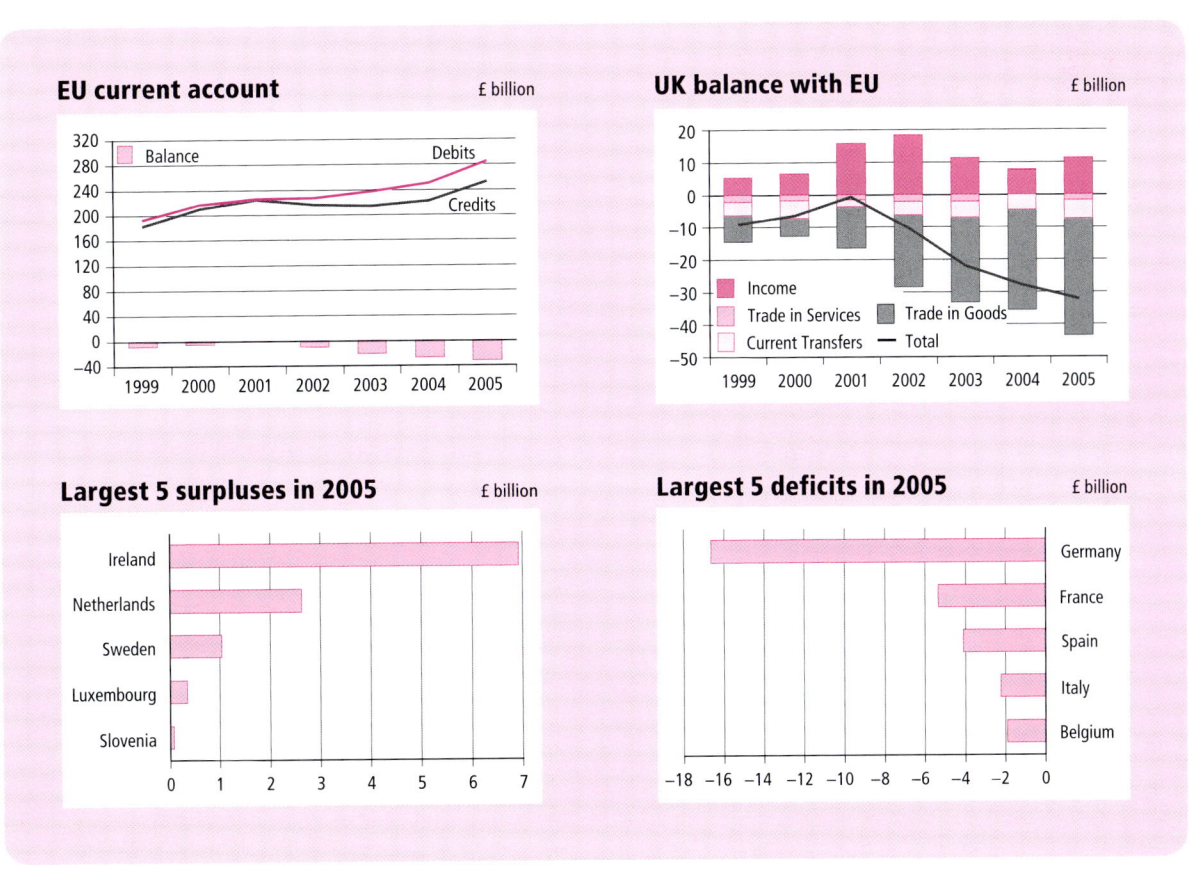

Chapter 9: Geographical breakdown of current account
The Pink Book: 2006 edition

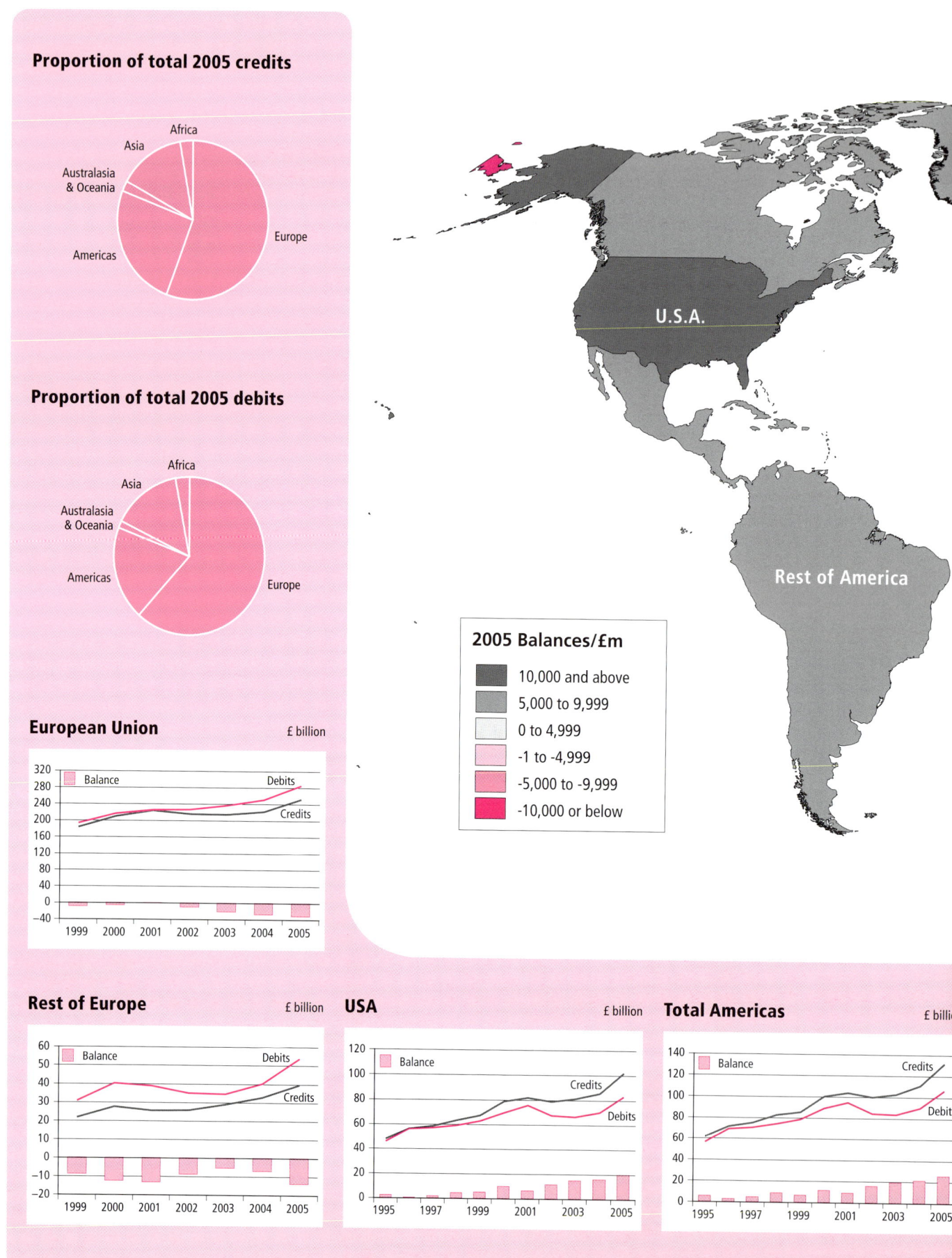

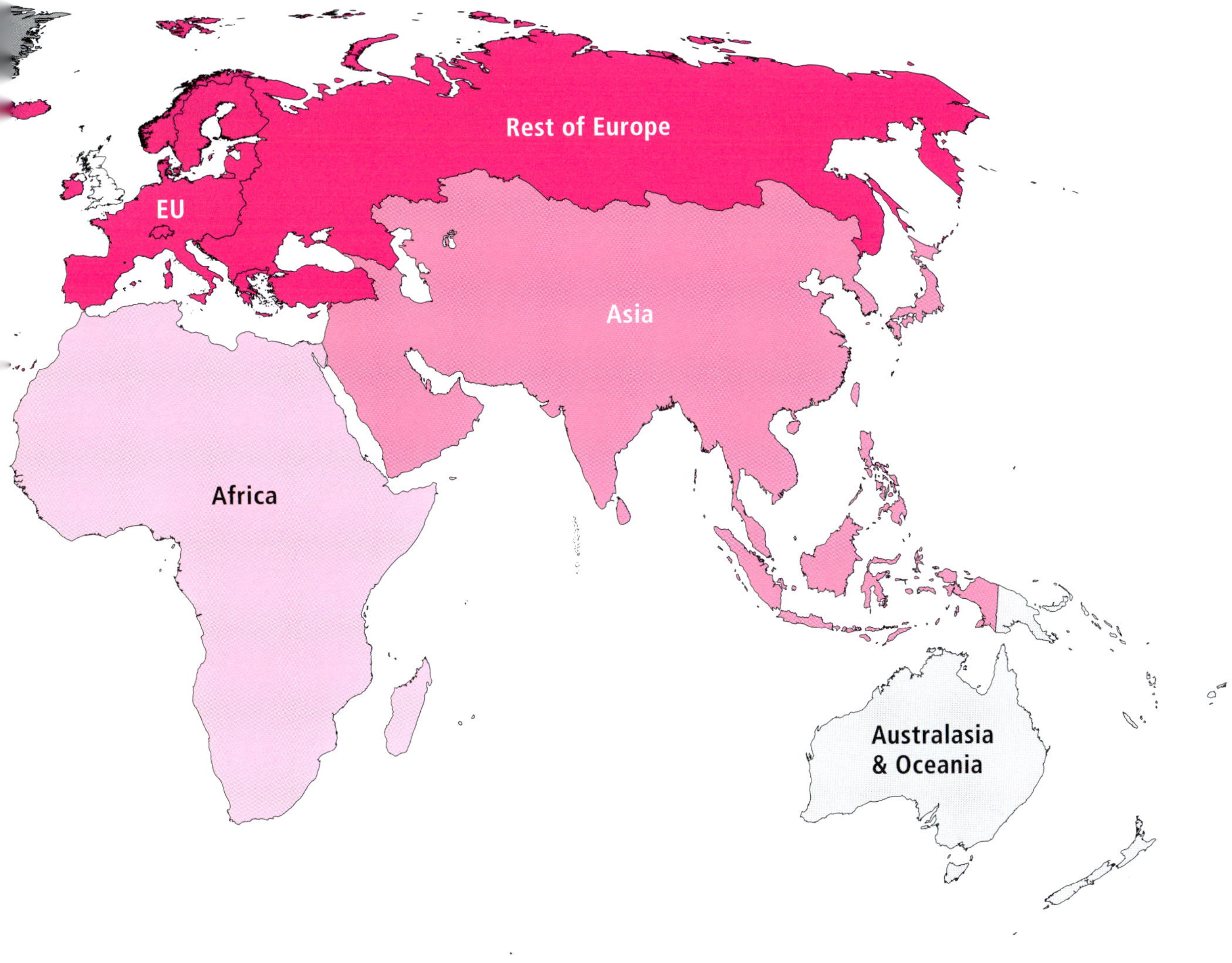

The Pink Book: 2006 edition — Chapter 9: Geographical breakdown of current account

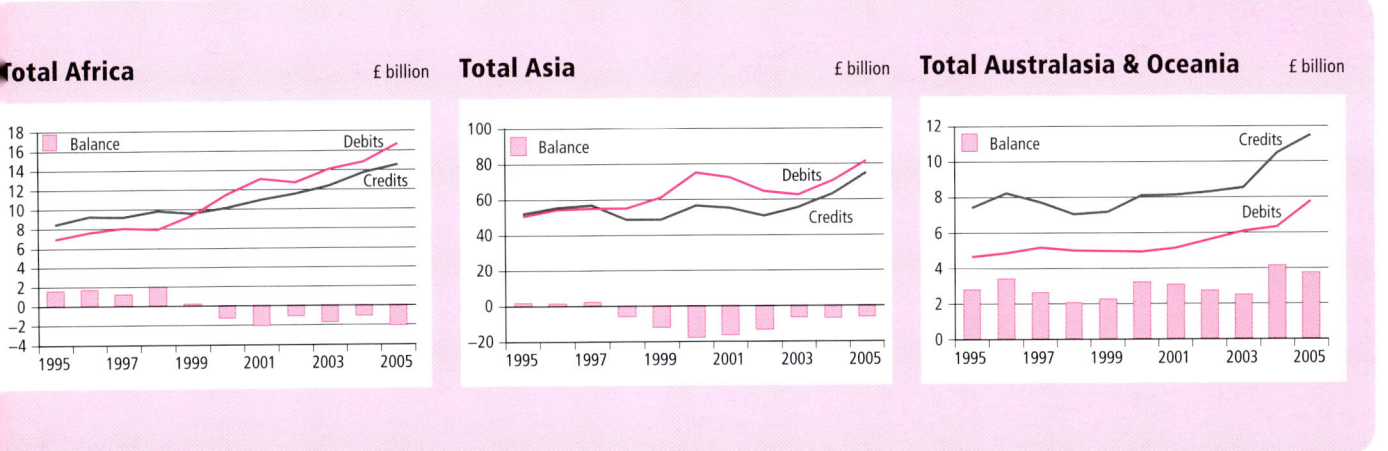

9.1 Current account
Summary transactions in 2005

£ million

Credits	Trade in goods	Trade in services	Income	Current transfers	Current account
Europe					
European Union (EU)					
Austria	1 329	485	994	40	2 848
Belgium	11 182	2 615	3 380	235	17 412
Cyprus	356	289	143	6	794
Czech Republic	1 020	335	250	15	1 620
Denmark	2 306	1 529	1 279	40	5 154
Estonia	114	38	15	–	167
Finland	1 505	866	624	18	3 013
France	19 873	5 949	11 306	278	37 406
Germany	23 021	7 760	11 746	472	42 999
Greece	1 364	787	913	47	3 111
Hungary	828	572	443	5	1 848
Ireland	16 273	5 768	8 010	147	30 198
Italy	8 769	3 442	5 443	154	17 808
Latvia	99	102	2	6	209
Lithuania	167	118	1	5	291
Luxembourg	211	721	4 556	20	5 508
Malta	238	75	259	3	575
Netherlands	12 656	6 589	17 641	359	37 245
Poland	1 603	798	504	81	2 986
Portugal	1 679	611	912	22	3 224
Slovak Republic	255	95	55	9	414
Slovenia	169	50	111	1	331
Spain	10 583	3 506	5 103	123	19 315
Sweden	4 583	1 688	2 619	98	8 988
European Central Bank	–	–	–	–	–
EU Institutions	–	596	547	7 771	8 914
Total EU25	120 183	45 384	76 856	9 955	252 378
European Free Trade Association (EFTA)					
Iceland	..	135	116	37	467
Liechtenstein	2	42	27	1	72
Norway	2 212	1 681	1 533	80	5 506
Switzerland	4 984	4 248	5 961	92	15 285
Total EFTA	7 377	6 106	7 637	210	21 330
Other Europe					
Albania	16	9	–	–	25
Belarus	57	8	–	1	66
Bulgaria	219	99	98	6	422
Croatia	118	41	70	2	231
Romania	641	136	134	7	918
Russia	1 868	1 081	1 533	24	4 506
Turkey	2 159	479	592	38	3 268
Ukraine	279	174	83	2	538
Serbia and Montenegro	58	36	16	2	112
Other	249	2 379	5 731	12	8 371
Total Europe	**133 224**	**55 932**	**92 750**	**10 259**	**292 165**
Americas					
Argentina	167	96	319	5	587
Brazil	836	353	934	11	2 134
Canada	3 278	1 519	2 726	224	7 747
Chile	150	90	607	8	855
Colombia	117	75	390	12	594
Mexico	638	298	974	27	1 937
United States of America	30 914	22 824	44 869	3 706	102 313
Uruguay	39	16	15	–	70
Venezuela	234	95	275	14	618
Other Central American Countries	676	3 128	10 172	213	14 189
Other	332	195	301	14	842
Total Americas	**37 381**	**28 689**	**61 582**	**4 234**	**131 886**
Asia					
China	2 811	1 339	749	14	4 913
Hong Kong	3 090	1 209	5 116	33	9 448
India	2 800	1 102	896	27	4 825
Indonesia	366	178	290	14	848
Iran	452	219	148	6	825
Israel	1 352	382	91	27	1 852
Japan	3 900	4 050	6 720	118	14 788
Malaysia	1 089	431	834	21	2 375
Pakistan	461	421	272	6	1 160
Philippines	279	150	246	8	683
Saudi Arabia	1 559	705	410	490	3 164
Singapore	2 080	2 479	2 576	26	7 161
South Korea	1 677	712	950	26	3 365
Taiwan	939	544	698	14	2 195
Thailand	638	244	284	10	1 176
Residual Gulf Arabian Countries	6 952	3 819	935	174	11 880
Other Near & Middle Eastern Countries	744	450	206	301	1 701
Other	580	1 073	846	40	2 539
Total Asia	**31 769**	**19 507**	**22 267**	**1 355**	**74 898**
Australasia & Oceania					
Australia	2 580	2 584	4 300	240	9 704
New Zealand	415	391	659	57	1 522
Other	81	117	55	7	260
Total Australasia & Oceania	**3 076**	**3 092**	**5 014**	**304**	**11 486**
Africa					
Egypt	543	339	313	10	1 205
Morocco	261	61	29	4	355
South Africa	2 074	1 083	3 046	90	6 293
Other North Africa	484	391	99	17	991
Other	2 363	1 938	1 381	40	5 722
Total Africa	**5 725**	**3 812**	**4 868**	**161**	**14 566**
International Organisations	–	91	544	–	635
World Total	**211 175**	**111 123**	**187 037**	**16 313**	**525 648**

9.1 Current account
Summary transactions in 2005
continued

£ million

	Trade in goods	Trade in services	Income	Current transfers	**Current account**
Debits					
Europe					
European Union (EU)					
Austria	2 426	853	820	38	4 137
Belgium	14 149	1 662	3 417	86	19 314
Cyprus	262	1 215	180	35	1 692
Czech Republic	1 663	453	82	16	2 214
Denmark	4 367	831	1 360	52	6 610
Estonia	360	47	2	4	413
Finland	2 427	359	471	17	3 274
France	22 088	8 518	11 801	342	42 749
Germany	39 018	6 866	13 325	433	59 642
Greece	701	1 992	732	58	3 483
Hungary	1 848	235	30	13	2 126
Ireland	10 395	3 165	9 109	610	23 279
Italy	12 590	4 119	3 145	190	20 044
Latvia	720	38	13	1	772
Lithuania	272	71	11	9	363
Luxembourg	919	240	3 991	14	1 692
Malta	175	365	34	24	598
Netherlands	20 353	3 375	10 669	218	34 615
Poland	2 305	508	251	63	3 127
Portugal	1 965	1 455	428	51	3 899
Slovak Republic	366	55	38	7	466
Slovenia	199	43	10	1	253
Spain	11 380	9 507	2 349	171	23 407
Sweden	5 283	1 260	1 374	36	7 953
European Central Bank	–	–	–	–	–
EU Institutions	–	4	1 939	13 081	15 024
Total EU25	156 231	47 236	65 581	15 570	284 618
European Free Trade Association (EFTA)					
Iceland	345	81	53	4	483
Liechtenstein	13	3	73	1	90
Norway	12 063	861	1 495	73	14 492
Switzerland	3 877	2 164	8 240	141	14 422
Total EFTA	16 298	3 109	9 861	219	29 487
Other Europe					
Albania	–	19	–	7	26
Belarus	270	13	6	2	291
Bulgaria	160	250	28	5	443
Croatia	54	134	40	14	242
Romania	758	153	31	14	956
Russia	4 982	542	1 105	64	6 693
Turkey	3 506	1 013	217	48	4 784
Ukraine	91	95	143	43	372
Serbia and Montenegro	42	45	10	33	130
Other	325	464	9 283	179	10 251
Total Europe	**182 717**	**53 073**	**86 305**	**16 198**	**338 293**
Americas					
Argentina	279	76	5	14	374
Brazil	1 731	263	160	42	2 196
Canada	4 110	1 128	2 338	364	7 940
Chile	477	77	38	12	604
Colombia	296	37	13	18	364
Mexico	442	317	104	29	892
United States of America	21 948	14 834	42 013	4 005	82 800
Uruguay	58	8	9	1	76
Venezuela	385	33	31	15	464
Other Central American Countries	1 183	1 778	6 166	467	9 594
Other	563	168	89	68	888
Total Americas	**31 472**	**18 719**	**50 966**	**5 035**	**106 192**
Asia					
China	12 897	668	367	143	14 075
Hong Kong	6 567	716	1 874	153	9 310
India	2 769	1 247	571	717	5 304
Indonesia	837	103	118	180	1 238
Iran	33	44	109	25	211
Israel	1 000	227	246	36	1 509
Japan	8 611	2 121	6 942	129	17 803
Malaysia	1 806	220	221	61	2 308
Pakistan	486	479	86	217	1 268
Philippines	707	141	33	38	919
Saudi Arabia	1 705	214	582	49	2 550
Singapore	3 810	687	2 221	116	6 834
South Korea	3 041	254	15	27	3 337
Taiwan	2 217	223	235	14	2 689
Thailand	1 712	613	107	50	2 482
Residual Gulf Arabian Countries	2 002	1 444	935	349	4 730
Other Near & Middle Eastern Countries	231	229	193	102	755
Other	2 067	864	214	875	4 020
Total Asia	**52 498**	**10 494**	**15 069**	**3 281**	**81 342**
Australasia & Oceania					
Australia	2 085	1 808	1 954	376	6 223
New Zealand	587	497	99	100	1 283
Other	130	64	40	30	264
Total Australasia & Oceania	**2 802**	**2 369**	**2 093**	**506**	**7 770**
Africa					
Egypt	348	504	244	40	1 136
Morocco	419	174	27	12	632
South Africa	3 919	1 002	943	357	6 221
Other North Africa	932	324	221	23	1 500
Other	3 366	1 365	7 174	1 994	7 174
Total Africa	**8 984**	**3 369**	**1 884**	**2 426**	**16 663**
International Organisations	–	43	837	1 046	1 926
World total	**278 473**	**88 067**	**157 166**	**28 492**	**552 198**

9.1 Current account
Summary transactions in 2005
continued

£ million

Balances	Trade in goods	Trade in services	Income	Current transfers	Current account
Europe					
European Union(EU)					
Austria	−1 097	−368	174	2	−1 289
Belgium	−2 967	953	−37	149	−1 902
Cyprus	94	−926	−37	−29	−898
Czech Republic	−643	−118	168	−1	−594
Denmark	−2 061	698	−81	−12	−1 456
Estonia	−246	−9	13	−4	−246
Finland	−922	507	153	1	−261
France	−2 215	−2 569	−495	−64	−5 343
Germany	−15 997	894	−1 579	39	−16 643
Greece	663	−1 205	181	−11	−372
Hungary	−1 020	337	413	−8	−278
Ireland	5 878	2 603	−1 099	−463	6 919
Italy	−3 821	−677	2 298	−36	−2 236
Latvia	−621	64	−11	5	−563
Lithuania	−105	47	−10	−4	−72
Luxembourg	−708	481	565	6	344
Malta	63	−290	225	−21	−23
Netherlands	−7 697	3 214	6 972	141	2 630
Poland	−702	290	253	18	−141
Portugal	−286	−844	484	−29	−675
Slovak Republic	−111	40	17	2	−52
Slovenia	−30	7	101	–	78
Spain	−797	−6 001	2 754	−48	−4 092
Sweden	−700	428	1 245	62	1 035
European Central Bank	–	–	–	–	–
EU Institutions	–	592	−1 392	−5 310	−6 110
Total EU25	−36 048	−1 852	11 275	−5 615	−32 240
European Free Trade Association (EFTA)					
Iceland	−166	54	63	33	−16
Liechtenstein	−11	39	−46	–	−18
Norway	−9 851	820	38	7	−8 986
Switzerland	1 107	2 084	−2 279	−49	863
Total EFTA	−8 921	2 997	−2 224	−9	−8 157
Other Europe					
Albania	16	−10	–	−7	−1
Belarus	−213	−5	−6	−1	−225
Bulgaria	59	−151	70	1	−21
Croatia	64	−93	30	−12	−11
Romania	−117	−17	103	−7	−38
Russia	−3 114	539	428	−40	−2 187
Turkey	−1 347	−534	375	−10	−1 516
Ukraine	188	79	−60	−41	166
Serbia and Montenegro	16	−9	6	−31	−18
Other	−76	1 915	−3 552	−167	−1 880
Total Europe	**−49 493**	**2 859**	**6 445**	**−5 939**	**−46 128**
Americas					
Argentina	−112	20	314	−9	213
Brazil	−895	90	774	−31	−62
Canada	−832	391	388	−140	−193
Chile	−327	13	569	−4	251
Colombia	−179	38	377	−6	230
Mexico	196	−19	870	−2	1 045
USA	8 966	7 990	2 856	−299	19 513
Uruguay	−19	8	6	−1	−6
Venezuela	−151	62	244	−1	154
Other Central American Countries	−507	1 350	4 006	−254	4 595
Other America	−231	27	212	−54	−46
Total Americas	**5 909**	**9 970**	**10 616**	**−801**	**25 694**
Asia					
China	−10 086	671	382	−129	−9 162
Hong Kong	−3 477	493	3 242	−120	138
India	31	−145	325	−690	−479
Indonesia	−471	75	172	−166	−390
Iran	419	175	39	−19	614
Israel	352	155	−155	−9	343
Japan	−4 711	1 929	−222	−11	−3 015
Malaysia	−717	211	613	−40	67
Pakistan	−25	−58	186	−211	−108
Philippines	−428	9	213	−30	−236
Saudi Arabia	−146	491	−172	441	614
Singapore	−1 730	1 792	355	−90	327
South Korea	−1 364	458	935	−1	28
Taiwan	−1 278	321	463	–	−494
Thailand	−1 074	−369	177	−40	−1 306
Residual Gulf Arabian Countries	4 950	2 375	–	−175	7 150
Other Near & Middle Eastern Countries	513	221	13	199	946
Other	−1 487	209	632	−835	−1 481
Total Asia	**−20 729**	**9 013**	**7 198**	**−1 926**	**−6 444**
Australasia & Oceania					
Australia	495	776	2 346	−136	3 481
New Zealand	−172	−106	560	−43	239
Other	−49	53	15	−23	−4
Total Australasia & Oceania	**274**	**723**	**2 921**	**−202**	**3 716**
Africa					
Egypt	195	−165	69	−30	69
Morocco	−158	−113	2	−8	−277
South Africa	−1 845	81	2 103	−267	72
Other North Africa	−448	67	−122	−6	−509
Other	−1 003	573	932	−1 954	−1 452
Total Africa	**−3 259**	**443**	**2 984**	**−2 265**	**−2 097**
International Organisations	–	48	−293	−1 046	−1 291
World total	**−67 298**	**23 056**	**29 871**	**−12 179**	**−26 550**

9.2 Current account

£ million

		1995	1996	1997	1998	1999	2000	2001	2002	2003	2004	2005
Credits												
Europe												
European Union (EU)												
Austria	CUGP	1 987	2 187	2 027	2 188	2 171	2 341	2 343	2 408	2 471	2 375	2 848
Belgium and Luxembourg	CTFH	13 001	13 708	13 854	14 469	15 354	19 180	19 283	19 253	19 571	21 180	22 920
of which Belgium	AA2Q	..	..	..	..	12 908	14 772	14 641	15 100	15 814	16 106	17 412
Luxembourg	AA2U	..	..	..	..	2 446	4 408	4 642	4 153	3 757	5 074	5 508
Cyprus[1]	AA2R	..	..	..	..	497	613	588	627	608	726	794
Czech Republic	LEPQ	732	955	967	1 013	1 022	1 223	1 457	1 363	1 470	1 537	1 620
Denmark	LEQR	3 293	3 791	3 730	4 199	4 008	4 643	4 677	4 688	4 430	4 555	5 154
Estonia	ZWVK	38	67	95	86	70	110	105	127	125	141	167
Finland	LEUD	2 795	2 952	2 655	2 449	2 441	3 039	3 155	2 876	2 975	2 857	3 013
France	LEUM	24 779	26 881	25 949	26 972	28 162	31 838	34 309	32 010	32 184	33 604	37 406
Germany	LEQI	34 360	34 875	33 921	34 437	34 920	40 126	42 372	39 682	37 200	39 646	42 999
Greece	LEUV	1 969	2 602	2 804	2 446	2 839	3 093	2 833	2 698	2 873	2 969	3 111
Hungary	BFKO	454	606	763	756	878	1 266	1 139	1 328	1 417	1 688	1 848
Ireland	BFLV	11 743	13 112	14 251	15 043	16 457	19 201	22 382	23 822	22 500	26 137	30 198
Italy	BFOD	14 943	15 388	15 612	18 077	15 912	17 670	17 773	16 438	16 584	16 604	17 808
Latvia	ZWVM	79	105	117	114	96	118	119	120	143	142	209
Lithuania	ZWVN	62	107	135	127	115	157	191	185	236	199	291
Malta[1]	AA2V	..	..	..	..	294	314	323	337	386	833	575
Netherlands	BFQF	22 178	24 329	25 797	25 412	27 322	32 846	36 338	34 744	34 335	28 825	37 245
Poland	BFRY	1 249	1 674	1 740	1 559	1 560	1 715	1 962	2 055	2 276	2 392	2 986
Portugal	BFSH	2 242	2 545	2 489	2 657	2 781	2 791	2 885	2 851	2 814	3 087	3 224
Slovak Republic	ZWVP	124	191	230	169	187	247	277	265	301	300	414
Slovenia	ZWVQ	133	147	174	163	218	229	293	294	281	301	331
Spain	LEST	9 252	10 775	10 701	11 456	11 826	13 209	13 105	13 258	14 197	15 800	19 315
Sweden	BFTI	6 560	7 217	7 144	7 309	7 116	8 028	7 988	7 570	7 363	8 461	8 988
European Central Bank	ZWVF	–	–	–	–	–	–	3	12	3	–	–
EU Institutions	CSFH	5 011	7 578	5 833	5 612	6 847	5 932	8 287	7 106	7 950	8 040	8 914
Total EU25	AA2S	..	..	..	..	183 093	209 932	224 196	216 108	214 690	222 399	252 378
European Free Trade Association (EFTA)												
Iceland	BFNH	216	266	268	274	257	334	286	263	304	345	467
Liechtenstein	BFPE	68	94	79	63	76	89	71	65	60	79	72
Norway	BFQO	3 514	3 844	4 268	4 585	3 935	4 132	3 871	3 907	4 059	4 355	5 506
Switzerland	LEOY	7 522	8 817	8 660	8 823	9 753	13 085	12 177	11 394	11 638	11 948	15 285
Total EFTA	CTFQ	11 320	13 021	13 275	13 745	14 021	17 640	16 404	15 629	16 061	16 727	21 330
Other Europe												
Albania	ZWVG	8	14	8	8	15	9	28	25	16	34	25
Belarus	ZWVH	28	33	42	33	30	64	37	43	49	59	66
Bulgaria	ZWVI	134	116	139	116	165	167	249	286	294	323	422
Croatia	ZWVJ	277	171	144	131	134	127	150	178	231	216	231
Romania	ZWVO	217	283	270	286	331	447	490	579	704	863	918
Russia	BFSQ	1 313	1 543	1 857	1 665	1 041	1 371	1 904	2 225	2 706	3 648	4 506
Turkey	BFUJ	1 528	2 028	2 291	2 199	1 976	2 671	1 985	2 007	2 433	2 683	3 268
Ukraine	ZWVR	151	199	205	247	173	200	234	501	395	325	538
Serbia and Montenegro	BFWC	48	104	125	63	72	138	95	93	111	86	112
Other[1]	LEVW	2 942	3 616	5 014	4 818	3 915	4 847	4 137	4 380	6 048	8 037	8 371
Total Europe	LERA	174 950	192 920	194 357	200 023	204 966	237 613	249 909	242 054	243 738	255 400	292 165
Americas												
Argentina	ZWVT	782	932	1 104	1 097	720	967	834	328	414	589	587
Brazil	LENO	2 001	2 261	2 143	2 120	1 471	1 670	1 873	1 873	1 659	1 890	2 134
Canada	LEOP	4 813	5 180	5 409	5 910	6 057	7 302	7 403	6 578	6 990	7 595	7 747
Chile	ZWVU	569	588	565	500	399	477	444	451	519	1 083	855
Colombia	ZWVV	288	388	350	309	279	491	501	421	492	624	594
Mexico	BFPN	1 065	961	1 165	1 115	1 364	1 413	1 299	1 622	1 392	1 743	1 937
United States of America	BFVB	48 050	55 956	58 530	63 323	67 320	78 979	81 759	78 738	81 186	85 592	102 313
Uruguay	ZWVW	87	106	116	102	82	80	66	55	99	41	70
Venezuela	ZWVX	400	457	425	287	278	225	614	602	343	572	618
Other Central American Countries	JISS	2 125	2 853	3 151	6 781	6 350	7 979	8 273	8 789	9 040	10 160	14 189
Other	LEVE	1 400	1 395	1 897	923	885	746	678	634	728	973	842
Total Americas	LESK	61 580	71 512	75 211	82 772	85 205	100 329	103 743	100 090	102 862	110 862	131 886
Asia												
China	LEPH	1 198	1 197	1 382	1 387	1 945	2 228	2 854	2 653	3 250	4 290	4 913
Hong Kong	BFJR	6 606	7 259	6 830	5 773	5 157	5 394	5 196	4 756	5 428	6 988	9 448
India	BFMY	2 416	2 389	2 452	2 154	2 368	3 111	2 958	3 041	3 667	3 846	4 825
Indonesia	BFKX	894	1 268	1 240	844	738	761	754	710	885	798	848
Iran	ZWWA	508	573	518	452	374	464	641	618	736	740	825
Israel	BFMP	1 532	1 754	1 664	1 481	1 922	2 039	1 963	1 931	1 830	1 876	1 852
Japan	BFOM	15 177	14 473	13 873	12 743	13 847	16 424	14 868	12 438	12 870	13 188	14 788
Malaysia	BFPW	2 097	2 141	2 218	1 604	1 771	1 806	1 828	1 881	2 112	2 271	2 375
Pakistan	BFRP	693	772	707	569	593	517	646	600	678	825	1 160
Philippines	BFRG	700	678	859	549	478	643	733	602	686	630	683
Saudi Arabia	BFSZ	3 620	4 762	6 031	5 214	4 357	4 688	4 400	2 696	3 136	2 884	3 164
Singapore	BFTR	5 191	5 339	5 233	3 972	4 274	5 514	5 404	4 493	4 829	6 159	7 161
South Korea	BFOV	1 867	2 226	2 288	1 535	1 683	2 174	2 202	2 670	2 599	3 211	3 365
Taiwan	BFUS	1 343	1 384	1 493	1 290	1 320	1 564	1 471	1 572	1 623	2 012	2 195
Thailand	BFUA	1 419	1 523	1 369	694	788	1 001	1 065	977	1 136	1 196	1 176
Residual Gulf Arabian Countries	JITT	3 739	4 055	4 613	5 264	4 532	5 362	5 145	6 227	7 181	8 473	11 880
Other Near & Middle Eastern Countries	ZWWC	794	954	1 097	1 020	980	1 103	1 310	1 159	1 275	1 573	1 701
Other	LEWF	2 284	2 548	2 820	1 918	1 433	1 810	1 746	1 779	1 634	1 914	2 539
Total Asia	LETC	52 078	55 338	56 731	48 553	48 560	56 603	55 183	50 802	55 555	62 874	74 898
Australasia & Oceania												
Australia	CWBG	5 840	6 534	6 272	5 573	5 929	6 955	6 890	6 919	7 039	8 994	9 704
New Zealand	BFQX	1 300	1 325	1 155	1 001	1 133	986	1 107	1 204	1 286	1 256	1 522
Other	LEVN	289	390	315	474	136	151	148	188	211	217	260
Total Australasia & Oceania	LETU	7 429	8 249	7 742	7 048	7 198	8 092	8 145	8 311	8 536	10 467	11 486
Africa												
Egypt	ZWWE	522	597	694	524	855	1 039	1 046	1 008	952	1 190	1 205
Morocco	ZWWF	357	390	413	424	435	499	482	438	425	413	355
South Africa	BFWU	3 365	3 553	3 360	3 324	3 507	3 670	4 093	4 716	4 851	6 084	6 293
Other North Africa	JIRU	607	623	801	844	571	790	797	675	1 010	916	991
Other	LEWO	3 605	4 117	3 953	4 709	4 198	4 158	4 550	4 726	5 139	5 116	5 722
Total Africa	LERS	8 456	9 280	9 221	9 825	9 566	10 156	10 968	11 562	12 377	13 719	14 566
International Organisations	CTEY	291	220	270	405	419	496	571	553	529	490	635
World total	HBOE	304 784	337 519	343 532	348 626	355 914	413 289	428 519	413 372	423 597	453 812	525 648

1 Cyprus and Malta are included within Other Europe before 1999.

9.2 Current account
continued

£ million

		1995	1996	1997	1998	1999	2000	2001	2002	2003	2004	2005
Debits												
Europe												
European Union (EU)												
Austria	CUGW	2 082	2 315	2 641	2 416	2 340	2 494	3 094	3 777	4 062	3 697	4 137
Belgium and Luxembourg	CTFI	14 197	14 732	14 565	15 652	16 236	17 792	19 493	20 382	20 315	21 915	24 478
of which Belgium	AA34	..	..	..	..	13 461	14 722	15 979	16 806	16 499	17 192	19 314
Luxembourg	AA38	..	..	..	..	2 775	3 070	3 514	3 576	3 816	4 723	5 164
Cyprus[1]	AA35	..	..	..	..	1 065	1 270	1 485	1 379	1 471	1 454	1 692
Czech Republic	LEPR	507	636	747	819	829	1 109	1 415	1 587	1 833	1 799	2 214
Denmark	LEQS	3 322	3 947	3 896	3 618	3 838	4 427	4 528	5 215	5 105	5 240	6 610
Estonia	ZWWN	166	194	212	214	232	371	321	348	302	430	413
Finland	LEUE	3 202	3 450	3 311	2 877	2 979	3 568	3 712	3 344	3 298	2 912	3 274
France	LEUN	27 407	27 669	28 602	29 907	32 613	34 013	34 989	34 379	36 538	38 497	42 749
Germany	LEQJ	40 057	41 157	37 977	36 938	39 811	45 303	44 778	45 045	49 649	53 517	59 642
Greece	LEUW	1 985	2 172	2 249	2 155	2 557	2 968	3 101	2 972	3 086	3 129	3 483
Hungary	BFKP	501	570	655	725	850	869	887	1 039	1 303	1 794	2 126
Ireland	BFLW	10 334	12 058	12 324	12 887	13 509	16 136	19 134	20 720	17 922	19 725	23 279
Italy	BFOE	12 430	14 401	15 457	15 764	15 090	15 442	16 075	15 746	17 192	18 185	20 044
Latvia	ZWWP	247	360	410	362	336	460	472	517	559	735	772
Lithuania	ZWWQ	227	244	212	209	216	288	278	322	311	301	363
Malta[1]	AA39	..	..	..	..	379	407	460	465	513	559	598
Netherlands	BFQG	18 853	20 289	18 670	19 703	22 856	31 075	31 157	26 907	27 255	29 413	34 615
Poland	BFRZ	925	882	1 054	1 043	1 026	1 347	1 582	1 615	1 918	2 311	3 127
Portugal	BFSI	2 613	2 910	3 017	2 901	3 080	3 051	3 079	3 276	3 713	3 743	3 899
Slovak Republic	ZWWS	200	164	168	191	194	214	211	254	290	315	466
Slovenia	ZWWT	169	162	165	196	184	192	200	229	208	217	253
Spain	LESU	9 914	11 586	11 529	13 009	13 450	14 047	16 255	18 521	20 122	20 015	23 407
Sweden	BFTJ	6 366	7 060	6 506	6 117	6 901	7 176	6 884	6 530	6 700	7 207	7 953
European Central Bank	ZWWI	–	–	–	–	1	1	1	1	1	–	–
EU Institutions	CSFI	9 886	9 959	9 014	11 153	11 752	12 211	11 324	11 785	12 995	13 111	15 024
Total EU25	AA36	..	..	..	..	192 324	216 231	224 915	226 355	236 660	250 221	284 618
European Free Trade Association (EFTA)												
Iceland	BFNI	279	301	274	309	321	453	335	348	364	465	483
Liechtenstein	BFPF	137	155	158	163	114	144	123	72	90	70	90
Norway	BFQP	5 306	5 855	6 033	4 100	4 261	6 750	6 861	7 211	7 843	9 983	14 492
Switzerland	LEOZ	13 160	13 558	12 781	12 316	14 731	18 810	16 306	12 897	11 426	10 999	14 422
Total EFTA	CTFR	18 882	19 869	19 246	16 888	19 427	26 157	23 625	20 528	19 723	21 517	29 487
Other Europe												
Albania	ZWWJ	58	53	63	115	48	41	28	26	24	27	26
Belarus	ZWWK	74	64	73	97	67	65	37	44	30	76	291
Bulgaria	ZWWL	217	207	218	196	164	189	183	249	242	339	443
Croatia	ZWWM	121	136	136	273	123	136	157	150	152	166	242
Romania	ZWWR	284	307	357	336	363	462	555	650	783	926	956
Russia	BFSR	1 342	1 661	1 873	1 712	1 678	2 196	2 708	2 540	3 088	4 437	6 693
Turkey	BFUK	1 564	1 819	1 791	1 859	1 857	2 134	2 430	2 991	3 507	4 180	4 784
Ukraine	ZWWU	92	95	133	150	141	137	169	200	202	274	372
Serbia and Montenegro	BFWD	137	188	210	99	94	141	111	100	110	109	130
Other[1]	LEVX	4 890	6 121	7 281	8 284	6 789	8 380	8 834	7 371	6 661	8 017	10 251
Total Europe	LERB	193 251	207 437	204 762	208 865	223 075	256 269	263 755	261 195	271 178	290 299	338 293
Americas												
Argentina	ZWWW	276	373	380	344	335	373	369	314	317	348	374
Brazil	LENP	1 302	1 521	1 357	1 308	1 272	1 513	1 715	1 748	1 772	1 882	2 196
Canada	LEOQ	4 420	4 635	4 512	4 721	5 450	6 889	6 338	6 530	6 346	7 631	7 940
Chile	ZWWX	516	594	566	458	406	506	561	542	484	582	604
Colombia	ZWWY	289	442	319	330	280	331	421	287	278	326	364
Mexico	BFPO	490	645	681	746	815	1 152	1 356	937	880	859	892
United States of America	BFVC	45 606	55 847	56 925	58 944	62 475	69 351	75 492	67 278	66 342	69 866	82 800
Uruguay	ZWWZ	81	99	103	45	55	62	54	74	56	53	76
Venezuela	ZWXA	265	277	216	186	214	282	256	258	178	264	464
Other Central American Countries	JIST	1 797	2 857	3 387	6 118	5 852	7 158	6 385	5 109	5 810	6 906	9 594
Other	LEVF	1 153	1 434	2 012	1 109	1 135	1 302	1 625	1 145	942	889	888
Total Americas	LESL	56 195	68 689	70 376	74 085	78 289	88 919	94 572	84 222	83 405	89 606	106 192
Asia												
China	LEPI	2 253	2 532	2 791	3 283	4 029	5 637	6 660	7 650	9 137	11 373	14 075
Hong Kong	BFJS	6 026	6 332	6 387	6 933	8 608	10 103	9 831	8 292	7 659	7 942	9 310
India	BFMZ	2 202	2 652	2 807	2 686	2 832	3 158	3 506	3 369	3 634	4 323	5 304
Indonesia	BFKY	1 163	1 328	1 189	1 099	1 274	1 382	1 474	1 323	1 355	1 205	1 238
Iran	ZWXD	317	351	202	168	165	316	352	176	174	206	211
Israel	BFMQ	1 080	1 248	1 298	1 281	1 444	1 548	1 546	1 298	1 269	1 315	1 509
Japan	BFON	17 480	17 230	16 532	17 399	18 397	23 073	20 723	16 786	13 628	14 942	17 803
Malaysia	BFPX	1 966	2 961	2 648	2 380	2 536	2 950	2 515	2 166	2 235	2 461	2 308
Pakistan	BFRQ	662	750	808	753	845	900	958	1 194	1 183	1 252	1 268
Philippines	BFRH	461	1 018	908	1 019	1 195	1 486	1 331	1 113	924	824	919
Saudi Arabia	BFTA	2 686	2 521	2 644	2 220	2 035	2 415	2 522	1 231	1 208	1 747	2 550
Singapore	BFTS	4 177	4 651	4 775	4 062	4 803	5 510	5 120	4 290	4 559	5 535	6 834
South Korea	BFOW	1 776	2 252	2 513	2 541	3 018	3 705	3 135	3 027	2 768	3 168	3 337
Taiwan	BFUT	1 947	2 287	2 504	2 444	2 859	3 854	3 141	2 696	2 492	2 707	2 689
Thailand	BFUB	1 401	1 549	1 562	1 671	1 695	2 149	2 279	2 187	2 247	2 497	2 482
Residual Gulf Arabian Countries	JITU	1 462	1 438	1 631	2 253	2 279	2 978	3 298	3 917	3 983	4 158	4 730
Other Near & Middle Eastern Countries	ZWXF	525	487	529	419	497	583	594	563	542	530	755
Other	LEWG	2 948	2 747	3 030	2 242	2 233	2 866	2 923	2 999	3 254	4 009	4 020
Total Asia	LETD	50 532	54 210	54 772	54 820	60 744	74 613	71 908	64 277	62 251	70 194	81 342
Australasia & Oceania												
Australia	CWBO	3 376	3 438	3 845	3 858	3 809	3 717	3 875	4 350	4 793	4 886	6 223
New Zealand	BFQY	990	1 107	1 054	1 004	951	1 000	1 043	1 045	1 071	1 193	1 283
Other	LEVO	294	302	244	137	188	187	186	194	205	237	264
Total Australasia & Oceania	LETV	4 660	4 847	5 143	4 999	4 948	4 904	5 104	5 589	6 069	6 316	7 770
Africa												
Egypt	ZWXH	1 227	1 186	973	718	778	1 040	1 063	909	926	1 005	1 136
Morocco	ZWXI	343	432	489	511	567	606	625	621	605	680	632
South Africa	BFWV	1 763	1 997	2 149	2 459	3 144	4 031	4 430	4 345	4 915	5 259	6 221
Other North Africa	JIRV	734	724	838	770	679	1 201	1 059	975	949	1 185	1 500
Other	LEWP	2 834	3 276	3 594	3 428	4 169	4 554	5 859	5 831	6 678	6 684	7 174
Total Africa	LERT	6 901	7 615	8 043	7 886	9 337	11 432	13 036	12 681	14 073	14 813	16 663
International Organisations	CTEZ	1 721	1 438	1 275	1 164	1 238	1 985	2 028	1 921	1 542	1 912	1 926
World total	HBOF	313 260	344 236	344 372	351 821	377 631	438 122	450 403	429 885	438 518	473 140	552 198

1 Cyprus and Malta are included in Other Europe before 1999.

9.2 Current account
continued

£ million

		1995	1996	1997	1998	1999	2000	2001	2002	2003	2004	2005
Balances												
Europe												
European Union (EU)												
Austria	CUGX	−95	−128	−614	−228	−169	−153	−751	−1 369	−1 591	−1 322	−1 289
Belgium and Luxembourg	CTFJ	−1 196	−1 024	−711	−1 183	−882	1 388	−210	−1 129	−744	−735	−1 558
of which Belgium	AA4H	..	..	..	..	−553	50	−1 338	−1 706	−685	−1 086	−1 902
Luxembourg	AA4L	..	..	..	..	−329	1 338	1 128	577	−59	351	344
Cyprus[1]	AA4I	..	..	..	..	−568	−657	−897	−752	−863	−728	−898
Czech Republic	LEPS	225	319	220	194	193	114	42	−224	−363	−262	−594
Denmark	LEQT	−29	−156	−166	581	170	216	149	−527	−675	−685	−1 456
Estonia	ZWXQ	−128	−127	−117	−128	−162	−261	−216	−221	−177	−289	−246
Finland	LEUF	−407	−498	−656	−428	−538	−529	−557	−468	−323	−55	−261
France	LEUO	−2 628	−788	−2 653	−2 935	−4 451	−2 175	−680	−2 369	−4 354	−4 893	−5 343
Germany	LEQK	−5 697	−6 282	−4 056	−2 501	−4 891	−5 177	−2 406	−5 363	−12 449	−13 871	−16 643
Greece	LEUX	−16	430	555	291	282	125	−268	−274	−213	−160	−372
Hungary	BFKQ	−47	36	108	31	28	397	252	289	114	−106	−278
Ireland	BFLX	1 409	1 054	1 927	2 156	2 948	3 065	3 248	3 102	4 578	6 412	6 919
Italy	BFOF	2 513	987	155	2 313	822	2 228	1 698	692	−608	−1 581	−2 236
Latvia	ZWXS	−168	−255	−293	−248	−240	−342	−353	−397	−416	−593	−563
Lithuania	ZWXT	−165	−137	−77	−82	−101	−131	−87	−137	−75	−102	−72
Malta[1]	AA4M	..	..	..	..	−85	−93	−137	−128	−127	274	−23
Netherlands	BFQH	3 325	4 040	7 127	5 709	4 466	1 771	5 181	7 837	7 080	−588	2 630
Poland	BFSA	324	792	686	516	534	368	380	440	358	81	−141
Portugal	BFSJ	−371	−365	−528	−244	−299	−260	−194	−425	−899	−656	−675
Slovak Republic	ZWXV	−76	27	62	−22	−7	33	66	11	11	−15	−52
Slovenia	ZWXW	−36	−15	9	−33	34	37	93	65	73	84	78
Spain	LESV	−662	−811	−828	−1 553	−1 624	−838	−3 150	−5 263	−5 925	−4 215	−4 092
Sweden	BFTK	194	157	638	1 192	215	852	1 104	1 040	663	1 254	1 035
European Central Bank	ZWXL	–	–	–	–	−1	2	11	2	–	–	–
EU Institutions	CSFJ	−4 875	−2 381	−3 181	−5 541	−4 905	−6 279	−3 037	−4 679	−5 045	−5 071	−6 110
Total EU25	AA4J	..	..	..	..	−9 231	−6 299	−719	−10 247	−21 970	−27 822	−32 240
European Free Trade Association (EFTA)												
Iceland	BFNJ	−63	−35	−6	−35	−64	−119	−49	−85	−60	−120	−16
Liechtenstein	BFPG	−69	−61	−79	−100	−38	−55	−52	−7	−30	9	−18
Norway	BFQQ	−1 792	−2 011	−1 765	485	−326	−2 618	−2 990	−3 304	−3 784	−5 628	−8 986
Switzerland	LEPA	−5 638	−4 741	−4 121	−3 493	−4 978	−5 725	−4 129	−1 503	212	949	863
Total EFTA	CTFS	−7 562	−6 848	−5 971	−3 143	−5 406	−8 517	−7 221	−4 899	−3 662	−4 790	−8 157
Other Europe												
Albania	ZWXM	−50	−39	−55	−107	−33	−32	–	−1	−8	7	−1
Belarus	ZWXN	−46	−31	−31	−64	−37	−1	–	−1	19	−17	−225
Bulgaria	ZWXO	−83	−91	−79	−80	1	−22	66	37	52	−16	−21
Croatia	ZWXP	156	35	8	−142	11	−9	−7	28	79	50	−11
Romania	ZWXU	−67	−24	−87	−50	−32	−15	−65	−71	−79	−63	−38
Russia	BFSS	−29	−118	−16	−47	−637	−825	−804	−315	−382	−789	−2 187
Turkey	BFUL	−36	209	500	340	119	537	−445	−984	−1 074	−1 497	−1 516
Ukraine	ZWXX	59	104	72	97	32	63	65	301	193	51	166
Serbia and Montenegro	BFWE	−89	−84	−85	−36	−22	−3	−16	−7	1	−23	−18
Other[1]	LEVY	−1 948	−2 505	−2 267	−3 466	−2 874	−3 533	−4 697	−2 991	−613	20	−1 880
Total Europe	LERC	−18 301	−14 517	−10 405	−8 842	−18 109	−18 656	−13 846	−19 141	−27 440	−34 899	−46 128
Americas												
Argentina	ZWXZ	506	559	724	753	385	594	465	14	97	241	213
Brazil	LENQ	699	740	786	812	199	157	158	125	−113	8	−62
Canada	LEOR	393	545	897	1 189	607	413	1 065	48	644	−36	−193
Chile	ZWYA	53	−6	−1	42	−7	−29	−117	−91	35	501	251
Colombia	ZWYB	−1	−54	31	−21	−1	160	80	134	214	298	230
Mexico	BFPP	575	316	484	369	549	261	−57	685	512	884	1 045
United States of America	BFVD	2 444	109	1 605	4 379	4 845	9 628	6 267	11 460	14 844	15 726	19 513
Uruguay	ZWYC	6	7	13	57	27	18	12	−19	43	−12	−6
Venezuela	ZWYD	135	180	209	101	64	−57	358	344	165	308	154
Other Central American Countries	JISU	328	−4	−236	663	498	821	1 888	3 680	3 230	3 254	4 595
Other	LEVG	247	−39	−115	−186	−250	−556	−947	−511	−214	84	−46
Total Americas	LESM	5 385	2 823	4 835	8 687	6 916	11 410	9 171	15 868	19 457	21 256	25 694
Asia												
China	LEPJ	−1 055	−1 335	−1 409	−1 896	−2 084	−3 409	−3 806	−4 997	−5 887	−7 083	−9 162
Hong Kong	BFJT	580	927	443	−1 160	−3 451	−4 709	−4 635	−3 536	−2 231	−954	138
India	BFNA	214	−263	−355	−532	−464	−47	−548	−328	33	−477	−479
Indonesia	BFKZ	−269	−60	51	−255	−536	−621	−720	−613	−470	−407	−390
Iran	ZWYG	191	222	316	284	209	148	289	442	562	534	614
Israel	BFMR	452	506	366	200	478	491	417	633	561	561	343
Japan	BFOO	−2 303	−2 757	−2 659	−4 656	−4 550	−6 649	−5 855	−4 348	−758	−1 754	−3 015
Malaysia	BFPY	131	−820	−430	−776	−765	−1 144	−687	−285	−123	−190	67
Pakistan	BFRR	31	22	−101	−184	−252	−383	−312	−594	−505	−427	−108
Philippines	BFRI	239	−340	−49	−470	−717	−843	−598	−511	−238	−194	−236
Saudi Arabia	BFTB	934	2 241	3 387	2 994	2 322	2 273	1 878	1 465	1 928	1 137	614
Singapore	BFTT	1 014	688	458	−90	−529	4	284	203	270	624	327
South Korea	BFOX	91	−26	−225	−1 006	−1 335	−1 531	−933	−357	−169	43	28
Taiwan	BFUU	−604	−903	−1 011	−1 154	−1 539	−2 290	−1 670	−1 124	−869	−695	−494
Thailand	BFUC	18	−26	−193	−977	−907	−1 148	−1 210	−1 111	−1 301	−1 306
Residual Gulf Arabian Countries	JITV	2 277	2 617	2 982	3 011	2 253	2 384	1 847	2 310	3 198	4 315	7 150
Other Near & Middle Eastern Countries	ZWYI	269	467	568	601	483	520	716	596	733	1 043	946
Other	LEWH	−664	−199	−210	−324	−800	−1 056	−1 177	−1 220	−1 620	−2 095	−1 481
Total Asia	LETE	1 546	1 128	1 959	−6 267	−12 184	−18 010	−16 725	−13 475	−6 696	−7 320	−6 444
Australasia & Oceania												
Australia	CWJK	2 464	3 096	2 427	1 715	2 120	3 238	3 015	2 569	2 246	4 108	3 481
New Zealand	BFQZ	310	218	101	−3	182	−14	64	159	215	63	239
Other	LEVP	−5	88	71	337	−52	−36	−38	−6	6	−20	−4
Total Australasia & Oceania	LETW	2 769	3 402	2 599	2 049	2 250	3 188	3 041	2 722	2 467	4 151	3 716
Africa												
Egypt	ZWYK	−705	−589	−279	−194	77	−1	−17	99	26	185	69
Morocco	ZWYL	14	−42	−76	−87	−132	−107	−143	−183	−180	−267	−277
South Africa	BFWW	1 602	1 556	1 211	865	363	−361	−337	371	−64	825	72
Other North Africa	JIRW	−127	−101	−37	74	−108	−411	−262	−300	61	−269	−509
Other	LEWQ	771	841	359	1 281	29	−396	−1 309	−1 105	−1 539	−1 568	−1 452
Total Africa	LERU	1 555	1 665	1 178	1 939	229	−1 276	−2 068	−1 119	−1 696	−1 094	−2 097
International Organisations	CTFA	−1 430	−1 218	−1 005	−759	−819	−1 489	−1 457	−1 368	−1 013	−1 422	−1 291
World total	HBOG	−8 476	−6 717	−840	−3 195	−21 717	−24 833	−21 884	−16 513	−14 921	−19 328	−26 550

1 Cyprus and Malta are included in Other Europe before 1999.

9.3 Trade in goods and services

£ million

		1995	1996	1997	1998	1999	2000	2001	2002	2003	2004	2005
Exports												
Europe												
European Union (EU)												
Austria	LGHT	1 450	1 617	1 458	1 533	1 551	1 539	1 610	1 668	1 658	1 507	1 814
Belgium and Luxembourg	LGHU	9 767	10 045	10 124	10 400	11 503	12 707	12 236	13 195	14 366	14 134	14 729
of which Belgium	A7RS	..	..	..	..	11 037	12 142	11 578	12 466	13 521	13 087	13 797
Luxembourg	A7RV	..	..	..	..	466	565	658	729	845	1 047	932
Cyprus[1]	A7RT	..	..	..	..	389	458	419	485	548	650	645
Czech Republic	LGIN	679	832	830	896	892	1 092	1 256	1 208	1 243	1 304	1 355
Denmark	LGHV	2 624	3 076	2 956	3 089	3 014	3 479	3 633	4 076	3 663	3 605	3 835
Estonia	ZWLX	34	63	79	73	65	108	97	121	117	135	152
Finland	LGHW	2 105	2 376	2 115	1 967	2 008	2 421	2 479	2 261	2 355	2 311	2 371
France	LGHX	18 329	20 452	20 136	20 629	21 603	23 901	25 118	24 545	24 824	24 891	25 822
Germany	LGHY	24 643	25 312	25 422	25 841	25 939	29 264	30 481	29 150	28 043	29 306	30 781
Greece	LGHZ	1 562	1 693	1 676	1 732	1 941	2 043	1 904	1 929	1 999	2 119	2 151
Hungary	XUXI	348	426	545	581	630	817	817	966	1 082	1 308	1 400
Ireland	LGIA	9 775	10 897	11 671	12 155	13 305	15 172	17 395	19 592	17 365	19 634	22 041
Italy	LGIB	9 589	9 979	10 293	10 978	10 530	11 051	11 177	11 401	11 460	11 689	12 211
Latvia	ZWMF	50	94	112	102	93	114	117	115	142	141	201
Lithuania	ZWME	62	102	126	123	111	148	183	181	232	198	285
Malta[1]	A7RW	..	..	..	..	254	263	267	288	332	335	313
Netherlands	LGIC	14 471	15 947	16 783	16 189	17 655	19 613	19 308	18 480	18 685	18 450	19 245
Poland	LGIO	1 125	1 546	1 592	1 463	1 445	1 568	1 629	1 704	1 868	1 919	2 401
Portugal	LGID	1 825	2 058	2 078	2 172	2 282	2 086	2 045	2 024	1 966	2 237	2 290
Slovak Republic	ZWMJ	100	140	161	127	145	196	246	231	273	270	350
Slovenia	ZWMI	133	147	166	141	162	181	192	221	192	203	219
Spain	LGIE	7 352	8 182	8 137	8 934	9 587	10 527	10 556	11 039	11 731	11 937	14 089
Sweden	LGIF	5 064	5 452	5 607	5 752	5 441	5 808	5 551	5 316	5 482	6 019	6 271
European Central Bank	ZWLL	–	–	–	–	–	3	12	3	–	–	–
EU Institutions	LGIG	381	247	248	228	232	248	544	544	511	578	596
Total EU25	A7RU	..	..	..	..	130 777	144 807	149 272	150 743	150 137	154 880	165 567
European Free Trade Association (EFTA)												
Iceland	LGII	179	200	212	206	189	266	213	195	213	256	314
Liechtenstein	LGIJ	15	19	21	25	34	42	30	42	39	55	44
Norway	LGIK	2 932	3 109	3 577	3 870	3 123	3 039	2 772	3 124	3 174	3 423	3 893
Switzerland	LGIL	4 007	5 105	4 983	5 274	5 183	5 593	6 608	6 653	6 896	6 773	9 232
Total EFTA	LGIM	7 133	8 433	8 793	9 375	8 529	8 940	9 622	10 014	10 322	10 507	13 483
Other Europe												
Albania	ZWLP	8	14	7	8	15	9	28	25	16	34	25
Belarus	ZWLS	28	33	42	33	30	64	36	36	42	59	65
Bulgaria	ZWLR	110	100	98	95	117	128	156	206	219	232	318
Croatia	ZWMC	244	155	124	118	118	110	122	147	186	166	159
Romania	ZWMH	195	241	245	262	287	427	412	503	607	743	777
Russia	LGIP	1 185	1 351	1 594	1 370	822	1 075	1 347	1 543	2 123	2 332	2 949
Turkey	LGIQ	1 368	1 810	2 065	1 895	1 592	2 145	1 537	1 651	2 033	2 263	2 638
Ukraine	ZWMK	140	191	204	231	170	199	231	483	367	276	453
Serbia and Montenegro	ZWMN	15	47	76	63	68	135	92	90	105	87	94
Other[1]	ZWLM	1 576	2 089	2 371	1 647	1 194	1 345	1 181	1 455	1 966	2 213	2 628
Total Europe	LGIS	123 470	135 147	137 934	140 202	143 719	159 384	164 036	166 896	168 123	173 792	189 156
Americas												
Argentina	ZWLQ	319	451	601	668	456	471	453	210	217	321	263
Brazil	LGIT	933	1 126	1 328	1 321	1 108	1 117	1 190	1 193	1 151	1 097	1 189
Canada	LGIU	2 750	2 944	3 286	3 546	3 892	4 857	4 791	4 597	4 772	5 075	4 797
Chile	ZWLT	214	216	267	295	221	211	235	202	202	219	240
Colombia	ZWLU	215	250	229	242	155	157	245	169	205	208	192
Mexico	LGIV	411	431	593	679	791	937	959	962	950	978	936
United States of America	LGIW	28 643	33 394	35 528	36 614	41 463	48 175	47 759	50 135	51 714	53 307	53 738
Uruguay	ZWML	63	74	87	79	75	64	53	35	90	42	55
Venezuela	ZWMM	216	225	254	314	273	305	424	405	245	289	329
Other Central American Countries	ZWLW	1 421	1 426	1 736	2 193	1 959	2 269	2 091	2 686	2 796	3 272	3 804
Other	ZWLZ	905	752	918	837	828	648	652	556	647	495	527
Total Americas	LGIY	36 090	41 289	44 827	46 788	51 221	59 211	58 851	61 149	62 989	65 303	66 070
Asia												
China	LGIZ	1 041	977	1 145	1 161	1 659	1 910	2 286	2 203	2 796	3 641	4 150
Hong Kong	LGJA	3 425	3 820	4 107	3 632	3 315	3 659	3 688	3 341	3 590	3 690	4 299
India	LGJB	2 106	2 102	2 151	1 726	1 960	2 606	2 436	2 378	2 989	3 216	3 902
Indonesia	LGJC	735	1 002	900	568	549	595	514	494	613	584	544
Iran	ZWMD	388	444	448	412	333	411	596	578	687	676	671
Israel	LGJD	1 458	1 587	1 550	1 387	1 818	1 957	1 861	1 850	1 718	1 790	1 734
Japan	LGJE	6 074	6 609	6 552	5 740	6 244	6 921	6 925	7 104	7 243	7 761	7 950
Malaysia	LGJF	1 676	1 669	1 708	1 219	1 322	1 292	1 398	1 275	1 443	1 537	1 520
Pakistan	LGJG	634	697	575	471	515	383	505	409	468	572	882
Philippines	LGJH	573	500	734	394	362	456	570	457	518	443	429
Saudi Arabia	LGJI	3 433	4 163	5 400	4 528	3 619	3 907	3 639	1 978	2 437	2 157	2 264
Singapore	LGJJ	2 557	2 613	2 551	2 155	2 690	2 659	2 796	2 512	3 117	3 819	4 559
South Korea	LGJK	1 583	1 775	1 705	1 153	1 353	1 738	1 680	1 956	1 930	2 385	2 389
Taiwan	LGJL	1 186	1 188	1 306	1 101	1 113	1 295	1 165	1 214	1 268	1 446	1 483
Thailand	LGJM	1 183	1 232	1 139	616	668	773	809	731	902	954	882
Residual Gulf Arabian Countries	ZWMA	3 257	3 483	3 941	3 632	3 251	3 654	3 800	5 245	6 294	7 449	10 771
Other Near & Middle Eastern Countries	ZWMB	501	564	645	641	607	620	813	760	884	1 113	1 194
Other	ZWLN	1 804	2 096	2 317	1 782	1 245	1 451	1 444	1 233	1 393	1 567	1 653
Total Asia	LGJO	33 614	36 521	38 874	32 318	32 623	36 287	36 924	35 717	40 290	44 800	51 276
Australasia & Oceania												
Australia	LGJP	3 389	3 931	3 961	3 638	3 744	4 273	4 236	3 908	4 269	4 716	5 164
New Zealand	LGJQ	720	727	687	656	677	606	609	584	646	799	806
Other	LGJR	84	156	145	89	99	127	137	152	179	176	198
Total Australasia & Oceania	LGJS	4 193	4 814	4 793	4 383	4 520	5 006	4 982	4 644	5 094	5 691	6 168
Africa												
Egypt	ZWLY	451	515	610	604	813	772	784	681	687	983	882
Morocco	ZWMG	309	324	384	405	406	468	450	396	406	403	322
South Africa	LGJT	2 577	2 609	2 473	2 451	2 396	2 381	2 651	2 594	2 838	2 941	3 157
Other North Africa	ZWLV	473	515	701	731	501	680	658	694	867	821	875
Other	ZWLO	2 844	3 350	3 346	4 107	3 548	3 367	3 745	3 662	4 040	3 921	4 301
Total Africa	LGJV	6 654	7 313	7 514	8 298	7 664	7 668	8 288	8 026	8 838	9 069	9 537
International Organisations	LGJW	130	74	77	45	35	46	57	78	63	39	91
World total	KTMW	204 151	225 158	234 019	232 034	239 782	267 602	273 140	276 511	285 397	298 694	322 298

1 Cyprus and Malta are included in Other Europe before 1999.

9.3 Trade in goods and services
continued

£ million

		1995	1996	1997	1998	1999	2000	2001	2002	2003	2004	2005
Imports												
Europe												
European Union (EU)												
Austria	LGJY	1 308	1 532	1 736	1 865	1 900	1 879	2 390	3 080	3 431	3 086	3 279
Belgium and Luxembourg	LGJZ	9 205	10 223	10 637	11 160	11 766	12 490	13 837	15 066	15 186	15 804	16 970
of which Belgium	A8EO	..	..	..	..	11 515	12 213	13 394	14 135	14 243	14 567	15 811
Luxembourg	A8ER	..	..	..	..	251	277	443	931	943	1 237	1 159
Cyprus[1]	A8EP	..	..	..	..	950	1 094	1 302	1 225	1 339	1 290	1 477
Czech Republic	LGKS	404	453	555	701	721	975	1 296	1 500	1 749	1 719	2 116
Denmark	LGKA	2 574	2 869	2 808	2 649	2 961	3 290	3 559	4 318	4 152	4 166	5 198
Estonia	ZWOD	109	145	149	155	191	340	310	345	299	424	407
Finland	LGKB	2 652	2 883	2 729	2 545	2 669	3 168	3 325	3 073	2 977	2 620	2 786
France	LGKC	20 417	21 141	22 755	23 661	25 057	26 137	27 623	28 610	28 607	28 546	30 606
Germany	LGKD	29 707	31 560	29 071	28 782	31 422	32 999	35 211	37 574	38 866	41 288	45 884
Greece	LGKE	1 474	1 314	1 341	1 414	1 825	2 060	2 367	2 459	2 589	2 581	2 693
Hungary	ZWOJ	400	468	537	617	750	781	836	975	1 271	1 762	2 083
Ireland	LGKF	8 143	8 881	9 033	9 546	10 821	12 474	14 428	15 422	12 413	13 045	13 560
Italy	LGKG	9 503	10 750	11 506	12 064	11 980	12 338	12 832	13 781	14 873	15 681	16 709
Latvia	ZWOM	166	297	336	295	284	425	459	504	554	730	758
Lithuania	ZWOL	165	181	143	147	164	252	251	305	310	291	343
Malta[1]	A8ES	..	..	..	..	322	346	400	406	464	504	540
Netherlands	LGKH	13 267	14 364	14 162	15 449	16 143	18 337	18 455	19 410	19 900	21 420	23 728
Poland	LGKT	732	664	726	797	843	1 107	1 366	1 453	1 768	2 132	2 813
Portugal	LGKI	2 126	2 445	2 591	2 615	2 827	2 755	2 739	2 964	3 327	3 325	3 420
Slovak Republic	ZWOQ	75	77	81	103	129	166	191	238	281	294	421
Slovenia	ZWOP	115	114	107	112	120	150	166	196	193	201	242
Spain	LGKJ	7 904	9 286	9 643	10 940	11 886	12 624	14 576	17 014	18 196	18 195	20 887
Sweden	LGKK	5 055	5 413	5 208	4 892	5 416	5 772	5 491	5 148	5 690	6 251	6 543
European Central Bank	ZWNR	–	–	–	–	1	1	1	1	–	–	–
EU Institutions	LGKL	1	2	1	10	6	3	5	14	17	4	4
Total EU25	A8EQ	..	..	..	..	141 154	151 963	163 416	175 081	178 452	185 359	203 467
European Free Trade Association (EFTA)												
Iceland	LGKN	267	286	263	297	307	440	325	331	342	412	426
Liechtenstein	LGKO	5	18	27	53	30	28	32	23	36	22	16
Norway	LGKP	4 865	5 307	5 262	3 963	4 044	6 099	6 170	6 523	7 129	9 243	12 924
Switzerland	LGKQ	6 062	6 453	5 738	6 148	6 787	7 048	6 062	6 221	5 502	5 198	6 041
Total EFTA	LGKR	11 199	12 064	11 290	10 461	11 168	13 615	12 589	13 098	13 009	14 875	19 407
Other Europe												
Albania	ZWNV	1	1	2	1	7	7	9	15	17	20	19
Belarus	ZWNY	20	16	15	25	21	36	22	34	26	73	283
Bulgaria	ZWNX	130	140	122	103	104	123	148	211	225	324	410
Croatia	ZWOI	62	75	67	77	60	68	106	115	123	139	188
Romania	ZWOO	186	203	220	244	300	397	519	598	765	897	911
Russia	LGKU	1 155	1 387	1 577	1 588	1 461	1 684	2 265	2 246	2 757	4 014	5 524
Turkey	LGKV	1 242	1 491	1 569	1 665	1 728	1 971	2 238	2 868	3 407	4 064	4 519
Ukraine	ZWOR	26	29	51	64	56	78	104	168	117	151	186
Serbia and Montenegro	ZWOU	8	24	44	41	45	45	61	63	75	76	87
Other[1]	ZWNS	1 570	2 055	2 210	2 234	907	1 298	1 233	1 253	1 203	1 076	789
Total Europe	LGKX	131 101	142 547	143 022	147 022	157 011	171 285	182 710	195 750	200 176	211 068	235 790
Americas												
Argentina	ZWNW	270	313	295	268	265	256	283	291	300	334	355
Brazil	LGKY	1 004	1 051	1 028	1 023	1 066	1 288	1 510	1 616	1 653	1 766	1 994
Canada	LGKZ	3 085	3 116	3 272	3 385	3 994	5 114	4 849	4 644	4 658	5 223	5 238
Chile	ZWNZ	302	383	394	372	370	486	512	508	457	546	554
Colombia	ZWOA	178	229	201	259	226	270	342	247	255	302	333
Mexico	LGLA	347	442	520	554	620	896	1 078	841	802	778	759
United States of America	LGLB	28 485	32 635	34 509	36 277	37 005	41 546	42 750	38 457	37 307	36 819	36 782
Uruguay	ZWOS	60	74	72	59	44	42	45	70	47	46	66
Venezuela	ZWOT	210	208	174	155	190	253	202	218	147	234	418
Other Central American Countries	ZWOC	1 108	1 436	1 523	1 745	1 817	2 492	2 001	2 055	2 366	2 959	2 961
Other	ZWOF	560	650	598	575	757	956	1 368	976	822	772	731
Total Americas	LGLD	35 609	40 537	42 586	44 672	46 354	53 599	54 940	49 923	48 814	49 779	50 191
Asia												
China	LGLE	1 998	2 268	2 544	3 035	3 650	5 118	6 116	7 222	8 805	10 972	13 565
Hong Kong	LGLF	3 850	4 382	4 729	4 892	5 512	6 476	6 348	6 172	6 200	6 473	7 283
India	LGLG	1 720	1 984	2 038	1 936	2 099	2 315	2 632	2 612	2 903	3 385	4 016
Indonesia	LGLH	983	1 030	989	958	1 081	1 206	1 249	1 142	1 172	1 017	940
Iran	ZWOK	121	115	46	57	54	73	65	62	74	102	77
Israel	LGLI	880	956	1 027	1 082	1 211	1 291	1 253	1 082	1 074	1 144	1 227
Japan	LGLJ	10 255	9 666	10 119	10 278	10 341	11 774	10 799	9 463	9 751	9 685	10 732
Malaysia	LGLK	1 615	2 479	2 175	2 145	2 147	2 515	2 130	1 924	2 037	2 234	2 026
Pakistan	LGLL	509	544	554	534	606	663	701	870	905	985	965
Philippines	LGLM	397	917	789	919	1 071	1 361	1 237	1 044	868	767	848
Saudi Arabia	LGLN	1 046	1 024	1 297	1 242	1 309	1 530	1 576	834	861	1 375	1 919
Singapore	LGLO	2 456	2 734	2 921	2 646	2 687	2 711	2 443	2 366	3 068	3 994	4 497
South Korea	LGLP	1 607	2 050	2 291	2 330	2 919	3 557	2 935	2 913	2 749	3 331	3 295
Taiwan	LGLQ	1 728	2 107	2 330	2 325	2 745	3 717	2 932	2 564	2 371	2 539	2 440
Thailand	LGLR	1 256	1 382	1 402	1 572	1 598	1 987	2 065	2 071	2 159	2 406	2 325
Residual Gulf Arabian Countries	ZWOG	815	823	973	1 219	1 254	1 509	1 675	2 965	3 176	3 137	3 446
Other Near & Middle Eastern Countries	ZWOH	176	181	160	159	219	243	288	340	328	316	460
Other	ZWNT	1 635	1 672	1 770	1 652	1 696	2 227	2 200	2 276	2 629	3 145	2 931
Total Asia	LGLT	33 047	36 314	38 154	38 981	42 199	50 273	48 644	47 922	51 130	57 007	62 992
Australasia & Oceania												
Australia	LGLU	1 779	2 052	2 209	2 324	2 380	2 733	3 027	3 118	3 260	3 256	3 893
New Zealand	LGLV	748	831	788	744	814	808	829	859	899	1 017	1 084
Other	LGLW	215	265	199	150	151	149	114	129	153	184	194
Total Australasia & Oceania	LGLX	2 742	3 148	3 196	3 218	3 345	3 690	3 970	4 106	4 312	4 457	5 171
Africa												
Egypt	ZWOE	383	429	386	394	484	701	773	729	782	843	852
Morocco	ZWON	290	355	424	468	520	568	586	594	584	660	593
South Africa	LGLY	1 410	1 597	1 742	1 838	2 116	3 081	3 439	3 369	3 870	4 201	4 921
Other North Africa	ZWOB	504	548	550	612	551	1 022	864	851	834	1 070	1 256
Other	ZWNU	2 047	2 129	2 128	1 904	2 600	2 664	3 940	4 102	4 311	4 547	4 731
Total Africa	LGMA	4 634	5 058	5 230	5 216	6 271	8 036	9 602	9 645	10 381	11 321	12 353
International Organisations	LGMB	84	72	67	66	56	80	62	40	29	37	43
World total	KTMX	207 217	227 676	232 255	239 175	255 236	286 963	299 929	307 386	314 842	333 669	366 540

1 Cyprus and Malta are included in Other Europe before 1999.

9.3 Trade in goods and services (continued)

£ million

		1995	1996	1997	1998	1999	2000	2001	2002	2003	2004	2005
Balances												
Europe												
European Union (EU)												
Austria	LGMD	142	85	−278	−332	−349	−340	−780	−1 412	−1 773	−1 579	−1 465
Belgium and Luxembourg	LGME	562	−178	−513	−760	−263	217	−1 601	−1 871	−820	−1 670	−2 241
of which Belgium	A8HC	..	..	..	..	−478	−71	−1 816	−1 669	−722	−1 480	−2 014
Luxembourg	A8HF	..	..	..	..	215	288	215	−202	−98	−190	−227
Cyprus[1]	A8HD	..	..	..	..	−561	−636	−883	−740	−791	−640	−832
Czech Republic	LGMX	275	379	275	195	171	117	−40	−292	−506	−415	−761
Denmark	LGMF	50	207	148	440	53	189	74	−242	−489	−561	−1 363
Estonia	ZWSQ	−75	−82	−70	−82	−126	−232	−213	−224	−182	−289	−255
Finland	LGMG	−547	−507	−614	−578	−661	−747	−846	−812	−622	−309	−415
France	LGMH	−2 088	−689	−2 619	−3 032	−3 454	−2 236	−2 505	−4 065	−3 783	−3 655	−4 784
Germany	LGMI	−5 064	−6 248	−3 649	−2 941	−5 483	−3 735	−4 730	−8 424	−10 823	−11 982	−15 103
Greece	LGMJ	88	379	335	318	116	−17	−463	−530	−590	−462	−542
Hungary	ZWSW	−52	−42	8	−36	−120	36	−19	−9	−189	−454	−683
Ireland	LGMK	1 632	2 016	2 638	2 609	2 484	2 698	2 967	4 170	4 952	6 589	8 481
Italy	LGML	86	−771	−1 213	−1 086	−1 450	−1 287	−1 655	−2 380	−3 413	−3 992	−4 498
Latvia	ZWSZ	−116	−203	−224	−193	−191	−311	−342	−389	−412	−589	−557
Lithuania	ZWSY	−103	−79	−17	−24	−53	−104	−68	−124	−78	−93	−58
Malta[1]	A8HG	..	..	..	..	−68	−83	−133	−118	−132	−169	−227
Netherlands	LGMM	1 204	1 583	2 621	740	1 512	1 276	853	−930	−1 215	−2 970	−4 483
Poland	LGMY	393	882	866	666	602	461	263	251	100	−213	−412
Portugal	LGMN	−301	−387	−513	−443	−545	−669	−694	−940	−1 361	−1 088	−1 130
Slovak Republic	ZWTD	25	63	80	24	16	30	55	−7	−8	−24	−71
Slovenia	ZWTC	18	33	59	29	42	31	26	25	−1	2	−23
Spain	LGMO	−552	−1 104	−1 506	−2 006	−2 299	−2 097	−4 020	−5 975	−6 465	−6 258	−6 798
Sweden	LGMP	9	39	399	860	25	36	60	168	−208	−232	−272
European Central Bank	ZWSE	−	−	−	−	−1	2	11	2	−	−	−
EU Institutions	LGMQ	380	245	247	218	226	245	539	530	494	574	592
Total EU25	A8HE	..	..	..	..	−10 377	−7 156	−14 144	−24 338	−28 315	−30 479	−37 900
European Free Trade Association (EFTA)												
Iceland	LGMS	−88	−86	−51	−91	−118	−174	−112	−136	−129	−156	−112
Liechtenstein	LGMT	10	1	−6	−28	4	14	−2	19	3	33	28
Norway	LGMU	−1 933	−2 198	−1 685	−93	−921	−3 060	−3 398	−3 399	−3 955	−5 820	−9 031
Switzerland	LGMV	−2 055	−1 348	−755	−874	−1 604	−1 455	546	432	1 394	1 575	3 191
Total EFTA	LGMW	−4 066	−3 631	−2 497	−1 086	−2 639	−4 675	−2 967	−3 084	−2 687	−4 368	−5 924
Other Europe												
Albania	ZWSI	7	13	5	7	8	2	19	10	−1	14	6
Belarus	ZWSL	8	17	27	8	9	28	14	2	16	−14	−218
Bulgaria	ZWSK	−20	−40	−24	−8	13	5	8	−5	−6	−92	−92
Croatia	ZWSV	182	80	57	41	58	42	16	32	63	27	−29
Romania	ZWTB	9	38	25	18	−13	30	−107	−95	−158	−154	−134
Russia	LGMZ	30	−36	17	−218	−639	−609	−918	−703	−634	−1 682	−2 575
Turkey	LGNA	126	319	496	230	−136	174	−701	−1 217	−1 374	−1 801	−1 881
Ukraine	ZWTE	114	162	153	167	114	121	127	315	250	125	267
Serbia and Montenegro	ZWTH	7	23	32	22	23	90	31	27	30	11	7
Other[1]	ZWSF	6	34	161	−587	287	47	−52	202	763	1 137	1 839
Total Europe	LGNC	−7 631	−7 400	−5 088	−6 820	−13 292	−11 901	−18 674	−28 854	−32 053	−37 276	−46 634
Americas												
Argentina	ZWSJ	49	138	306	400	191	215	170	−81	−83	−13	−92
Brazil	LGND	−71	75	300	298	42	−171	−320	−423	−502	−669	−805
Canada	LGNE	−335	−172	14	161	−102	−257	−58	−47	114	−148	−441
Chile	ZWSM	−88	−167	−127	−77	−149	−275	−277	−306	−255	−327	−314
Colombia	ZWSN	37	5	21	28	−17	−71	−113	−97	−78	−50	−94	−141
Mexico	LGNF	64	−11	73	125	171	41	−119	121	148	200	177
United States of America	LGNG	158	759	1 019	337	4 458	6 629	5 009	11 678	14 407	16 488	16 956
Uruguay	ZWTF	3	−	15	20	31	22	8	−35	43	−4	−11
Venezuela	ZWTG	6	17	80	159	83	52	222	187	98	55	−89
Other Central American Countries	ZWSP	313	−10	213	448	142	−223	90	631	430	313	843
Other	ZWSS	345	102	320	262	71	−308	−716	−420	−175	−277	−204
Total Americas	LGNI	481	752	2 241	2 116	4 867	5 612	3 911	11 226	14 175	15 524	15 879
Asia												
China	LGNJ	−957	−1 291	−1 399	−1 874	−1 991	−3 208	−3 830	−5 019	−6 009	−7 331	−9 415
Hong Kong	LGNK	−425	−562	−622	−1 260	−2 197	−2 817	−2 660	−2 831	−2 610	−2 783	−2 984
India	LGNL	386	118	113	−210	−139	291	−196	−234	86	−169	−114
Indonesia	LGNM	−248	−28	−89	−390	−532	−611	−735	−648	−559	−433	−396
Iran	ZWSX	267	329	402	355	279	338	531	516	613	574	594
Israel	LGNN	578	631	523	305	607	666	608	768	644	646	507
Japan	LGNO	−4 181	−3 057	−3 567	−4 538	−4 097	−4 853	−3 874	−2 359	−2 508	−1 924	−2 782
Malaysia	LGNP	61	−810	−467	−926	−825	−1 223	−732	−649	−594	−697	−506
Pakistan	LGNQ	125	153	21	−63	−91	−280	−196	−461	−437	−413	−83
Philippines	LGNR	176	−417	−55	−525	−709	−905	−667	−587	−350	−324	−419
Saudi Arabia	LGNS	2 387	3 139	4 103	3 286	2 310	2 377	2 063	1 144	1 576	782	345
Singapore	LGNT	101	−121	−370	−491	3	−52	353	146	49	−175	62
South Korea	LGNU	−24	−275	−586	−1 177	−1 566	−1 819	−1 255	−957	−819	−946	−906
Taiwan	LGNV	−542	−919	−1 024	−1 224	−1 632	−2 422	−1 767	−1 350	−1 103	−1 093	−957
Thailand	LGNW	−73	−150	−263	−956	−930	−1 214	−1 256	−1 340	−1 257	−1 452	−1 443
Residual Gulf Arabian Countries	ZWST	2 442	2 660	2 968	2 413	1 997	2 145	2 125	2 280	3 118	4 312	7 325
Other Near & Middle Eastern Countries	ZWSU	325	383	485	482	388	377	525	420	556	797	734
Other	ZWSG	169	424	547	130	−451	−776	−756	−1 043	−1 236	−1 578	−1 278
Total Asia	LGNY	567	207	720	−6 663	−9 576	−13 986	−11 720	−12 205	−10 840	−12 207	−11 716
Australasia & Oceania												
Australia	LGNZ	1 610	1 879	1 752	1 314	1 364	1 540	1 209	790	1 009	1 460	1 271
New Zealand	LGOA	−28	−104	−101	−88	−137	−202	−220	−275	−253	−218	−278
Other	LGOB	−131	−109	−54	−61	−52	−22	23	23	26	−8	4
Total Australasia & Oceania	LGOC	1 451	1 666	1 597	1 165	1 175	1 316	1 012	538	782	1 234	997
Africa												
Egypt	ZWSR	68	86	224	210	329	71	11	−48	−95	140	30
Morocco	ZWTA	19	−31	−40	−63	−114	−100	−136	−198	−178	−257	−271
South Africa	LGOD	1 167	1 012	731	613	280	−700	−788	−775	−1 032	−1 260	−1 764
Other North Africa	ZWSO	−31	−33	151	119	−50	−342	−206	−157	33	−249	−381
Other	ZWSH	797	1 221	1 218	2 203	948	703	−195	−440	−271	−626	−430
Total Africa	LGOF	2 020	2 255	2 284	3 082	1 393	−368	−1 314	−1 619	−1 543	−2 252	−2 816
International Organisations	LGOG	46	2	10	−21	−21	−34	−5	38	34	2	48
World total	KTMY	−3 066	−2 518	1 764	−7 141	−15 454	−19 361	−26 789	−30 875	−29 445	−34 975	−44 242

1 Cyprus and Malta are included in Other Europe before 1999.

9.4 Trade in goods

£ million

		1995	1996	1997	1998	1999	2000	2001	2002	2003	2004	2005
Exports												
Europe												
European Union (EU)												
Austria	QBRY	1 122	1 263	1 159	1 190	1 168	1 146	1 224	1 265	1 264	1 094	1 329
Belgium and Luxembourg	QBSB	8 298	8 522	8 451	8 445	9 241	10 322	9 893	10 552	11 374	10 511	11 393
of which Belgium	QDOH	..	..	..	..	9 117	10 102	9 609	10 182	11 073	10 250	11 182
Luxembourg	QDOK	..	..	..	..	124	220	284	370	301	261	211
Cyprus[1]	QDNZ	..	..	..	..	259	311	291	272	317	322	356
Czech Republic	QDLF	574	719	709	698	742	926	1 075	1 031	1 003	975	1 020
Denmark	QBSE	2 108	2 214	2 093	2 057	2 054	2 315	2 267	2 729	2 180	2 042	2 306
Estonia	QAMN	30	56	64	68	51	96	83	100	95	106	114
Finland	QBSH	1 716	1 810	1 570	1 434	1 354	1 471	1 442	1 493	1 493	1 362	1 505
France	QDJA	15 265	17 093	16 601	16 449	16 907	18 577	19 249	18 757	18 885	18 564	19 873
Germany	QDJD	20 242	20 715	20 685	20 590	20 464	22 789	23 655	22 064	20 805	21 671	23 021
Greece	QDJG	1 038	1 147	1 047	1 045	1 153	1 229	1 124	1 199	1 252	1 413	1 364
Hungary	QDLI	299	351	435	486	492	611	612	750	856	933	828
Ireland	QDJJ	7 794	8 661	9 357	9 604	10 783	12 372	13 835	15 422	12 224	14 133	16 273
Italy	QDJM	7 883	8 027	8 214	8 608	7 831	8 429	8 404	8 506	8 603	8 401	8 769
Latvia	QAMO	41	82	85	86	69	84	84	77	113	92	99
Lithuania	QAMP	51	84	107	116	96	131	137	149	189	142	167
Malta[1]	QDOC	..	..	..	..	189	206	215	228	260	258	238
Netherlands	QDJP	12 346	13 484	13 923	12 983	13 632	15 167	14 599	14 011	13 597	12 030	12 656
Poland	QDLL	953	1 358	1 354	1 178	1 179	1 286	1 297	1 318	1 462	1 413	1 603
Portugal	QDJT	1 469	1 677	1 752	1 722	1 712	1 660	1 579	1 518	1 453	1 580	1 679
Slovak Republic	QAMR	78	106	132	103	114	157	203	201	237	224	255
Slovenia	QAMS	123	131	149	136	140	157	160	182	161	162	169
Spain	QDJW	6 098	6 725	6 745	7 171	7 526	8 302	8 363	8 490	8 943	9 100	10 583
Sweden	QDJZ	4 157	4 420	4 451	4 392	4 035	4 211	3 951	3 873	3 823	4 355	4 583
European Central Bank	QARP	–	–	–	–	–	–	–	–	–	–	–
EU Institutions	EOAY	–	–	–	–	–	–	–	–	–	–	–
Total EU25	LGCJ	..	..	..	..	101 191	111 955	113 911	114 136	110 589	110 883	120 183
European Free Trade Association (EFTA)												
Iceland	QDKW	137	152	157	158	159	193	150	131	141	167	179
Liechtenstein	EPOW	13	14	10	4	2	6	3	2	3	6	2
Norway	QDKZ	1 993	2 039	2 609	2 658	1 999	2 018	1 813	1 696	1 886	1 937	2 212
Switzerland	QDLC	2 727	3 166	2 955	2 892	2 768	3 061	3 496	3 080	2 786	2 840	4 984
Total EFTA	EPOT	4 870	5 371	5 731	5 712	4 928	5 278	5 461	4 909	4 816	4 950	7 377
Other Europe												
Albania	QAMC	8	14	7	8	12	7	23	19	10	12	16
Belarus	QAME	23	27	35	32	27	37	33	32	38	53	57
Bulgaria	QAMF	103	88	78	81	76	86	121	131	152	154	219
Croatia	QAMM	232	137	105	106	80	72	88	94	138	125	118
Romania	QAMQ	179	213	213	233	243	383	339	427	510	606	641
Russia	QDLO	874	1 018	1 233	929	532	668	893	981	1 420	1 466	1 868
Turkey	QDLR	1 150	1 545	1 734	1 562	1 198	1 800	1 150	1 287	1 638	1 903	2 159
Ukraine	QAMT	113	145	166	166	147	156	202	182	245	224	279
Serbia and Montenegro	QAMW	8	34	37	43	29	32	51	62	65	66	58
Other[1]	BOQE	714	693	682	604	160	231	220	177	211	214	249
Total Europe	EPLM	99 959	107 930	109 104	108 037	108 623	120 705	122 492	122 437	119 832	120 656	133 224
Americas												
Argentina	QAOM	235	335	486	458	293	288	264	127	134	178	167
Brazil	QDLU	677	853	1 030	899	739	775	808	880	825	790	836
Canada	QATH	1 806	1 963	2 146	2 147	2 532	3 487	3 203	3 107	3 239	3 339	3 278
Chile	QAMG	172	168	211	171	115	115	132	115	123	135	150
Colombia	QAML	146	182	170	175	107	101	105	83	108	118	117
Mexico	QDLX	276	316	428	516	577	675	681	704	687	629	638
United States of America	QAMH	17 899	19 753	20 853	21 082	24 040	29 276	29 244	28 197	28 672	28 576	30 914
Uruguay	QAMU	56	66	78	68	66	57	48	30	30	32	39
Venezuela	QAMV	178	182	206	242	205	222	314	306	143	190	234
Other Central American Countries	BOQQ	798	669	758	785	922	979	684	690	712	620	676
Other	BOQT	724	555	655	589	596	431	437	409	491	352	332
Total Americas	EPLO	22 967	25 042	27 021	27 132	30 192	36 406	35 919	34 647	35 164	34 959	37 381
Asia												
China	QDMA	830	741	922	860	1 211	1 468	1 709	1 493	1 924	2 372	2 811
Hong Kong	QDMD	2 666	2 943	3 215	2 671	2 312	2 673	2 683	2 411	2 481	2 632	3 090
India	QDMG	1 691	1 718	1 576	1 242	1 450	2 058	1 772	1 755	2 284	2 235	2 800
Indonesia	QDMJ	518	809	674	369	385	404	313	324	452	398	366
Iran	QAON	330	385	380	320	237	292	434	397	471	442	452
Israel	QDMM	1 114	1 277	1 178	1 079	1 295	1 516	1 357	1 428	1 359	1 389	1 352
Japan	QAMJ	3 818	4 296	4 180	3 127	3 300	3 672	3 673	3 583	3 710	3 862	3 900
Malaysia	QDMP	1 195	1 169	1 206	677	934	907	1 029	877	1 028	995	1 089
Pakistan	QDMS	343	347	271	228	221	207	229	240	291	344	461
Philippines	QDMV	435	398	601	301	239	273	392	352	377	315	279
Saudi Arabia	QDMY	1 625	2 425	3 656	2 605	1 481	1 557	1 525	1 388	1 819	1 611	1 559
Singapore	QDNB	2 077	2 158	2 047	1 598	1 597	1 625	1 592	1 445	1 582	1 710	2 080
South Korea	QDNE	1 162	1 314	1 222	666	949	1 350	1 262	1 461	1 468	1 482	1 677
Taiwan	QDNH	966	945	1 036	867	865	1 015	875	848	897	951	939
Thailand	QDNK	839	981	863	386	463	582	594	529	572	637	638
Residual Gulf Arabian Countries	BOQW	2 570	2 756	3 127	2 640	2 258	2 586	2 749	2 620	3 353	4 027	6 952
Other Near & Middle Eastern Countries	QARJ	394	456	498	466	406	393	481	499	632	776	744
Other	BORB	699	1 005	1 186	771	592	644	564	510	600	608	580
Total Asia	EPLP	23 272	26 123	27 838	20 873	20 195	23 222	23 232	22 159	25 300	26 786	31 769
Australasia & Oceania												
Australia	QDNN	2 141	2 492	2 454	2 188	2 155	2 699	2 298	2 114	2 289	2 455	2 580
New Zealand	QDNQ	440	474	409	336	324	305	309	311	348	418	415
Other	EGIZ	55	54	84	42	38	43	42	55	64	43	81
Total Australasia & Oceania	EPLQ	2 636	3 020	2 947	2 566	2 517	3 047	2 649	2 480	2 701	2 916	3 076
Africa												
Egypt	QDNT	385	434	501	505	539	498	452	463	458	667	543
Morocco	QAOO	273	284	356	348	359	407	368	346	358	341	261
South Africa	QDNW	1 840	1 894	1 646	1 520	1 281	1 413	1 534	1 597	1 766	1 877	2 074
Other North Africa	BORU	372	390	451	440	386	419	445	478	587	540	484
Other	BOQH	1 873	2 079	2 059	2 635	2 074	1 819	2 000	1 917	2 154	2 135	2 363
Total Africa	EPLN	4 743	5 081	5 013	5 448	4 639	4 556	4 799	4 800	5 323	5 560	5 725
International Organisations	EPLR	–	–	–	–	–	–	–	–	–	–	–
World total	LQAD	153 577	167 196	171 923	164 056	166 166	187 936	189 093	186 524	188 320	190 877	211 175

1 Cyprus and Malta are included in Other Europe before 1999.

9.4 Trade in goods
continued

£ million

Imports

		1995	1996	1997	1998	1999	2000	2001	2002	2003	2004	2005
Europe												
European Union (EU)												
Austria	QBRZ	925	1 172	1 393	1 411	1 453	1 410	1 888	2 397	2 776	2 354	2 426
Belgium and Luxembourg	QBSC	8 130	9 026	9 390	9 831	10 156	10 927	12 159	13 202	13 207	13 848	15 068
of which Belgium	QDOI	..	..	..	..	10 079	10 795	11 859	12 449	12 481	12 909	14 149
Luxembourg	QDOL	..	..	..	..	77	132	300	753	726	939	919
Cyprus[1]	QDOA	..	..	..	..	184	209	243	248	251	206	262
Czech Republic	QDLG	313	353	450	555	575	807	1 097	1 250	1 412	1 290	1 663
Denmark	QBSF	2 199	2 393	2 316	2 156	2 342	2 631	2 922	3 595	3 400	3 357	4 367
Estonia	QAND	105	139	147	150	185	307	283	327	264	378	360
Finland	QBTG	2 500	2 682	2 544	2 328	2 364	2 765	2 965	2 791	2 663	2 336	2 427
France	QDJB	16 457	16 873	18 020	17 956	18 415	18 642	20 130	20 800	20 391	20 139	22 088
Germany	QDJE	26 234	27 597	25 632	25 095	26 817	28 461	30 195	32 446	33 668	35 390	39 018
Greece	QDJH	429	401	396	363	399	443	481	552	612	635	701
Hungary	QDLJ	358	402	465	535	661	686	710	846	1 120	1 578	1 848
Ireland	QDJK	7 045	7 342	7 391	7 802	8 708	10 261	12 142	13 175	9 920	10 134	10 395
Italy	QDJN	8 264	8 900	9 548	9 744	9 385	9 516	9 862	10 676	11 483	12 186	12 590
Latvia	QANE	163	294	331	290	272	400	439	485	525	692	720
Lithuania	QANF	163	178	140	140	159	246	235	268	285	271	272
Malta[1]	QDOD	..	..	..	..	126	123	144	168	185	184	175
Netherlands	QDJQ	11 516	12 596	12 328	13 408	13 772	15 379	15 396	16 144	16 693	18 200	20 353
Poland	QDLM	615	570	597	653	668	923	1 166	1 266	1 544	1 833	2 305
Portugal	QDJU	1 467	1 686	1 763	1 790	1 822	1 734	1 625	1 760	1 966	1 928	1 965
Slovak Republic	QANH	64	62	71	71	101	136	177	211	259	261	366
Slovenia	QANI	109	104	95	100	106	121	149	173	169	167	199
Spain	QDJX	4 356	5 120	5 102	5 738	5 967	6 140	7 361	9 191	9 249	9 123	11 380
Sweden	QDKA	4 537	4 840	4 693	4 361	4 649	4 950	4 671	4 330	4 570	5 120	5 283
European Central Bank	QARQ	–	–	–	–	–	–	–	–	–	–	–
EU Institutions	EOBS	–	–	–	–	–	–	–	–	–	–	–
Total EU25	LGDB	..	..	..	..	109 286	117 217	126 440	136 301	136 612	141 610	156 231
European Free Trade Association (EFTA)												
Iceland	QDKX	243	255	229	251	282	365	281	289	296	355	345
Liechtenstein	EPOX	4	17	26	20	23	22	25	22	25	18	13
Norway	QDLA	4 182	4 751	4 666	3 440	3 546	5 563	5 523	5 258	6 423	8 479	12 063
Switzerland	QDLD	4 983	5 183	4 636	4 755	5 341	5 485	4 544	4 595	3 759	3 439	3 877
Total EFTA	EPOU	9 412	10 206	9 557	8 466	9 192	11 435	10 373	10 164	10 503	12 291	16 298
Other Europe												
Albania	QAMX	–	–	2	–	1	2	–	2	3	–	–
Belarus	QAMY	18	13	14	20	20	34	18	31	22	72	270
Bulgaria	QAMZ	112	111	88	74	68	85	96	115	121	148	160
Croatia	QANC	36	36	34	40	39	41	51	68	50	55	54
Romania	QANG	165	174	195	222	249	333	442	514	674	772	758
Russia	QDLP	917	1 222	1 418	1 406	1 324	1 496	2 047	1 950	2 454	3 511	4 982
Turkey	QDLS	770	892	990	1 103	1 204	1 450	1 669	2 164	2 619	3 246	3 506
Ukraine	QANJ	21	22	37	50	47	64	71	143	94	108	91
Serbia and Montenegro	QANM	–	12	30	30	14	23	23	30	34	38	42
Other[1]	BOQF	317	378	345	407	191	186	266	262	280	265	325
Total Europe	EPMM	107 717	115 796	115 522	116 295	121 635	132 366	141 496	151 744	153 466	162 116	182 717
Americas												
Argentina	QAOP	240	273	258	198	190	181	209	236	252	265	279
Brazil	QDLV	937	942	911	883	910	1 114	1 279	1 365	1 477	1 547	1 731
Canada	QATI	2 304	2 409	2 480	2 519	3 026	4 009	3 664	3 563	3 664	4 187	4 110
Chile	QANA	285	359	374	329	328	451	464	460	414	474	477
Colombia	QANB	165	203	178	199	191	231	311	212	223	278	296
Mexico	QDLY	290	326	371	366	395	613	680	505	490	411	442
United States of America	QAMI	19 620	22 287	24 329	24 785	24 360	28 416	29 345	25 149	22 857	22 067	21 948
Uruguay	QANK	59	73	70	53	40	36	36	47	43	41	58
Venezuela	QANL	194	181	150	115	144	207	160	183	113	208	385
Other Central American Countries	BOQR	546	678	680	712	871	1 044	616	766	1 000	1 101	1 183
Other	BOQU	333	413	388	410	542	700	1 220	852	692	661	563
Total Americas	EPMO	24 973	28 144	30 189	30 569	30 997	37 002	37 984	33 338	31 225	31 240	31 472
Asia												
China	QDMB	1 841	2 110	2 379	2 816	3 384	4 826	5 741	6 726	8 342	10 405	12 897
Hong Kong	QDME	3 358	3 904	4 146	4 391	4 909	5 917	5 754	5 561	5 500	5 771	6 567
India	QDMH	1 362	1 542	1 546	1 382	1 426	1 651	1 816	1 804	2 093	2 290	2 769
Indonesia	QDMK	814	852	862	854	931	1 081	1 128	1 006	875	917	837
Iran	QAOQ	113	102	29	32	33	30	28	33	29	41	33
Israel	QDMN	658	796	839	875	996	1 025	939	880	861	923	1 000
Japan	QAMK	9 288	8 545	9 031	9 124	9 118	10 214	9 080	8 079	8 085	8 106	8 611
Malaysia	QDMQ	1 414	2 280	1 931	1 892	1 961	2 288	1 939	1 731	1 867	2 024	1 806
Pakistan	QDMT	347	375	362	340	318	363	421	472	519	554	486
Philippines	QDMW	334	858	726	855	983	1 155	1 155	944	713	657	707
Saudi Arabia	QDMZ	645	654	841	791	783	977	933	677	715	1 158	1 705
Singapore	QDNC	2 097	2 465	2 585	2 343	2 348	2 395	2 067	1 959	2 672	3 382	3 810
South Korea	QDNF	1 506	1 935	2 147	2 201	2 784	3 416	2 756	2 728	2 563	3 083	3 041
Taiwan	QDNI	1 640	2 001	2 230	2 217	2 626	3 561	2 784	2 385	2 198	2 344	2 217
Thailand	QDNL	987	1 140	1 166	1 264	1 291	1 602	1 607	1 550	1 646	1 762	1 712
Residual Gulf Arabian Countries	BOQX	496	597	734	847	833	1 109	1 138	1 225	1 516	1 724	2 002
Other Near & Middle Eastern Countries	QARK	124	123	87	81	135	118	133	189	177	111	231
Other	BORD	779	1 065	1 138	1 117	1 217	1 596	1 665	1 719	1 984	2 244	2 067
Total Asia	EPMP	27 803	31 344	32 779	33 422	36 076	43 324	41 084	39 668	42 355	47 496	52 498
Australasia & Oceania												
Australia	QDNO	1 070	1 230	1 320	1 363	1 338	1 543	1 776	1 688	1 789	1 868	2 085
New Zealand	QDNR	556	602	555	517	565	544	542	522	552	584	587
Other	HFKF	188	203	164	124	122	124	94	96	125	130	130
Total Australasia & Oceania	EPMQ	1 814	2 035	2 039	2 004	2 025	2 211	2 412	2 306	2 466	2 582	2 802
Africa												
Egypt	QDNU	236	271	256	277	255	411	406	416	432	496	348
Morocco	QAOR	241	290	331	349	383	454	439	453	443	510	419
South Africa	QDNX	1 057	1 170	1 323	1 351	1 636	2 553	2 841	2 685	2 949	3 277	3 919
Other North Africa	BORW	391	367	323	280	333	699	515	588	568	736	932
Other	BOQJ	1 368	1 501	1 503	1 322	1 877	1 892	3 127	3 031	3 023	3 317	3 366
Total Africa	EPMN	3 293	3 599	3 736	3 579	4 484	6 009	7 328	7 173	7 415	8 336	8 984
International Organisations	EPMR	–	–	–	–	–	–	–	–	–	–	–
World total	LQBL	165 600	180 918	184 265	185 869	195 217	220 912	230 305	234 229	236 927	251 770	278 473

1 Cyprus and Malta are included in Other Europe before 1999.

9.4 Trade in goods
continued

£ million

		1995	1996	1997	1998	1999	2000	2001	2002	2003	2004	2005
Balances												
Europe												
European Union (EU)												
Austria	QBSA	197	91	−234	−221	−285	−264	−664	−1 132	−1 512	−1 260	−1 097
Belgium and Luxembourg	QBSD	168	−504	−939	−1 386	−915	−605	−2 266	−2 650	−1 833	−3 337	−3 675
of which Belgium	QDOJ	..	..	..	..	−962	−693	−2 250	−2 267	−1 408	−2 659	−2 967
Luxembourg	QDOM	..	..	..	..	47	88	−16	−383	−425	−678	−708
Cyprus[1]	QDOB	..	..	..	..	75	102	48	24	66	116	94
Czech Republic	QDLH	261	366	259	143	167	119	−22	−219	−409	−315	−643
Denmark	QBSG	−91	−179	−223	−99	−288	−316	−655	−866	−1 220	−1 315	−2 061
Estonia	QANT	−75	−83	−83	−82	−134	−211	−200	−227	−169	−272	−246
Finland	QBTL	−784	−872	−974	−894	−1 010	−1 294	−1 354	−1 349	−1 170	−974	−922
France	QDJC	−1 192	220	−1 419	−1 507	−1 508	−65	−881	−2 043	−1 506	−1 575	−2 215
Germany	QDJF	−5 992	−6 882	−4 947	−4 505	−6 353	−5 672	−6 540	−10 382	−12 863	−13 719	−15 997
Greece	QDJI	609	746	651	682	754	786	643	647	640	778	663
Hungary	QDLK	−59	−51	−30	−49	−169	−75	−98	−96	−264	−645	−1 020
Ireland	QDJL	749	1 319	1 966	1 802	2 075	2 111	1 693	2 247	2 304	3 999	5 878
Italy	QDJO	−381	−873	−1 334	−1 136	−1 554	−1 087	−1 458	−2 170	−2 880	−3 785	−3 821
Latvia	QANU	−122	−212	−246	−204	−203	−316	−355	−408	−412	−600	−621
Lithuania	QANV	−112	−94	−33	−24	−63	−115	−98	−119	−96	−129	−105
Malta[1]	QDOE	..	..	..	..	63	83	71	60	75	74	63
Netherlands	QDJR	830	888	1 595	−425	−140	−212	−797	−2 133	−3 096	−6 170	−7 697
Poland	QDLN	338	788	757	525	511	363	131	52	−82	−420	−702
Portugal	QDJV	2	−9	−11	−68	−110	−74	−46	−242	−513	−348	−286
Slovak Republic	QAOG	14	44	61	32	13	21	26	−10	−22	−37	−111
Slovenia	QAOH	14	27	54	36	34	36	11	9	−8	−5	−30
Spain	QDJY	1 742	1 605	1 643	1 433	1 559	2 162	1 002	−701	−306	−23	−797
Sweden	QDKV	−380	−420	−242	31	−614	−739	−720	−457	−747	−765	−700
European Central Bank	QARR	−	−	−	−	−	−	−	−	−	−	−
EU Institutions	EOCM	−	−	−	−	−	−	−	−	−	−	−
Total EU25	LGCF	..	..	..	..	−8 095	−5 262	−12 529	−22 165	−26 023	−30 727	−36 048
European Free Trade Association (EFTA)												
Iceland	QDKY	−106	−103	−72	−93	−123	−172	−131	−158	−155	−188	−166
Liechtenstein	EPOY	9	−3	−16	−16	−21	−16	−22	−20	−22	−12	−11
Norway	QDLB	−2 189	−2 712	−2 057	−782	−1 547	−3 545	−3 710	−3 562	−4 537	−6 542	−9 851
Switzerland	QDLE	−2 256	−2 017	−1 681	−1 863	−2 573	−2 424	−1 048	−1 515	−973	−599	1 107
Total EFTA	EPOV	−4 542	−4 835	−3 826	−2 754	−4 264	−6 157	−4 912	−5 255	−5 687	−7 341	−8 921
Other Europe												
Albania	QANN	8	14	5	8	11	5	23	17	7	12	16
Belarus	QANO	5	14	21	12	7	3	15	1	16	−19	−213
Bulgaria	QANP	−9	−23	−10	7	8	1	25	16	31	6	59
Croatia	QANS	196	101	71	66	41	31	37	26	88	70	64
Romania	QAOD	14	39	18	11	−6	50	−103	−87	−164	−166	−117
Russia	QDLQ	−43	−204	−185	−477	−792	−828	−1 154	−969	−1 034	−2 045	−3 114
Turkey	QDLT	380	653	744	459	−6	350	−519	−877	−981	−1 343	−1 347
Ukraine	QAOI	92	123	129	116	100	92	131	39	151	116	188
Serbia and Montenegro	QAOL	8	22	7	13	15	9	28	32	31	28	16
Other[1]	BOQG	397	315	337	197	−31	45	−46	−85	−69	−51	−76
Total Europe	EPNM	−7 758	−7 866	−6 418	−8 258	−13 012	−11 661	−19 004	−29 307	−33 634	−41 460	−49 493
Americas												
Argentina	QAOS	−5	62	228	260	103	107	55	−109	−118	−87	−112
Brazil	QDLW	−260	−89	119	16	−171	−339	−471	−485	−652	−757	−895
Canada	QBRV	−498	−446	−334	−372	−494	−522	−461	−456	−425	−848	−832
Chile	QANQ	−113	−191	−163	−158	−213	−336	−332	−345	−291	−339	−327
Colombia	QANR	−19	−21	−8	−24	−84	−130	−206	−127	−115	−160	−179
Mexico	QDLZ	−14	−10	57	150	182	62	1	199	197	218	196
United States of America	QBRP	−1 721	−2 534	−3 476	−3 703	−320	860	−101	3 048	5 815	6 509	8 966
Uruguay	QAOJ	−3	−7	8	15	26	21	12	−17	−13	−9	−19
Venezuela	QAOK	−16	1	56	127	61	15	154	123	30	−18	−151
Other Central American Countries	BOQS	252	−9	78	73	51	−65	68	−76	−288	−481	−507
Other	BOQV	391	142	267	179	54	−269	−783	−443	−201	−309	−231
Total Americas	EPNO	−2 006	−3 102	−3 168	−3 437	−805	−596	−2 065	1 309	3 939	3 719	5 909
Asia												
China	QDMC	−1 011	−1 369	−1 457	−1 956	−2 173	−3 358	−4 032	−5 233	−6 418	−8 033	−10 086
Hong Kong	QDMF	−692	−961	−931	−1 720	−2 597	−3 244	−3 071	−3 150	−3 019	−3 139	−3 477
India	QDMI	329	176	30	−140	24	407	−44	−49	191	−55	31
Indonesia	QDML	−296	−43	−188	−485	−546	−677	−815	−682	−423	−519	−471
Iran	QAOT	217	283	351	288	204	262	406	364	442	401	419
Israel	QDMO	456	481	339	204	299	491	418	548	498	466	352
Japan	QBRR	−5 470	−4 249	−4 851	−5 997	−5 818	−6 542	−5 407	−4 496	−4 375	−4 244	−4 711
Malaysia	QDMR	−219	−1 111	−725	−1 215	−1 027	−1 381	−910	−854	−839	−1 029	−717
Pakistan	QDMU	−4	−28	−91	−112	−97	−156	−192	−232	−228	−210	−25
Philippines	QDMX	101	−460	−125	−554	−744	−882	−763	−592	−336	−342	−428
Saudi Arabia	QDNA	980	1 771	2 815	1 814	698	580	592	711	1 104	453	−146
Singapore	QDND	−20	−307	−538	−745	−751	−770	−475	−514	−1 090	−1 672	−1 730
South Korea	QDNG	−344	−621	−925	−1 535	−1 535	−2 066	−1 494	−1 267	−1 095	−1 601	−1 364
Taiwan	QDNJ	−674	−1 056	−1 194	−1 350	−1 761	−2 546	−1 909	−1 537	−1 301	−1 393	−1 278
Thailand	QDNM	−148	−159	−303	−878	−1 020	−1 013	−1 021	−1 074	−1 125	−1 074
Residual Gulf Arabian Countries	BORA	2 074	2 159	2 393	1 793	1 425	1 477	1 611	1 395	1 837	2 303	4 950
Other Near & Middle Eastern Countries	QARL	270	333	411	385	271	275	348	310	455	665	513
Other	BORE	−80	−60	48	−346	−625	−952	−1 101	−1 209	−1 384	−1 636	−1 487
Total Asia	EPNP	−4 531	−5 221	−4 941	−12 549	−15 881	−20 102	−17 852	−17 509	−17 055	−20 710	−20 729
Australasia & Oceania												
Australia	QDNP	1 071	1 262	1 134	825	817	1 156	522	426	500	587	495
New Zealand	QDNS	−116	−128	−146	−181	−241	−239	−233	−211	−204	−166	−172
Other	HFKK	−133	−149	−80	−82	−84	−81	−52	−41	−61	−87	−49
Total Australasia & Oceania	EPNQ	822	985	908	562	492	836	237	174	235	334	274
Africa												
Egypt	QDNV	149	163	245	228	284	87	46	47	26	171	195
Morocco	QAOU	32	−6	25	−1	−24	−47	−71	−107	−85	−169	−158
South Africa	QDNY	783	724	323	169	−355	−1 140	−1 307	−1 088	−1 183	−1 400	−1 845
Other North Africa	BORX	−19	23	128	160	53	−280	−70	−110	19	−196	−448
Other	BOQK	505	578	556	1 313	197	−73	−1 127	−1 114	−869	−1 182	−1 003
Total Africa	EPNN	1 450	1 482	1 277	1 869	155	−1 453	−2 529	−2 373	−2 092	−2 776	−3 259
International Organisations	EPNR	−	−	−	−	−	−	−	−	−	−	−
World total	LQCT	−12 023	−13 722	−12 342	−21 813	−29 051	−32 976	−41 212	−47 705	−48 607	−60 893	−67 298

1 Cyprus and Malta are included in Other Europe before 1999.

9.5 Trade in services

£ million

		1995	1996	1997	1998	1999	2000	2001	2002	2003	2004	2005
Exports												
Europe												
European Union (EU)												
Austria	FYVC	328	354	299	343	383	393	386	403	394	413	485
Belgium and Luxembourg	FYVD	1 469	1 523	1 673	1 955	2 262	2 385	2 343	2 643	2 992	3 623	3 336
of which Belgium	A7RX	..	..	..	..	1 920	2 040	1 969	2 284	2 448	2 837	2 615
Luxembourg	A7S2	..	..	..	..	342	345	374	359	544	786	721
Cyprus[1]	A7RY	..	174	156	126	130	147	128	213	231	328	289
Czech Republic	FYVW	105	113	121	198	150	166	181	177	240	329	335
Denmark	FYVE	516	862	863	1 032	960	1 164	1 366	1 347	1 483	1 563	1 529
Estonia	ZWKU	4	7	15	5	14	12	14	21	22	29	38
Finland	FYVF	389	566	545	533	654	950	868	819	862	949	866
France	FYVG	3 064	3 359	3 535	4 180	4 696	5 324	5 869	5 788	5 939	6 327	5 949
Germany	FYVH	4 401	4 597	4 737	5 251	5 475	6 475	6 826	7 086	7 238	7 635	7 760
Greece	FYVI	524	546	629	687	788	814	780	730	747	706	787
Hungary	GYWV	49	75	110	95	138	206	205	216	226	375	572
Ireland	FYVJ	1 981	2 236	2 314	2 551	2 522	2 800	3 560	4 170	5 141	5 501	5 768
Italy	FYVK	1 706	1 952	2 079	2 370	2 699	2 622	2 773	2 895	2 857	3 288	3 442
Latvia	ZWLC	9	12	27	16	24	30	33	38	29	49	102
Lithuania	ZWLB	11	18	19	7	15	17	46	32	43	56	118
Malta[1]	A7S3	..	105	125	67	65	57	52	60	72	77	75
Netherlands	FYVL	2 125	2 463	2 860	3 206	4 023	4 446	4 709	4 469	5 088	6 420	6 589
Poland	FYVX	172	188	238	285	266	282	332	386	406	506	798
Portugal	FYVM	356	381	326	450	570	426	466	506	513	657	611
Slovak Republic	ZWLG	22	34	29	24	31	39	43	30	36	46	95
Slovenia	ZWLF	10	16	17	5	22	24	32	39	31	41	50
Spain	FYVN	1 254	1 457	1 392	1 763	2 061	2 225	2 193	2 549	2 788	2 837	3 506
Sweden	FYVO	907	1 032	1 156	1 360	1 406	1 597	1 600	1 443	1 659	1 664	1 688
European Central Bank	KNWZ	–	–	–	–	–	–	3	12	3	–	–
EU Institutions	FYVP	381	247	248	228	232	248	544	544	511	578	596
Total EU25	A7RZ	..	22 317	23 513	26 737	29 586	32 852	35 361	36 607	39 548	43 997	45 384
European Free Trade Association (EFTA)												
Iceland	FYVR	42	48	55	48	30	73	63	64	72	89	135
Liechtenstein	FYVS	2	5	11	21	32	36	27	40	36	49	42
Norway	FYVT	939	1 070	968	1 212	1 124	1 021	959	1 428	1 288	1 486	1 681
Switzerland	FYVU	1 280	1 939	2 028	2 382	2 415	2 532	3 112	3 573	4 110	3 933	4 248
Total EFTA	FYVV	2 263	3 062	3 062	3 663	3 601	3 662	4 161	5 105	5 506	5 557	6 106
Other Europe												
Albania	ZWKM	–	–	–	–	3	2	5	6	6	22	9
Belarus	ZWKP	5	6	7	1	3	27	3	4	4	6	8
Bulgaria	ZWKO	7	12	20	14	41	42	35	75	67	78	99
Croatia	ZWKZ	12	18	19	12	38	38	34	53	48	41	41
Romania	ZWLE	16	28	32	29	44	44	73	76	97	137	136
Russia	FYVY	311	333	361	441	290	407	454	562	703	866	1 081
Turkey	FYVZ	218	265	331	333	394	345	387	364	395	360	479
Ukraine	ZWLH	27	46	38	65	23	43	29	301	122	52	174
Serbia and Montenegro	ZWLK	7	13	39	20	39	103	41	28	40	21	36
Other[1]	ZWKJ	862	1 117	1 408	850	1 034	1 114	961	1 278	1 755	1 999	2 379
Total Europe	FYWB	23 511	27 217	28 830	32 165	35 096	38 679	41 544	44 459	48 291	53 136	55 932
Americas												
Argentina	ZWKN	84	116	115	210	163	183	189	83	83	143	96
Brazil	FYWC	256	273	298	422	369	342	382	313	326	307	353
Canada	FYWD	944	981	1 140	1 399	1 360	1 370	1 588	1 490	1 533	1 736	1 519
Chile	ZWKQ	42	48	56	124	106	96	103	87	79	84	90
Colombia	ZWKR	69	68	59	67	48	56	140	86	97	90	75
Mexico	FYWE	135	115	165	163	214	262	278	258	263	349	298
United States of America	FYWF	10 744	13 641	14 675	15 532	17 423	18 899	18 515	21 938	23 042	24 731	22 824
Uruguay	ZWLI	7	8	9	11	9	7	5	5	60	10	16
Venezuela	ZWLJ	38	43	48	72	68	83	110	99	102	99	95
Other Central American Countries	ZWKT	623	757	978	1 408	1 037	1 290	1 407	1 996	2 084	2 652	3 128
Other	ZWKW	181	197	263	248	232	217	215	147	156	143	195
Total Americas	FYWH	13 123	16 247	17 806	19 656	21 029	22 805	22 932	26 502	27 825	30 344	28 689
Asia												
China	FYWI	211	236	223	301	448	442	577	710	872	1 269	1 339
Hong Kong	FYWJ	759	877	892	961	1 003	986	1 005	930	1 109	1 058	1 209
India	FYWK	415	384	575	484	510	548	664	623	705	981	1 102
Indonesia	FYWL	217	193	226	199	164	191	201	170	161	186	178
Iran	ZWLA	58	59	68	92	96	119	162	181	216	234	219
Israel	FYWM	344	310	372	308	523	441	504	422	359	401	382
Japan	FYWN	2 256	2 313	2 372	2 613	2 944	3 249	3 252	3 521	3 533	3 899	4 050
Malaysia	FYWO	481	500	502	542	388	385	369	398	415	542	431
Pakistan	FYWP	291	350	304	243	294	176	276	169	177	228	421
Philippines	FYWQ	138	102	133	93	123	183	178	105	141	128	150
Saudi Arabia	FYWR	1 808	1 738	1 744	1 923	2 138	2 350	2 114	590	618	546	705
Singapore	FYWS	480	455	504	557	1 093	1 034	1 204	1 067	1 535	2 109	2 479
South Korea	FYWT	421	461	483	487	404	388	418	495	462	903	712
Taiwan	FYWU	220	243	270	234	248	280	290	366	371	495	544
Thailand	FYWV	344	251	276	230	205	191	215	202	330	317	244
Residual Gulf Arabian Countries	ZWKX	687	727	814	992	993	1 068	1 051	2 625	2 941	3 422	3 819
Other Near & Middle Eastern Countries	ZWKY	107	108	147	175	201	227	332	261	252	337	450
Other Asian Countries	ZWKK	1 105	1 091	1 131	1 011	653	807	880	723	793	959	1 073
Total Asia	FYWX	10 342	10 398	11 036	11 445	12 428	13 065	13 692	13 558	14 990	18 014	19 507
Australasia & Oceania												
Australia	FYWY	1 248	1 439	1 507	1 450	1 589	1 574	1 938	1 794	1 980	2 261	2 584
New Zealand	FYWZ	280	253	278	320	353	301	300	273	298	381	391
Other	FYXA	29	102	61	47	61	84	95	97	115	133	117
Total Australasia & Oceania	FYXB	1 557	1 794	1 846	1 817	2 003	1 959	2 333	2 164	2 393	2 775	3 092
Africa												
Egypt	ZWKV	66	81	109	99	274	274	332	218	229	316	339
Morocco	ZWLD	36	40	28	57	47	61	82	50	48	62	61
South Africa	FYXC	737	715	827	931	1 115	968	1 117	997	1 072	1 064	1 083
Other North Africa	ZWKS	101	125	250	291	115	261	213	216	280	281	391
Other	ZWKL	971	1 271	1 287	1 472	1 474	1 548	1 745	1 745	1 886	1 786	1 938
Total Africa	FYXE	1 911	2 232	2 501	2 850	3 025	3 112	3 489	3 226	3 515	3 509	3 812
International Organisations	FYXF	130	74	77	45	35	46	57	78	63	39	91
World total	KTMQ	50 574	57 962	62 096	67 978	73 616	79 666	84 047	89 987	97 077	107 817	111 123

1 Cyprus and Malta are included in Other Europe before 1999.

9.5 Trade in services
continued

£ million

		1995	1996	1997	1998	1999	2000	2001	2002	2003	2004	2005
Imports												
Europe												
European Union (EU)												
Austria	GGOR	383	360	343	454	447	469	502	683	655	732	853
Belgium and Luxembourg	GGOS	1 075	1 197	1 247	1 329	1 610	1 563	1 678	1 864	1 979	1 956	1 902
of which Belgium	A8ET	..	..	..	..	1 436	1 418	1 535	1 686	1 762	1 658	1 662
Luxembourg	A8EW	..	..	..	..	174	145	143	178	217	298	240
Cyprus[1]	A8EU	..	708	771	802	766	885	1 059	977	1 088	1 084	1 215
Czech Republic	GGPL	91	100	105	146	146	168	199	250	337	429	453
Denmark	GGOT	375	476	492	493	619	659	637	723	752	809	831
Estonia	ZWNA	4	6	2	5	6	33	27	18	35	46	47
Finland	GGOU	152	201	185	217	305	403	360	282	314	284	359
France	GGOV	3 960	4 268	4 735	5 705	6 642	7 495	7 493	7 810	8 216	8 407	8 518
Germany	GGOW	3 473	3 963	3 439	3 687	4 605	4 538	5 016	5 128	5 198	5 898	6 866
Greece	GGOX	1 045	913	945	1 051	1 426	1 617	1 886	1 907	1 977	1 946	1 992
Hungary	GYXH	42	66	72	82	89	95	126	129	151	184	235
Ireland	GGOY	1 098	1 539	1 642	1 744	2 113	2 213	2 286	2 247	2 493	2 911	3 165
Italy	GGOZ	1 239	1 850	1 958	2 320	2 595	2 822	2 970	3 105	3 390	3 495	4 119
Latvia	ZWNI	3	3	5	5	12	25	20	19	29	38	38
Lithuania	ZWNH	2	3	3	7	5	6	16	37	25	20	71
Malta[1]	A8EX	..	175	186	198	196	223	256	238	279	320	365
Netherlands	GGPA	1 751	1 768	1 834	2 041	2 371	2 958	3 059	3 266	3 207	3 220	3 375
Poland	GGPM	117	94	129	144	175	184	200	187	224	299	508
Portugal	GGPB	659	759	828	825	1 005	1 021	1 114	1 204	1 361	1 397	1 455
Slovak Republic	ZWNM	11	15	10	32	28	30	14	27	22	33	55
Slovenia	ZWNL	6	10	12	12	14	29	17	23	24	34	43
Spain	GGPC	3 548	4 166	4 541	5 202	5 919	6 484	7 215	7 823	8 947	9 072	9 507
Sweden	GGPD	518	573	515	531	767	822	820	818	1 120	1 131	1 260
European Central Bank	KOFJ	–	–	–	–	1	1	1	1	–	–	–
EU Institutions	GGPE	1	2	1	10	6	3	5	14	17	4	4
Total EU25	A8EV	..	23 215	24 000	27 042	31 868	34 746	36 976	38 780	41 840	43 749	47 236
European Free Trade Association (EFTA)												
Iceland	GGPG	24	31	34	46	25	75	44	42	46	57	81
Liechtenstein	GGPH	1	1	1	33	7	6	7	1	11	4	3
Norway	GGPI	683	556	596	523	498	536	647	1 265	706	764	861
Switzerland	GGPJ	1 079	1 270	1 102	1 393	1 446	1 563	1 518	1 626	1 743	1 759	2 164
Total EFTA	GGPK	1 787	1 858	1 733	1 995	1 976	2 180	2 216	2 934	2 506	2 584	3 109
Other Europe												
Albania	ZWMS	1	1	–	1	6	5	9	13	14	20	19
Belarus	ZWMV	2	3	1	5	1	2	4	3	4	1	13
Bulgaria	ZWMU	18	29	34	29	36	38	52	96	104	176	250
Croatia	ZWNF	26	39	33	37	21	27	55	47	73	84	134
Romania	ZWNK	21	29	25	22	51	64	77	84	91	125	153
Russia	GGPN	238	165	159	182	137	188	218	296	303	503	542
Turkey	GGPO	472	599	579	562	524	521	569	704	788	818	1 013
Ukraine	ZWNN	5	7	14	14	9	14	33	25	23	43	95
Serbia and Montenegro	ZWNQ	8	12	14	11	31	22	38	33	41	38	45
Other[1]	ZWMP	1 253	794	908	827	716	1 112	967	991	923	811	464
Total Europe	GGPQ	23 384	26 751	27 500	30 727	35 376	38 919	41 214	44 006	46 710	48 952	53 073
Americas												
Argentina	ZWMT	30	40	37	70	75	75	74	55	48	69	76
Brazil	GGPR	67	109	117	140	156	174	231	251	176	219	263
Canada	GGPS	781	707	792	866	968	1 105	1 185	1 081	994	1 036	1 128
Chile	ZWMW	17	24	20	43	42	35	48	48	43	72	77
Colombia	ZWMX	13	26	23	60	35	39	31	35	32	24	37
Mexico	GGPT	57	116	149	188	225	283	398	336	312	367	317
United States of America	GGPU	8 865	10 348	10 180	11 492	12 645	13 130	13 405	13 308	14 450	14 752	14 834
Uruguay	ZWNO	1	1	2	6	4	6	9	23	4	5	8
Venezuela	ZWNP	16	27	24	40	46	46	42	35	34	26	33
Other Central American Countries	ZWMZ	562	758	843	1 033	946	1 448	1 385	1 289	1 366	1 858	1 778
Other	ZWNC	227	237	210	165	215	256	148	124	130	111	168
Total Americas	GGPW	10 636	12 393	12 397	14 103	15 357	16 597	16 956	16 585	17 589	18 539	18 719
Asia												
China	GGPX	157	158	165	219	266	292	375	496	463	567	668
Hong Kong	GGPY	492	478	583	501	603	559	594	611	700	702	716
India	GGPZ	358	442	492	554	673	664	816	808	810	1 095	1 247
Indonesia	GGQA	169	178	127	104	150	125	121	136	297	100	103
Iran	ZWNG	8	13	17	25	21	43	37	29	45	61	44
Israel	GGQB	222	160	188	207	215	266	314	202	213	221	227
Japan	GGQC	967	1 121	1 088	1 154	1 223	1 560	1 719	1 384	1 666	1 579	2 121
Malaysia	GGQD	201	199	244	253	186	227	191	193	170	210	220
Pakistan	GGQE	162	169	192	194	288	300	280	398	386	431	479
Philippines	GGQF	63	59	63	64	88	206	82	100	155	110	141
Saudi Arabia	GGQG	401	370	456	451	526	553	643	157	146	217	214
Singapore	GGQH	359	269	336	303	339	316	376	407	396	612	687
South Korea	GGQI	101	115	144	129	135	141	179	185	186	248	254
Taiwan	GGQJ	88	106	100	108	119	156	148	179	173	195	223
Thailand	GGQK	269	242	236	308	307	385	458	521	513	644	613
Residual Gulf Arabian Countries	ZWND	319	226	239	372	421	400	537	1 740	1 660	1 413	1 444
Other Near & Middle Eastern Countries	ZWNE	52	58	73	78	84	125	155	151	151	205	229
Other	ZWMQ	856	607	632	535	479	631	535	557	645	901	864
Total Asia	GGQM	5 244	4 970	5 375	5 559	6 123	6 949	7 560	8 254	8 775	9 511	10 494
Australasia & Oceania												
Australia	GGQN	709	822	889	961	1 042	1 190	1 251	1 430	1 471	1 388	1 808
New Zealand	GGQO	192	229	233	227	249	264	287	337	347	433	497
Other	GGQP	27	62	35	26	29	25	20	33	28	54	64
Total Australasia & Oceania	GGQQ	928	1 113	1 157	1 214	1 320	1 479	1 558	1 800	1 846	1 875	2 369
Africa												
Egypt	ZWNB	147	158	130	117	229	290	367	313	350	347	504
Morocco	ZWNJ	49	65	93	119	137	114	147	141	141	150	174
South Africa	GGQR	353	427	419	487	480	528	598	684	921	924	1 002
Other North Africa	ZWMY	113	181	227	332	218	323	349	263	266	334	324
Other	ZWMR	679	628	625	582	723	772	813	1 071	1 288	1 230	1 365
Total Africa	GGQT	1 341	1 459	1 494	1 637	1 787	2 027	2 274	2 472	2 966	2 985	3 369
International Organisations	GGQU	84	72	67	66	56	80	62	40	29	37	43
World total	KTMR	41 617	46 758	47 990	53 306	60 019	66 051	69 624	73 157	77 915	81 899	88 067

1 Cyprus and Malta are included in Other Europe before 1999.

9.5 Trade in services
continued

£ million

		1995	1996	1997	1998	1999	2000	2001	2002	2003	2004	2005
Balances												
Europe												
European Union (EU)												
Austria	GGQW	−55	−6	−44	−111	−64	−76	−116	−280	−261	−319	−368
Belgium and Luxembourg	GGQX	394	326	426	626	652	822	665	779	1 013	1 667	1 434
of which Belgium	A8HH	..	..	..	..	484	622	434	598	686	1 179	953
Luxembourg	A8HK	..	..	..	..	168	200	231	181	327	488	481
Cyprus[1]	A8HI	..	−534	−615	−676	−636	−738	−931	−764	−857	−756	−926
Czech Republic	GGRQ	14	13	16	52	4	−2	−18	−73	−97	−100	−118
Denmark	GGQY	141	386	371	539	341	505	729	624	731	754	698
Estonia	ZWTU	−	1	13	−	8	−21	−13	3	−13	−17	−9
Finland	GGQZ	237	365	360	316	349	547	508	537	548	665	507
France	GGRA	−896	−909	−1 200	−1 525	−1 946	−2 171	−1 624	−2 022	−2 277	−2 080	−2 569
Germany	GGRB	928	634	1 298	1 564	870	1 937	1 810	1 958	2 040	1 737	894
Greece	GGRC	−521	−367	−316	−364	−638	−803	−1 106	−1 177	−1 230	−1 240	−1 205
Hungary	GYXT	7	9	38	13	49	111	79	87	75	191	337
Ireland	GGRD	883	697	672	807	409	587	1 274	1 923	2 648	2 590	2 603
Italy	GGRE	467	102	121	50	104	−200	−197	−210	−533	−207	−677
Latvia	ZWUC	6	9	22	11	12	5	13	19	−	11	64
Lithuania	ZWUB	9	15	16	−	10	11	30	−5	18	36	47
Malta[1]	A8HL	..	−70	−61	−131	−131	−166	−204	−178	−207	−243	−290
Netherlands	GGRF	374	695	1 026	1 165	1 652	1 488	1 650	1 203	1 881	3 200	3 214
Poland	GGRR	55	94	109	141	91	98	132	199	182	207	290
Portugal	GGRG	−303	−378	−502	−375	−435	−595	−648	−698	−848	−740	−844
Slovak Republic	ZWUG	11	19	19	−8	3	9	29	3	14	13	40
Slovenia	ZWUF	4	6	5	−7	8	−5	15	16	7	7	7
Spain	GGRH	−2 294	−2 709	−3 149	−3 439	−3 858	−4 259	−5 022	−5 274	−6 159	−6 235	−6 001
Sweden	GGRI	389	459	641	829	639	775	780	625	539	533	428
European Central Bank	ZWTI	−	−	−	−	−1	9	11	2	−	−	−
EU Institutions	GGRJ	380	245	247	218	226	245	539	530	494	574	592
Total EU25	A8HJ	..	−898	−487	−305	−2 282	−1 894	−1 615	−2 173	−2 292	248	−1 852
European Free Trade Association (EFTA)												
Iceland	GGRL	18	17	21	2	5	−2	19	22	26	32	54
Liechtenstein	GGRM	1	4	10	−12	25	30	20	39	25	45	39
Norway	GGRN	256	514	372	689	626	485	312	163	582	722	820
Switzerland	GGRO	201	669	926	989	969	969	1 594	1 947	2 367	2 174	2 084
Total EFTA	GGRP	476	1 204	1 329	1 668	1 625	1 482	1 945	2 171	3 000	2 973	2 997
Other Europe												
Albania	ZWTM	−1	−1	−	−1	−3	−3	−4	−7	−8	2	−10
Belarus	ZWTP	3	3	6	−4	2	25	−1	1	−	5	−5
Bulgaria	ZWTO	−11	−17	−14	−15	5	4	−17	−21	−37	−98	−151
Croatia	ZWTZ	−14	−21	−14	−25	17	11	−21	6	−25	−43	−93
Romania	ZWUE	−5	−1	7	7	−7	−20	−4	−8	6	12	−17
Russia	GGRS	73	168	202	259	153	219	236	266	400	363	539
Turkey	GGRT	−254	−334	−248	−229	−130	−176	−182	−340	−393	−458	−534
Ukraine	ZWUH	22	39	24	51	14	29	−4	276	99	9	79
Serbia and Montenegro	ZWUK	−1	1	25	9	8	81	3	−5	−1	−17	−9
Other[1]	ZWTJ	−391	323	500	23	318	2	−6	287	832	1 188	1 915
Total Europe	GGRV	127	466	1 330	1 438	−280	−240	330	453	1 581	4 184	2 859
Americas												
Argentina	ZWTN	54	76	78	140	88	108	115	28	35	74	20
Brazil	GGRW	189	164	181	282	213	168	151	62	150	88	90
Canada	GGRX	163	274	348	533	392	265	403	409	539	700	391
Chile	ZWTQ	25	24	36	81	64	61	55	39	36	12	13
Colombia	ZWTR	56	42	36	7	13	17	109	51	65	66	38
Mexico	GGRY	78	−1	16	−25	−11	−21	−120	−78	−49	−18	−19
United States of America	GGRZ	1 879	3 293	4 495	4 040	4 778	5 769	5 110	8 630	8 592	9 979	7 990
Uruguay	ZWUI	6	7	7	5	5	1	−4	−18	56	5	8
Venezuela	ZWUJ	22	16	24	32	22	37	68	64	68	73	62
Other Central American Countries	ZWTT	61	−1	135	375	91	−158	22	707	718	794	1 350
Other	ZWTW	−46	−40	53	83	17	−39	67	23	26	32	27
Total Americas	GGSB	2 487	3 854	5 409	5 553	5 672	6 208	5 976	9 917	10 236	11 805	9 970
Asia												
China	GGSC	54	78	58	82	182	150	202	214	409	702	671
Hong Kong	GGSD	267	399	309	460	400	427	411	319	409	356	493
India	GGSE	57	−58	83	−70	−163	−116	−152	−185	−105	−114	−145
Indonesia	GGSF	48	15	99	95	14	66	80	34	−136	86	75
Iran	ZWUA	50	46	51	67	75	76	125	152	171	173	175
Israel	GGSG	122	150	184	101	308	175	190	220	146	180	155
Japan	GGSH	1 289	1 192	1 284	1 459	1 721	1 689	1 533	2 137	1 867	2 320	1 929
Malaysia	GGSI	280	301	258	289	202	158	178	205	245	332	211
Pakistan	GGSJ	129	181	112	49	6	−124	−4	−229	−209	−203	−58
Philippines	GGSK	75	43	70	29	35	−23	96	5	−14	18	9
Saudi Arabia	GGSL	1 407	1 368	1 288	1 472	1 612	1 797	1 471	433	472	329	491
Singapore	GGSM	121	186	168	254	754	718	828	660	1 139	1 497	1 792
South Korea	GGSN	320	346	339	358	269	247	239	310	276	655	458
Taiwan	GGSO	132	137	170	126	129	124	142	187	198	300	321
Thailand	GGSP	75	9	40	−78	−102	−194	−243	−319	−183	−327	−369
Residual Gulf Arabian Countries	ZWTX	368	501	575	620	572	668	514	885	1 281	2 009	2 375
Other Near & Middle Eastern Countries	ZWTY	55	50	74	97	117	102	177	110	101	132	221
Other	ZWTK	249	484	499	476	174	176	345	166	148	58	209
Total Asia	GGSR	5 098	5 428	5 661	5 886	6 305	6 116	6 132	5 304	6 215	8 503	9 013
Australasia & Oceania												
Australia	GGSS	539	617	618	489	547	384	687	364	509	873	776
New Zealand	GGST	88	24	45	93	104	37	13	−64	−49	−52	−106
Other	GGSU	2	40	26	21	32	59	75	64	87	79	53
Total Australasia & Oceania	GGSV	629	681	689	603	683	480	775	364	547	900	723
Africa												
Egypt	ZWTV	−81	−77	−21	−18	45	−16	−35	−95	−121	−31	−165
Morocco	ZWUD	−13	−25	−65	−62	−90	−53	−65	−91	−93	−88	−113
South Africa	GGSW	384	288	408	444	635	440	519	313	151	140	81
Other North Africa	ZWTS	−12	−56	23	−41	−103	−62	−136	−47	14	−53	67
Other	ZWTL	292	643	662	890	751	776	932	674	598	556	573
Total Africa	GGSY	570	773	1 007	1 213	1 238	1 085	1 215	754	549	524	443
International Organisations	GGSZ	46	2	10	−21	−21	−34	−5	38	34	2	48
World total	KTMS	8 957	11 204	14 106	14 672	13 597	13 615	14 423	16 830	19 162	25 918	23 056

1 Cyprus and Malta are included in Other Europe before 1999.

9.6 Income

£ million

Credits

		1995	1996	1997	1998	1999	2000	2001	2002	2003	2004	2005
Europe												
European Union (EU)												
Austria	CUGY	531	528	534	621	585	772	701	707	762	833	994
Belgium and Luxembourg	CTFK	3 113	3 395	3 508	3 820	3 605	6 249	6 824	5 811	4 972	6 807	7 936
of which Belgium	AA2K	..	..	..	..	1 646	2 421	2 857	2 404	2 075	2 798	3 380
Luxembourg	AA2O	..	..	..	..	1 959	3 828	3 967	3 407	2 897	4 009	4 556
Cyprus[1]	AA2L	..	..	..	..	106	153	166	139	56	72	143
Czech Republic	LEPT	49	109	130	113	125	125	198	152	195	214	250
Denmark	LEQU	612	623	722	1 064	952	1 136	1 008	571	738	913	1 279
Estonia	ZWYT	2	3	13	13	5	2	8	6	8	6	15
Finland	LEUG	656	540	518	464	415	594	659	599	527	530	624
France	LEUP	6 017	5 904	5 472	5 992	6 227	7 656	8 836	7 129	7 162	8 417	11 306
Germany	LEQL	9 225	8 917	8 031	8 130	8 525	10 469	11 475	10 088	8 773	9 904	11 746
Greece	LEUY	348	799	1 069	655	845	1 010	882	721	833	804	913
Hungary	BFKR	106	180	218	175	248	448	321	354	334	379	443
Ireland	BFLY	1 707	1 909	2 389	2 698	2 998	3 913	4 809	4 075	5 019	6 349	8 010
Italy	BFOG	5 187	5 067	5 139	6 929	5 229	6 516	6 459	4 914	5 020	4 756	5 443
Latvia	ZWYU	2	–	1	8	–	2	–	–	–	–	2
Lithuania	ZWYV	–	5	8	4	4	9	7	4	4	1	1
Malta[1]	AA2P	..	..	..	..	39	51	55	48	53	496	259
Netherlands	BFQI	7 097	7 962	8 679	8 904	9 343	12 909	16 705	15 890	15 356	10 038	17 641
Poland	BFSB	114	111	119	91	110	144	329	337	401	444	504
Portugal	BFSK	394	460	391	466	480	677	824	806	833	829	912
Slovak Republic	ZWYX	24	51	69	37	38	51	29	31	28	30	55
Slovenia	ZWYY	–	–	8	22	56	48	101	68	89	97	111
Spain	LESW	1 806	2 428	2 444	2 402	2 124	2 589	2 444	2 114	2 379	3 746	5 103
Sweden	BFTL	1 417	1 594	1 424	1 449	1 571	2 131	2 339	2 157	1 796	2 341	2 619
European Central Bank	ZWYO	–	–	–	–	–	–	–	–	–	–	–
EU Institutions	CSFK	211	143	163	281	224	370	468	657	751	514	547
Total EU25	AA2M	..	..	..	..	43 854	58 024	65 647	57 378	56 089	58 520	76 856
European Free Trade Association (EFTA)												
Iceland	BFNQ	32	25	20	32	32	35	38	33	27	55	116
Liechtenstein	BFPH	51	64	52	34	39	45	39	21	20	23	27
Norway	BFQR	531	586	605	632	734	1 032	1 030	703	817	861	1 533
Switzerland	LEPB	3 289	3 517	3 569	3 448	4 477	7 432	5 490	4 661	4 676	5 096	5 961
Total EFTA	CTFT	3 903	4 192	4 246	4 146	5 282	8 544	6 597	5 418	5 540	6 035	7 637
Other Europe												
Albania	ZWYP	–	–	–	–	–	–	–	–	–	–	–
Belarus	ZWYQ	–	–	–	–	–	–	–	7	–	–	–
Bulgaria	ZWYR	24	16	41	21	48	37	93	72	74	90	98
Croatia	ZWYS	4	4	14	9	13	15	26	29	43	48	70
Romania	ZWYW	22	42	25	22	42	20	78	65	96	118	134
Russia	BFST	105	130	234	269	196	286	541	666	571	1 299	1 533
Turkey	BFUM	136	158	190	268	349	498	415	322	369	385	592
Ukraine	ZWYZ	11	8	–	16	3	–	3	18	21	34	83
Serbia and Montenegro	BFWF	33	57	49	–	1	1	1	1	1	–2	16
Other[1]	LEVZ	1 242	1 476	2 617	3 146	2 705	3 492	2 924	2 908	4 065	5 816	5 731
Total Europe	LERD	44 098	46 811	48 464	52 234	52 493	70 917	76 325	66 884	66 869	72 343	92 750
Americas												
Argentina	ZWZB	449	437	484	410	249	488	368	107	190	261	319
Brazil	LENR	1 053	1 079	790	776	342	543	665	664	496	782	934
Canada	LEOS	1 817	1 609	1 820	2 078	1 891	2 268	2 371	1 772	2 050	2 299	2 726
Chile	ZWZC	345	339	283	190	166	259	196	240	310	856	607
Colombia	ZWZD	51	68	92	41	100	320	224	235	261	403	390
Mexico	BFPQ	630	455	535	404	543	458	315	632	420	741	974
United States of America	BFVE	16 639	18 645	21 125	25 048	23 984	29 591	31 518	26 909	28 181	30 881	44 869
Uruguay	ZWZE	24	32	29	23	7	16	13	20	9	–1	15
Venezuela	ZWZF	169	184	149	–48	–14	–88	176	185	89	271	275
Other Central American Countries	JISP	515	819	1 150	4 331	4 163	5 566	5 989	5 922	6 099	6 697	10 172
Other	LEVH	481	599	960	67	41	90	12	65	67	466	301
Total Americas	LESN	22 185	24 211	27 431	33 212	31 472	39 511	41 847	36 751	38 172	43 656	61 582
Asia												
China	LEPK	132	174	217	206	268	309	554	437	444	637	749
Hong Kong	BFJU	3 034	3 253	2 589	2 065	1 771	1 685	1 445	1 375	1 808	3 264	5 116
India	BFNB	252	251	284	411	387	490	496	638	664	606	896
Indonesia	BFLP	138	181	303	238	159	149	215	192	227	198	290
Iran	ZWZG	113	118	64	36	38	50	42	37	47	61	148
Israel	BFMS	38	56	59	42	56	52	63	46	78	57	91
Japan	BFOP	8 869	7 516	7 171	6 872	7 453	9 404	7 808	5 204	5 528	5 313	6 720
Malaysia	BFPZ	366	418	481	356	423	497	404	585	620	714	834
Pakistan	BFRS	57	62	123	93	72	131	136	187	181	249	272
Philippines	BFRJ	115	133	103	136	101	179	150	134	161	179	246
Saudi Arabia	BFTC	154	127	144	195	232	296	264	230	199	247	410
Singapore	BFTU	2 602	2 679	2 661	1 797	1 565	2 846	2 589	1 966	1 698	2 319	2 576
South Korea	BFOY	267	407	564	363	310	427	506	694	658	806	950
Taiwan	BFUV	141	163	172	174	195	262	293	343	332	555	698
Thailand	BFUD	225	268	221	70	112	223	249	237	228	235	284
Residual Gulf Arabian Countries	JITQ	359	368	493	1 454	1 102	1 546	1 174	810	718	851	935
Other Near & Middle Eastern Countries	ZWZH	131	117	158	85	67	189	197	104	97	168	206
Other	LEWI	406	328	396	71	125	305	236	502	202	307	846
Total Asia	LETF	17 399	16 646	16 222	14 714	14 436	19 040	16 821	13 721	13 890	16 766	22 267
Australasia & Oceania												
Australia	CXAT	2 059	2 227	2 075	1 724	1 983	2 523	2 466	2 808	2 600	4 052	4 300
New Zealand	BFRA	472	493	390	276	392	327	436	562	588	396	659
Other	LEVQ	200	222	164	381	33	21	8	33	30	36	55
Total Australasia & Oceania	LETX	2 731	2 942	2 629	2 381	2 408	2 871	2 910	3 403	3 218	4 484	5 014
Africa												
Egypt	ZWZJ	57	69	79	–81	37	264	259	323	239	201	313
Morocco	ZWZK	36	55	23	15	26	29	30	40	17	8	29
South Africa	BFWX	661	703	761	748	995	1 207	1 339	2 039	1 941	3 055	3 046
Other North Africa	JIRR	106	83	87	98	58	100	128	–32	122	82	99
Other	LEWR	648	666	553	547	603	757	762	1 022	1 063	1 155	1 381
Total Africa	LERV	1 508	1 576	1 503	1 327	1 719	2 357	2 518	3 392	3 382	4 501	4 868
International Organisations	CTFB	161	146	193	360	384	450	514	475	466	451	544
World total	HMBQ	88 082	92 332	96 442	104 228	102 912	135 146	140 935	124 626	125 997	142 201	187 037

1 Cyprus and Malta are included in Other Europe before 1999.

9.6 Income
continued

£ million

Debits

		1995	1996	1997	1998	1999	2000	2001	2002	2003	2004	2005
Europe												
European Union (EU)												
Austria	CUGZ	743	763	882	535	422	592	682	673	606	583	820
Belgium and Luxembourg	CTFL	4 873	4 381	3 820	4 416	4 390	5 224	5 557	5 218	5 023	5 991	7 408
of which Belgium	AA2W	..	..	..	..	1 879	2 439	2 499	2 584	2 161	2 518	3 417
Luxembourg	AA32	..	..	..	..	2 511	2 785	3 058	2 634	2 862	3 473	3 991
Cyprus[1]	AA2X	..	..	..	..	104	158	163	127	106	133	180
Czech Republic	LEPU	87	163	177	104	86	125	109	76	72	66	82
Denmark	LEQV	692	989	1 034	920	827	1 100	931	861	922	1 035	1 360
Estonia	ZWZR	57	49	59	59	41	31	11	3	2	2	2
Finland	LEUH	535	540	569	320	298	378	371	257	308	277	471
France	LEUQ	6 657	6 070	5 590	5 999	7 214	7 606	7 051	5 436	7 659	9 638	11 801
Germany	LEQM	9 904	9 084	8 556	7 845	8 072	12 008	9 255	7 105	10 462	11 829	13 325
Greece	LEUZ	471	758	860	692	688	858	682	467	450	491	732
Hungary	BFKS	92	89	101	90	83	72	39	45	22	21	30
Ireland	BFLZ	1 683	2 411	2 541	2 543	2 275	3 190	4 248	4 822	5 000	6 122	9 109
Italy	BFOH	2 755	3 265	3 707	3 446	2 926	2 969	3 084	1 815	2 184	2 335	3 145
Latvia	ZWZS	57	50	67	63	45	33	11	11	4	4	13
Lithuania	ZWZT	60	60	66	62	44	35	12	3	1	2	11
Malta[1]	AA33	..	..	..	..	48	46	43	41	30	33	34
Netherlands	BFQJ	5 051	5 702	4 360	4 118	6 539	12 550	12 548	7 319	7 226	7 814	10 669
Poland	BFSC	158	163	266	190	147	182	169	112	98	122	251
Portugal	BFSL	437	436	398	260	221	248	303	272	345	371	428
Slovak Republic	ZWZV	122	82	79	78	58	43	17	10	6	20	38
Slovenia	ZWZW	53	46	56	83	56	42	29	29	15	15	10
Spain	LESX	1 816	2 128	1 754	1 925	1 390	1 297	1 546	1 366	1 774	1 648	2 349
Sweden	BFTM	1 230	1 525	1 230	1 169	1 419	1 370	1 351	1 343	974	915	1 374
European Central Bank	ZWZM	–	–	–	–	–	–	–	–	–	–	–
EU Institutions	CSFL	693	680	745	878	1 222	1 489	1 762	1 674	1 493	1 602	1 939
Total EU25	AA2Y	..	..	..	..	38 615	51 646	49 974	39 085	44 782	51 069	65 581
European Free Trade Association (EFTA)												
Iceland	BFNR	4	4	5	8	11	9	7	11	20	51	53
Liechtenstein	BFPI	129	126	125	106	81	114	89	47	53	47	73
Norway	BFQS	363	409	687	63	121	594	623	625	658	670	1 495
Switzerland	LEPC	6 934	6 910	6 937	6 071	7 840	11 667	10 139	6 552	5 831	5 685	8 240
Total EFTA	CTFU	7 430	7 449	7 754	6 248	8 053	12 384	10 858	7 235	6 562	6 453	9 861
Other Europe												
Albania	ZWZN	56	50	59	112	39	28	10	–1	–	–	–
Belarus	ZWZO	53	46	56	70	42	29	9	–1	1	1	6
Bulgaria	ZWZP	87	67	96	93	52	54	27	24	10	10	28
Croatia	ZWZQ	58	49	56	188	58	60	45	29	24	21	40
Romania	ZWZU	78	91	124	78	53	51	25	32	8	16	31
Russia	BFSU	136	159	196	20	141	422	384	271	281	363	1 105
Turkey	BFUN	297	269	184	156	92	135	155	82	67	74	217
Ukraine	ZWZX	53	46	58	63	43	29	11	17	57	82	143
Serbia and Montenegro	BFWG	127	161	165	58	40	29	11	6	5	5	10
Other[1]	LEWA	3 218	3 935	4 935	5 909	5 756	6 956	7 451	5 953	5 264	6 782	9 283
Total Europe	LERE	49 819	51 756	50 600	48 790	52 984	71 823	68 960	52 732	57 061	64 876	86 305
Americas												
Argentina	ZWZZ	–17	14	64	55	53	101	66	5	3	–	5
Brazil	LENS	274	402	290	244	163	189	156	75	83	78	160
Canada	LEOT	709	857	773	874	1 108	1 507	1 165	1 571	1 404	2 060	2 338
Chile	ZXAA	188	173	151	67	19	8	35	21	17	24	38
Colombia	ZXAB	83	137	81	38	21	38	42	18	8	5	13
Mexico	BFPR	114	122	119	152	157	231	250	65	56	56	104
United States of America	BFVF	15 133	18 489	19 886	20 327	23 365	26 353	30 102	26 929	27 512	31 338	42 013
Uruguay	ZXAC	20	24	30	–15	10	19	8	3	8	6	9
Venezuela	ZXAD	35	25	23	12	8	20	39	27	21	17	31
Other Central American Countries	JISQ	236	542	1 292	3 756	3 508	4 233	3 935	2 628	3 069	3 511	6 166
Other	LEVI	494	674	1 345	375	286	266	157	73	55	48	89
Total Americas	LESO	17 269	21 598	24 263	25 990	28 698	32 965	35 955	31 415	32 236	37 143	50 966
Asia												
China	LEPL	204	145	155	139	276	437	454	303	254	263	367
Hong Kong	BFJV	2 055	1 799	1 559	1 949	2 960	3 509	3 353	1 987	1 332	1 326	1 874
India	BFNC	347	305	370	343	307	354	337	200	227	313	571
Indonesia	BFLQ	113	119	96	39	108	122	174	130	115	105	118
Iran	ZXAE	179	213	133	91	91	223	269	95	83	83	109
Israel	BFMT	156	180	220	149	186	223	249	174	159	134	246
Japan	BFOQ	7 055	7 067	6 220	6 931	7 895	11 202	9 794	7 191	3 774	5 135	6 942
Malaysia	BFQA	311	412	430	192	322	390	338	193	151	175	221
Pakistan	BFRT	86	57	71	61	87	93	94	91	95	69	86
Philippines	BFRK	31	31	63	48	84	95	58	34	23	22	33
Saudi Arabia	BFTD	1 603	1 452	1 324	957	684	844	901	346	298	320	582
Singapore	BFTV	1 687	1 856	1 814	1 374	2 034	2 723	2 597	1 836	1 400	1 433	2 221
South Korea	BFOZ	155	157	202	189	76	137	183	93	8	–184	15
Taiwan	BFUW	208	147	158	104	97	130	195	117	111	157	235
Thailand	BFUE	132	137	140	81	72	133	182	76	57	55	107
Residual Gulf Arabian Countries	JITR	533	494	561	945	926	1 395	1 534	864	613	692	935
Other Near & Middle Eastern Countries	ZXAF	303	262	318	214	221	273	228	146	141	122	193
Other	LEWJ	916	789	944	307	231	235	181	114	95	108	214
Total Asia	LETG	16 074	15 636	14 814	14 110	16 657	22 518	21 121	13 990	8 936	10 328	15 069
Australasia & Oceania												
Australia	CXCM	902	926	1 297	1 200	1 180	766	589	957	1 264	1 300	1 954
New Zealand	BFRB	105	163	181	178	78	126	135	106	92	86	99
Other	LEVR	2	5	13	–38	5	27	61	42	30	27	40
Total Australasia & Oceania	LETY	1 009	1 094	1 491	1 340	1 263	919	785	1 105	1 386	1 413	2 093
Africa												
Egypt	ZXAH	826	746	582	324	265	315	266	153	118	127	244
Morocco	ZXAI	40	58	54	34	36	30	31	19	12	11	27
South Africa	BFWY	58	112	216	421	778	692	690	667	729	716	943
Other North Africa	JIRS	159	136	254	115	103	169	181	110	88	95	221
Other	LEWS	336	292	510	539	518	621	592	366	279	293	449
Total Africa	LERW	1 419	1 344	1 616	1 433	1 700	1 827	1 760	1 315	1 226	1 242	1 884
International Organisations	CTFC	328	348	343	243	340	554	690	626	506	603	837
World total	HMBR	85 918	91 776	93 128	91 908	101 642	130 606	129 271	101 183	101 351	115 605	157 166

1 Cyprus and Malta are included in Other Europe before 1999.

9.6 Income continued

£ million

Balances

		1995	1996	1997	1998	1999	2000	2001	2002	2003	2004	2005
Europe												
European Union (EU)												
Austria	CUHA	−212	−235	−348	86	163	180	19	34	156	250	174
Belgium and Luxembourg	CTFM	−1 760	−986	−312	−596	−785	1 025	1 267	593	−51	816	528
of which Belgium	AA3A	..	..	..	..	−233	−18	358	−180	−86	280	−37
Luxembourg	AA3E	..	..	..	..	−552	1 043	909	773	35	536	565
Cyprus[1]	AA3B	..	..	..	..	2	−5	3	12	−50	−61	−37
Czech Republic	LEPV	−38	−54	−47	9	39	−	89	76	123	148	168
Denmark	LEQW	−80	−366	−312	144	125	36	77	−290	−184	−122	−81
Estonia	ZXAP	−55	−46	−46	−46	−36	−29	−3	3	6	4	13
Finland	LEUI	121	−	−51	144	117	216	288	342	219	253	153
France	LEUR	−640	−166	−118	−7	−987	50	1 785	1 693	−497	−1 221	−495
Germany	LEQN	−679	−167	−525	285	453	−1 539	2 220	2 983	−1 689	−1 925	−1 579
Greece	LEVA	−123	41	209	−37	157	152	200	254	383	313	181
Hungary	BFKT	14	91	117	85	165	376	282	309	312	358	413
Ireland	BFML	24	−502	−152	155	723	723	561	−747	19	227	−1 099
Italy	BFOI	2 432	1 802	1 432	3 483	2 303	3 547	3 375	3 099	2 836	2 421	2 298
Latvia	ZXAQ	−55	−50	−66	−55	−45	−31	−11	−11	−4	−4	−11
Lithuania	ZXAR	−60	−55	−58	−58	−40	−26	−5	1	3	−1	−10
Malta[1]	AA3F	..	..	..	..	−9	5	12	7	23	463	225
Netherlands	BFQK	2 046	2 260	4 319	4 786	2 804	359	4 157	8 571	8 130	2 224	6 972
Poland	BFSD	−44	−52	−147	−99	−37	−38	160	225	303	322	253
Portugal	BFSM	−43	24	−7	206	259	429	521	534	488	458	484
Slovak Republic	ZXAT	−98	−31	−10	−41	−20	8	12	21	22	10	17
Slovenia	ZXAU	−53	−46	−48	−61	−	6	72	39	74	82	101
Spain	LESY	−10	300	690	477	734	1 292	898	748	605	2 098	2 754
Sweden	BFTN	187	69	194	280	152	761	988	814	822	1 426	1 245
European Central Bank	ZXAK	−	−	−	−	−	−	−	−	−	−	−
EU Institutions	CSFM	−482	−537	−582	−597	−998	−1 119	−1 294	−1 017	−742	−1 088	−1 392
Total EU25	AA3C	..	..	..	..	5 239	6 378	15 673	18 293	11 307	7 451	11 275
European Free Trade Association (EFTA)												
Iceland	BFNU	28	21	15	24	21	26	31	22	7	4	63
Liechtenstein	BFPJ	−78	−62	−73	−72	−42	−69	−50	−26	−33	−24	−46
Norway	BFQT	168	177	−82	569	613	438	407	78	159	191	38
Switzerland	LEPD	−3 645	−3 393	−3 368	−2 623	−3 363	−4 235	−4 649	−1 891	−1 155	−589	−2 279
Total EFTA	CTFV	−3 527	−3 257	−3 508	−2 102	−2 771	−3 840	−4 261	−1 817	−1 022	−418	−2 224
Other Europe												
Albania	ZXAL	−56	−50	−59	−112	−39	−28	−10	1	−	−	−
Belarus	ZXAM	−53	−46	−56	−70	−42	−29	−9	8	−1	−1	−6
Bulgaria	ZXAN	−63	−51	−55	−72	−4	−17	66	48	64	80	70
Croatia	ZXAO	−54	−45	−42	−179	−45	−45	−19	−	19	27	30
Romania	ZXAS	−56	−49	−99	−56	−11	−31	53	33	88	102	103
Russia	BFSV	−31	−29	38	249	55	−136	157	395	290	936	428
Turkey	BFUO	−161	−111	6	112	257	363	260	240	302	311	375
Ukraine	ZXAV	−42	−38	−58	−47	−40	−29	−8	1	−36	−48	−60
Serbia and Montenegro	BFWH	−94	−104	−116	−58	−39	−28	−10	−5	−4	−7	6
Other[1]	LEWB	−1 976	−2 459	−2 318	−2 763	−3 051	−3 464	−4 527	−3 045	−1 199	−966	−3 552
Total Europe	LERF	−5 721	−4 945	−2 136	3 444	−491	−906	7 365	14 152	9 808	7 467	6 445
Americas												
Argentina	ZXAX	466	423	420	355	196	387	302	102	187	261	314
Brazil	LENT	779	677	500	532	179	354	509	589	413	704	774
Canada	LEOU	1 108	752	1 047	1 204	783	761	1 206	201	646	239	388
Chile	ZXAY	157	166	132	123	147	251	161	219	293	832	569
Colombia	ZXAZ	−32	−69	11	3	79	282	182	217	253	398	377
Mexico	BFPS	516	333	416	252	386	227	65	567	364	685	870
United States of America	BFVG	1 506	156	1 239	4 721	619	3 238	1 416	−20	669	−457	2 856
Uruguay	ZXBA	4	8	−1	38	−3	−3	5	17	1	−7	6
Venezuela	ZXBB	134	159	126	−60	−22	−108	137	158	68	254	244
Other Central American Countries	JISR	279	277	−142	575	655	1 333	2 054	3 294	3 030	3 186	4 006
Other	LEVJ	−13	−75	−385	−308	−245	−176	−145	−8	12	418	212
Total Americas	LESP	4 916	2 613	3 168	7 222	2 774	6 546	5 892	5 336	5 936	6 513	10 616
Asia												
China	LEPM	−72	29	62	67	−8	−128	100	134	190	374	382
Hong Kong	BFJW	979	1 454	1 030	116	−1 189	−1 824	−1 908	−612	476	1 938	3 242
India	BFND	−95	−54	−86	68	80	136	159	438	437	293	325
Indonesia	BFLR	25	62	207	199	51	27	41	62	112	93	172
Iran	ZXBC	−66	−95	−69	−55	−53	−173	−227	−58	−36	−22	39
Israel	BFMU	−118	−124	−161	−107	−130	−171	−186	−128	−81	−77	−155
Japan	BFOR	1 814	449	951	−59	−442	−1 798	−1 986	−1 987	1 754	178	−222
Malaysia	BFQB	55	6	51	164	101	107	66	392	469	539	613
Pakistan	BFRU	−29	5	52	32	−15	38	42	96	86	180	186
Philippines	BFRL	84	102	40	88	17	84	92	100	138	157	213
Saudi Arabia	BFTE	−1 449	−1 325	−1 180	−762	−452	−548	−637	−116	−99	−73	−172
Singapore	BFTW	915	823	847	423	−469	123	−8	130	298	886	355
South Korea	BFPA	112	250	362	174	234	290	323	601	650	990	935
Taiwan	BFUX	−67	16	14	70	98	132	98	226	221	398	463
Thailand	BFUF	93	131	81	−11	40	90	67	161	171	180	177
Residual Gulf Arabian Countries	JITS	−174	−126	−68	509	176	151	−360	−54	105	159	−
Other Near & Middle Eastern Countries	ZXBD	−172	−145	−160	−129	−154	−84	−31	−42	−44	46	13
Other	LEWK	−510	−461	−548	−236	−106	70	55	388	107	199	632
Total Asia	LETH	1 325	1 010	1 408	604	−2 221	−3 478	−4 300	−269	4 954	6 438	7 198
Australasia & Oceania												
Australia	CYAA	1 157	1 301	778	524	803	1 757	1 877	1 851	1 336	2 752	2 346
New Zealand	BFRC	367	330	209	98	314	201	301	456	496	310	560
Other	LEVS	198	217	151	419	28	−6	−53	−9	−	9	15
Total Australasia & Oceania	LETZ	1 722	1 848	1 138	1 041	1 145	1 952	2 125	2 298	1 832	3 071	2 921
Africa												
Egypt	ZXBF	−769	−677	−503	−405	−228	−51	−7	170	121	74	69
Morocco	ZXBG	−4	−3	−31	−19	−10	−1	−1	21	5	−3	2
South Africa	BFWZ	603	591	545	327	217	515	649	1 372	1 212	2 339	2 103
Other North Africa	JIRT	−53	−53	−167	−17	−45	−69	−53	−142	34	−13	−122
Other	LEWT	312	374	43	8	85	136	170	656	784	862	932
Total Africa	LERX	89	232	−113	−106	19	530	758	2 077	2 156	3 259	2 984
International Organisations	CTFD	−167	−202	−150	117	44	−104	−176	−151	−40	−152	−293
World total	HMBP	2 164	556	3 314	12 320	1 270	4 540	11 664	23 443	24 646	26 596	29 871

1 Cyprus and Malta are included in Other Europe before 1999.

9.7 Current transfers

£ million

		1995	1996	1997	1998	1999	2000	2001	2002	2003	2004	2005
Credits												
Europe												
European Union (EU)												
Austria	GXVQ	6	42	35	34	35	30	32	33	51	35	40
Belgium and Luxembourg	GXVR	121	268	222	249	246	224	223	247	233	239	255
of which Belgium	A7PL	..	..	..	..	225	209	206	230	218	221	235
Luxembourg	A7PO	..	..	..	..	21	15	17	17	15	18	20
Cyprus[1]	A7PM	..	..	..	..	2	2	3	3	4	4	6
Czech Republic	GXWK	4	14	7	4	5	6	3	3	32	19	15
Denmark	GXVS	57	92	52	46	42	28	36	41	29	37	40
Estonia	LWMG	2	1	3	–	–	–	–	–	–	–	–
Finland	GXVT	34	36	22	18	18	24	17	16	93	16	18
France	GXVU	433	525	341	351	332	281	355	336	198	296	278
Germany	GXVV	492	646	468	466	456	393	416	444	384	436	472
Greece	GXVW	59	110	59	59	53	40	47	48	41	46	47
Hungary	HZXT	–	–	–	–	–	1	1	8	1	1	5
Ireland	GXVX	261	306	191	190	154	116	178	155	116	154	147
Italy	GXVY	167	342	180	170	153	103	137	123	104	159	154
Latvia	LWWC	27	11	4	4	3	2	2	5	1	1	6
Lithuania	LYTR	–	–	1	–	–	–	1	–	–	–	5
Malta[1]	A7PP	..	..	..	..	1	–	1	1	1	2	3
Netherlands	GXVZ	610	420	335	319	324	324	325	374	294	337	359
Poland	GXWL	10	17	29	5	5	3	4	14	7	29	81
Portugal	GXWA	23	27	20	19	19	28	16	21	15	21	22
Slovak Republic	HZXX	–	–	–	5	4	–	2	3	–	–	9
Slovenia	HZXY	–	–	–	–	–	–	–	5	–	1	1
Spain	GXWB	94	165	120	120	115	93	105	105	87	117	123
Sweden	GXWC	79	171	113	108	104	89	98	97	85	101	98
European Central Bank	KNWK	–	–	–	–	–	–	–	–	–	–	–
EU Institutions	GXWD	4 419	7 188	5 422	5 103	6 391	5 314	7 275	5 905	6 688	6 948	7 771
Total EU25	A7PN	..	..	..	..	8 462	7 101	9 277	7 987	8 464	8 999	9 955
European Free Trade Association (EFTA)												
Iceland	GXWF	5	41	36	36	36	33	35	35	64	34	37
Liechtenstein	GXWG	2	11	6	4	3	2	2	2	1	1	1
Norway	GXWH	51	149	86	83	78	61	69	80	68	71	80
Switzerland	GXWI	226	195	108	101	93	60	79	80	66	79	92
Total EFTA	GXWJ	284	396	236	224	210	156	185	197	199	185	210
Other Europe												
Albania	HZXP	–	–	1	–	–	–	–	–	–	–	–
Belarus	HZXQ	–	–	–	–	–	–	1	–	7	–	1
Bulgaria	KOLZ	–	–	–	–	–	2	–	8	1	1	6
Croatia	HZXR	29	12	6	4	3	2	2	2	2	2	2
Romania	HZXV	–	–	–	2	2	–	–	11	1	2	7
Russia	GXWM	23	62	29	26	23	10	16	16	12	17	24
Turkey	GXWN	24	60	36	36	35	28	33	34	31	35	38
Ukraine	HZYA	–	–	1	–	–	1	–	–	7	15	2
Serbia and Montenegro	LTVE	–	–	–	–	3	2	2	2	5	1	2
Other[1]	HKJF	124	51	26	25	16	10	30	16	17	8	12
Total Europe	GXWP	7 382	10 962	7 959	7 587	8 754	7 312	9 546	8 273	8 746	9 265	10 259
Americas												
Argentina	HZYJ	14	44	19	19	15	8	13	11	7	7	5
Brazil	GXWQ	15	56	25	23	21	10	18	16	12	11	11
Canada	GXWR	246	636	312	300	274	177	241	209	168	221	224
Chile	HZYL	10	33	15	15	12	7	13	9	7	8	8
Colombia	HZYM	22	70	29	26	24	14	32	17	26	13	12
Mexico	GXWS	24	75	37	32	30	18	25	28	22	24	27
United States of America	GXWT	2 756	4 398	2 210	2 060	1 873	1 213	2 482	1 694	1 291	1 404	3 706
Uruguay	HZYN	–	–	–	–	–	–	–	–	–	–	–
Venezuela	HZYO	15	48	22	21	19	8	14	12	9	12	14
Other Central American Countries	HZYG	189	608	265	257	228	144	193	181	145	191	213
Other	HZYI	14	44	19	19	16	8	14	13	14	12	14
Total Americas	GXWV	3 305	6 012	2 953	2 772	2 512	1 607	3 045	2 190	1 701	1 903	4 234
Asia												
China	GXWW	25	46	20	20	18	9	14	13	10	12	14
Hong Kong	GXWX	147	186	134	76	71	50	63	40	30	34	33
India	GXWY	58	36	17	17	21	15	26	25	14	24	27
Indonesia	GXWZ	21	85	37	38	30	17	25	24	45	16	14
Iran	HZYQ	7	11	6	4	3	3	3	3	2	3	6
Israel	GXXA	36	111	55	52	48	30	39	35	34	29	27
Japan	GXXB	234	364	175	171	150	99	135	130	99	114	118
Malaysia	GXXC	55	54	29	29	26	17	26	21	49	20	21
Pakistan	GXXD	2	13	9	5	6	3	5	4	29	4	6
Philippines	GXXE	12	45	22	19	15	8	13	11	7	8	8
Saudi Arabia	GXXF	33	472	487	491	506	485	497	488	500	480	490
Singapore	GXXG	32	47	21	20	19	9	19	15	14	21	26
South Korea	GXXH	17	44	19	19	20	9	16	20	11	20	26
Taiwan	GXXI	16	33	15	15	12	7	13	15	23	11	14
Thailand	GXXJ	11	23	9	8	8	5	7	9	6	7	10
Residual Gulf Arabian Countries	HZYS	123	204	179	178	179	162	171	172	169	173	174
Other Near & Middle Eastern Countries	HZYU	162	273	294	294	306	294	300	295	294	292	301
Other	HZVR	74	124	107	65	63	54	66	44	39	40	40
Total Asia	GXXL	1 065	2 171	1 635	1 521	1 501	1 276	1 438	1 364	1 375	1 308	1 355
Australasia & Oceania												
Australia	GXXM	392	376	236	211	202	159	188	203	170	226	240
New Zealand	GXXN	108	105	78	69	64	53	62	58	52	61	57
Other	GXXO	5	12	6	4	4	3	3	3	2	5	7
Total Australasia & Oceania	GXXP	505	493	320	284	270	215	253	264	224	292	304
Africa												
Egypt	LZDN	14	13	5	1	5	3	3	4	26	6	10
Morocco	HICY	12	11	6	4	3	2	2	2	2	2	4
South Africa	GXXQ	127	241	126	125	116	82	103	83	72	88	90
Other North Africa	HICX	28	25	13	15	12	10	11	13	21	13	17
Other	HZUI	113	101	54	55	47	34	43	42	36	40	40
Total Africa	GXXS	294	391	204	200	183	131	162	144	157	149	161
International Organisations	GXXT	–	–	–	–	–	–	–	–	–	–	–
World total	KTND	12 551	20 029	13 071	12 364	13 220	10 541	14 444	12 235	12 203	12 917	16 313

1 Cyprus and Malta are included in Other Europe before 1999.

9.7 Current transfers continued

£ million

Debits		1995	1996	1997	1998	1999	2000	2001	2002	2003	2004	2005
Europe												
European Union (EU)												
Austria	GXXV	31	20	23	16	18	23	22	24	25	28	38
Belgium and Luxembourg	GXXW	119	128	108	76	80	78	99	98	106	120	100
of which Belgium	A8BV	..	..	..	..	67	70	86	87	95	107	86
Luxembourg	A8BY	..	..	..	..	13	8	13	11	11	13	14
Cyprus[1]	A8BW	..	..	..	..	11	18	20	27	26	31	35
Czech Republic	GXYP	16	20	15	14	22	9	10	11	12	14	16
Denmark	GXXX	56	89	54	49	50	37	38	36	31	39	52
Estonia	LWQY	–	–	4	–	–	–	–	–	1	4	4
Finland	GXXY	15	27	13	12	12	22	16	14	13	15	17
France	GXXZ	333	458	257	247	342	270	315	333	272	313	342
Germany	GXYA	446	513	350	311	317	296	312	366	321	400	433
Greece	GXYB	40	100	48	49	44	50	52	46	47	57	58
Hungary	HIEC	9	13	17	18	17	16	12	19	10	11	13
Ireland	GXYC	508	766	750	798	413	472	458	476	509	558	610
Italy	GXYD	172	386	244	254	184	135	159	150	135	169	190
Latvia	LYON	24	13	7	4	7	2	2	2	1	1	1
Lithuania	LYYJ	2	3	3	–	8	1	15	14	–	8	9
Malta[1]	A8BZ	..	..	..	..	9	15	17	18	19	22	24
Netherlands	GXYE	535	223	148	136	174	188	154	178	129	179	218
Poland	GXYQ	35	55	62	56	36	58	47	50	52	57	63
Portugal	GXYF	50	29	28	26	32	48	37	40	41	47	51
Slovak Republic	HIEG	3	5	8	10	7	5	3	6	3	1	7
Slovenia	HIEH	1	2	2	1	8	–	5	4	–	1	1
Spain	GXYG	194	172	132	144	174	126	133	141	152	172	171
Sweden	GXYH	81	122	68	56	66	34	42	39	36	41	36
European Central Bank	KOEJ	–	–	–	–	–	–	–	–	–	–	–
EU Institutions	GXYI	9 192	9 277	8 268	10 265	10 524	10 719	9 557	10 097	11 485	11 505	13 081
Total EU25	A8BX	..	..	..	..	12 555	12 622	11 525	12 189	13 426	13 793	15 570
European Free Trade Association (EFTA)												
Iceland	GXYK	8	11	6	4	3	4	3	6	2	2	4
Liechtenstein	GXYL	3	11	6	4	3	2	2	2	1	1	1
Norway	GXYM	78	139	84	74	96	57	68	63	56	70	73
Switzerland	GXYN	164	195	106	97	104	95	105	124	93	116	141
Total EFTA	GXYO	253	356	202	179	206	158	178	195	152	189	219
Other Europe												
Albania	HIDY	1	2	2	2	2	6	9	12	7	7	7
Belarus	HIDZ	1	2	2	2	4	–	6	11	3	2	2
Bulgaria	LTQA	–	–	–	–	8	12	8	14	7	5	5
Croatia	HIEA	1	12	13	8	5	8	6	6	5	6	14
Romania	HIEE	20	13	13	14	10	14	11	20	10	13	14
Russia	GXYR	51	115	100	104	76	90	59	23	50	60	64
Turkey	GXYS	25	59	38	38	37	28	37	41	33	42	48
Ukraine	HIEJ	13	20	24	23	42	30	54	15	28	41	43
Serbia and Montenegro	LWHC	2	3	1	–	9	67	39	31	30	28	33
Other[1]	HZWJ	102	131	136	141	126	126	152	156	190	169	179
Total Europe	GXYU	12 331	13 134	11 140	13 053	13 080	13 161	12 084	12 713	13 941	14 355	16 198
Americas												
Argentina	HIES	23	46	21	21	17	16	20	18	14	14	14
Brazil	GXYV	24	68	39	41	43	36	49	57	36	38	42
Canada	GXYW	626	688	370	356	348	268	324	315	284	348	364
Chile	HIEU	26	38	21	19	17	12	14	13	10	12	12
Colombia	HIEV	28	76	37	33	33	23	37	22	15	19	18
Mexico	GXYX	29	81	42	40	38	25	28	31	22	25	29
United States of America	GXYY	1 988	4 523	2 336	2 117	2 105	1 452	2 640	1 892	1 523	1 709	4 005
Uruguay	HIEW	1	1	1	1	1	1	1	1	1	1	1
Venezuela	HIEX	20	44	19	19	16	9	15	13	10	13	15
Other Central American Countries	HIEP	453	879	572	617	527	433	449	426	375	436	467
Other	HIER	99	110	69	159	92	80	100	96	65	69	68
Total Americas	GXZA	3 317	6 554	3 527	3 423	3 237	2 355	3 677	2 884	2 355	2 684	5 035
Asia												
China	GXZB	51	119	92	109	103	82	90	125	78	138	143
Hong Kong	GXZC	121	151	99	92	136	118	130	133	127	143	153
India	GXZD	135	363	399	407	426	489	537	557	504	625	717
Indonesia	GXZE	67	179	104	102	85	54	51	51	68	83	180
Iran	HIEZ	17	23	23	20	20	20	18	19	17	21	25
Israel	GXZF	44	112	51	50	47	34	44	42	36	37	36
Japan	GXZG	170	359	171	160	161	97	130	132	103	122	129
Malaysia	GXZH	40	70	43	43	67	45	47	49	47	52	61
Pakistan	GXZI	67	149	183	158	152	144	163	233	183	198	217
Philippines	GXZJ	33	70	56	52	40	30	36	35	33	35	38
Saudi Arabia	GXZK	37	45	23	21	42	41	45	51	49	52	49
Singapore	GXZL	34	61	40	42	82	76	80	88	91	108	116
South Korea	GXZM	14	45	20	22	23	11	17	21	11	21	27
Taiwan	GXZN	11	33	16	15	17	7	14	15	10	11	14
Thailand	GXZO	13	30	20	18	25	29	32	40	31	36	50
Residual Gulf Arabian Countries	HIFB	114	121	97	89	99	74	89	88	194	329	349
Other Near & Middle Eastern Countries	HIFD	46	44	51	46	57	67	78	77	73	92	102
Other	HZWN	397	286	316	283	306	404	542	609	530	756	875
Total Asia	GXZQ	1 411	2 260	1 804	1 729	1 888	1 822	2 143	2 365	2 185	2 859	3 281
Australasia & Oceania												
Australia	GXZR	695	460	339	334	249	218	259	275	269	330	376
New Zealand	GXZS	137	113	85	82	59	66	79	80	80	90	100
Other	GXZT	77	32	32	25	32	11	11	23	22	26	30
Total Australasia & Oceania	GXZU	909	605	456	441	340	295	349	378	371	446	506
Africa												
Egypt	LZIF	18	11	5	–	29	24	24	27	26	35	40
Morocco	HIYZ	13	19	11	9	11	8	8	8	9	9	12
South Africa	GXZV	295	288	191	200	250	258	301	309	316	342	357
Other North Africa	HIYX	71	40	34	43	25	10	14	14	27	20	23
Other	HZUA	451	855	956	985	1 051	1 269	1 327	1 363	2 088	1 844	1 994
Total Africa	GXZX	848	1 213	1 197	1 237	1 366	1 569	1 674	1 721	2 466	2 250	2 426
International Organisations	GXZY	1 309	1 018	865	855	842	1 351	1 276	1 255	1 007	1 272	1 046
World total	KTNE	20 125	24 784	18 989	20 738	20 753	20 553	21 203	21 316	22 325	23 866	28 492

1 Cyprus and Malta are included in Other Europe before 1999.

9.7 Current transfers
continued

£ million

Balances

		1995	1996	1997	1998	1999	2000	2001	2002	2003	2004	2005
Europe												
European Union (EU)												
Austria	GZDU	−25	22	12	18	17	7	10	9	26	7	2
Belgium and Luxembourg	GZDV	2	140	114	173	166	146	124	149	127	119	155
of which Belgium	A8H2	..	..	..	..	158	139	120	143	123	114	149
Luxembourg	A8H5	..	..	..	..	8	7	4	6	4	5	6
Cyprus[1]	A8H3	..	..	..	..	−9	−16	−17	−24	−22	−27	−29
Czech Republic	GZCJ	−12	−6	−8	10	−17	−3	−7	−8	20	5	−1
Denmark	GZDW	1	3	−2	−3	−8	−9	−2	5	−2	−2	−12
Estonia	ZWRN	2	1	−1	−	−	−	−	−	−1	−4	−4
Finland	GZDX	19	9	9	6	6	2	1	2	80	1	1
France	GZDY	100	67	84	104	−10	11	40	3	−74	−17	−64
Germany	GZDZ	46	133	118	155	139	97	104	78	63	36	39
Greece	GZEA	19	10	11	10	9	−10	−5	2	−6	−11	−11
Hungary	GYWH	−9	−13	−17	−18	−17	−15	−11	−11	−9	−10	−8
Ireland	GZEB	−247	−460	−559	−608	−259	−356	−280	−321	−393	−404	−463
Italy	GZEC	−5	−44	−64	−84	−31	−32	−22	−27	−31	−10	−36
Latvia	ZWRV	3	−2	−3	−	−4	−	−	3	−	−	5
Lithuania	ZWRU	−2	−3	−2	−	−8	−1	−14	−14	−	−8	−4
Malta[1]	A8H6	..	..	..	..	−8	−15	−16	−17	−18	−20	−21
Netherlands	GZED	75	197	187	183	150	136	171	196	165	158	141
Poland	GZCK	−25	−38	−33	−51	−31	−55	−43	−36	−45	−28	18
Portugal	GZEE	−27	−2	−8	−7	−13	−20	−21	−19	−26	−26	−29
Slovak Republic	ZWRZ	−3	−5	−8	−5	−3	−5	−1	−3	−3	−1	2
Slovenia	ZWRY	−1	−2	−2	−1	−8	−	−5	1	−	−	−
Spain	GZEF	−100	−7	−12	−24	−59	−33	−28	−36	−65	−55	−48
Sweden	GYRO	−2	49	45	52	38	55	56	58	49	60	62
European Central Bank	ZWRB	−	−	−	−	−	−	−	−	−	−	−
EU Institutions	GYRP	−4 773	−2 089	−2 846	−5 162	−4 133	−5 405	−2 282	−4 192	−4 797	−4 557	−5 310
Total EU25	A8H4	..	..	..	..	−4 093	−5 521	−2 248	−4 202	−4 962	−4 794	−5 615
European Free Trade Association (EFTA)												
Iceland	GXEL	−3	30	30	32	33	29	32	29	62	32	33
Liechtenstein	GXEM	−1	−	−	−	−	−	−	−	−	−	−
Norway	GXEN	−27	10	2	9	−18	4	1	17	12	1	7
Switzerland	GZCH	62	−	2	4	−11	−35	−26	−44	−27	−37	−49
Total EFTA	GZCI	31	40	34	45	4	−2	7	2	47	−4	−9
Other Europe												
Albania	ZWRF	−1	−2	−1	−2	−2	−6	−9	−12	−7	−7	−7
Belarus	ZWRI	−1	−2	−2	−2	−4	−	−5	−11	4	−2	−1
Bulgaria	ZWRH	−	−	−	−	−8	−10	−8	−6	−6	−4	1
Croatia	ZWRS	28	−	−7	−4	−2	−6	−4	−4	−3	−4	−12
Romania	ZWRX	−20	−13	−13	−12	−8	−14	−11	−9	−9	−11	−7
Russia	GZCL	−28	−53	−71	−78	−53	−80	−43	−7	−38	−43	−40
Turkey	GZCM	−1	1	−2	−2	−2	−	−4	−7	−2	−7	−10
Ukraine	ZWSA	−13	−20	−23	−23	−42	−29	−54	−15	−21	−26	−41
Serbia and Montenegro	ZWSD	−2	−3	−1	−	−6	−65	−37	−29	−25	−27	−31
Other[1]	ZWRC	22	−80	−110	−116	−110	−116	−122	−140	−173	−161	−167
Total Europe	GZCO	−4 949	−2 172	−3 181	−5 466	−4 326	−5 849	−2 538	−4 440	−5 195	−5 090	−5 939
Americas												
Argentina	ZWRG	−9	−2	−2	−2	−2	−8	−7	−7	−7	−7	−9
Brazil	GZCP	−9	−12	−14	−18	−22	−26	−31	−41	−24	−27	−31
Canada	GZCQ	−380	−52	−58	−56	−74	−91	−83	−106	−116	−127	−140
Chile	ZWRJ	−16	−5	−6	−4	−5	−5	−1	−4	−3	−4	−4
Colombia	ZWRK	−6	−6	−8	−7	−9	−9	−5	−5	11	−6	−6
Mexico	GZCR	−5	−6	−5	−8	−8	−7	−3	−3	−	−1	−2
United States of America	GZCS	768	−125	−126	−57	−232	−239	−158	−198	−232	−305	−299
Uruguay	ZWSB	−1	−1	−1	−1	−1	−1	−1	−1	−1	−1	−1
Venezuela	ZWSC	−5	4	3	2	3	−1	−1	−1	−1	−1	−1
Other Central American Countries	ZWRM	−264	−271	−307	−360	−299	−289	−256	−245	−230	−245	−254
Other	ZWRP	−85	−66	−50	−140	−76	−72	−86	−83	−51	−57	−54
Total Americas	GZCU	−12	−542	−574	−651	−725	−748	−632	−694	−654	−781	−801
Asia												
China	GZCV	−26	−73	−72	−89	−85	−73	−76	−112	−68	−126	−129
Hong Kong	GZCW	26	35	35	−16	−65	−68	−67	−93	−97	−109	−120
India	GZCX	−77	−327	−382	−390	−405	−474	−511	−532	−490	−601	−690
Indonesia	GZCY	−46	−94	−67	−64	−55	−37	−26	−27	−23	−67	−166
Iran	ZWRT	−10	−12	−17	−16	−17	−17	−15	−16	−15	−18	−19
Israel	GZCZ	−8	−1	4	2	1	−4	−5	−7	−2	−8	−9
Japan	GZDA	64	5	4	11	−11	2	5	−2	−4	−8	−11
Malaysia	GZDB	15	−16	−14	−14	−41	−28	−21	−28	2	−32	−40
Pakistan	GZDC	−65	−136	−174	−153	−146	−141	−158	−229	−154	−194	−211
Philippines	GZDD	−21	−25	−34	−33	−25	−22	−23	−24	−26	−27	−30
Saudi Arabia	GZDE	−4	427	464	470	464	444	452	437	451	428	441
Singapore	GZDF	−2	−14	−19	−22	−63	−67	−61	−73	−77	−87	−90
South Korea	GZDG	3	−1	−1	−3	−3	−2	−1	−1	−	−1	−1
Taiwan	GZDH	5	−	−1	−	−5	−	−1	−	13	−	−
Thailand	GZDI	−2	−7	−11	−10	−17	−24	−25	−31	−25	−29	−40
Residual Gulf Arabian Countries	ZWRQ	9	83	82	89	80	88	82	84	−25	−156	−175
Other Near & Middle Eastern Countries	ZWRR	116	229	243	248	249	227	222	218	221	200	199
Other	ZWRD	−323	−162	−209	−218	−243	−350	−476	−565	−491	−716	−835
Total Asia	GZDK	−346	−89	−169	−208	−387	−546	−705	−1 001	−810	−1 551	−1 926
Australasia & Oceania												
Australia	GZDL	−303	−84	−103	−123	−47	−59	−71	−72	−99	−104	−136
New Zealand	GZDM	−29	−8	−7	−13	5	−13	−17	−22	−28	−29	−43
Other	GZDN	−72	−20	−26	−21	−28	−8	−8	−20	−20	−21	−23
Total Australasia & Oceania	GZDO	−404	−112	−136	−157	−70	−80	−96	−114	−147	−154	−202
Africa												
Egypt	ZWRO	−4	2	−	1	−24	−21	−21	−23	−	−29	−30
Morocco	ZWRW	−1	−8	−5	−5	−8	−6	−6	−6	−7	−7	−8
South Africa	GZDP	−168	−47	−65	−75	−134	−176	−198	−226	−244	−254	−267
Other North Africa	ZWRL	−43	−15	−21	−28	−13	−	−3	−1	−6	−7	−6
Other	ZWRE	−338	−754	−902	−930	−1 004	−1 235	−1 284	−1 321	−2 052	−1 804	−1 954
Total Africa	GZDR	−554	−822	−993	−1 037	−1 183	−1 438	−1 512	−1 577	−2 309	−2 101	−2 265
International Organisations	GZDS	−1 309	−1 018	−865	−855	−842	−1 351	−1 276	−1 255	−1 007	−1 272	−1 046
World total	KTNF	−7 574	−4 755	−5 918	−8 374	−7 533	−10 012	−6 759	−9 081	−10 122	−10 949	−12 179

1 Cyprus and Malta are included in Other Europe before 1999.

9.8 Current account
Transactions with Europe and USA[1][2]

£ million

		2000	2001	2002	2003	2004	2005
Credits							
Exports of goods							
EMU members	QATL	101 464	103 536	103 226	99 893	99 859	108 445
EU members	LGCJ	111 955	113 911	114 136	110 589	110 883	120 183
Total Europe	EPLM	120 705	122 492	122 437	119 832	120 656	133 224
USA	QAMH	29 276	29 244	28 197	28 672	28 576	30 914
Exports of services							
EMU members	LJHS	28 863	30 785	32 061	34 559	38 356	39 099
EU members	A7RZ	32 852	35 361	36 607	39 548	43 997	45 384
Total Europe	FYWB	38 679	41 544	44 459	48 291	53 136	55 932
USA	FYWF	18 899	18 515	21 938	23 042	24 731	22 824
Income							
EMU members	BDJR	53 354	60 618	52 854	51 636	53 013	70 628
EU members	AA2M	58 024	65 647	57 378	56 089	58 520	76 856
Total Europe	LERD	70 917	76 325	66 884	66 869	72 343	92 750
USA	BFVE	29 591	31 518	26 909	28 181	30 881	44 869
Current transfers							
EMU members	XNSJ	1 656	1 851	1 902	1 616	1 856	1 915
EU members	A7PN	7 101	9 277	7 987	8 464	8 999	9 955
Total Europe	GXWP	7 312	9 546	8 273	8 746	9 265	10 259
USA	GXWT	1 213	2 482	1 694	1 291	1 404	3 706
TOTAL CREDITS							
EMU members	BDJV	185 337	196 790	190 043	187 704	193 084	220 087
EU members	AA2S	209 932	224 196	216 108	214 690	222 399	252 378
Total Europe	LERA	237 613	249 909	242 054	243 738	255 400	292 165
USA	BFVB	78 979	81 759	78 738	81 186	85 592	102 313
Debits							
Imports of goods							
EMU members	QBRM	105 678	114 204	123 134	122 628	126 273	138 411
EU members	LGDB	117 217	126 440	136 301	136 612	141 610	156 231
Total Europe	EPMM	132 366	141 496	151 744	153 466	162 116	182 717
USA	QAMI	28 416	29 345	25 149	22 857	22 067	21 948
Imports of services							
EMU members	XWLM	31 584	33 580	35 320	37 737	39 318	42 111
EU members	A8EV	34 746	36 976	38 780	41 840	43 749	47 236
Total Europe	GGPQ	38 919	41 214	44 006	46 710	48 952	53 073
USA	GGPU	13 130	13 405	13 308	14 450	14 752	14 834
Income							
EMU members	BDJT	46 920	45 327	34 750	41 037	47 099	60 257
EU members	AA2Y	51 646	49 974	39 085	44 782	51 069	65 581
Total Europe	LERE	71 823	68 960	52 732	57 061	64 876	86 305
USA	BFVF	26 353	30 102	26 929	27 512	31 338	42 013
Current transfers							
EMU members	XNSK	1 708	1 757	1 866	1 750	2 058	2 228
EU members	A8BX	12 622	11 525	12 189	13 426	13 793	15 570
Total Europe	GXYU	13 161	12 084	12 713	13 941	14 355	16 198
USA	GXYY	1 452	2 640	1 892	1 523	1 709	4 005
TOTAL DEBITS							
EMU members	LTLV	185 890	194 868	195 070	203 152	214 748	243 007
EU members	AA36	216 231	224 915	226 355	236 660	250 221	284 618
Total Europe	LERB	256 269	263 755	261 195	271 178	290 299	338 293
USA	BFVC	69 351	75 492	67 278	66 342	69 866	82 800
Balances							
Trade in goods							
EMU members	QBRX	−4 214	−10 668	−19 908	−22 735	−26 414	−29 966
EU members	LGCF	−5 262	−12 529	−22 165	−26 023	−30 727	−36 048
Total Europe	EPNM	−11 661	−19 004	−29 307	−33 634	−41 460	−49 493
USA	QBRP	860	−101	3 048	5 815	6 509	8 966
Trade in services							
EMU members	XQXB	−2 721	−2 795	−3 259	−3 178	−962	−3 012
EU members	A8HJ	−1 894	−1 615	−2 173	−2 292	248	−1 852
Total Europe	GGRV	−240	330	453	1 581	4 184	2 859
USA	GGRZ	5 769	5 110	8 630	8 592	9 979	7 990
Income							
EMU members	BDJU	6 434	15 291	18 104	10 599	5 914	10 371
EU members	AA3C	6 378	15 673	18 293	11 307	7 451	11 275
Total Europe	LERF	−906	7 365	14 152	9 808	7 467	6 445
USA	BFVG	3 238	1 416	−20	669	−457	2 856
Current transfers							
EMU members	XUGX	−52	94	36	−134	−202	−313
EU members	A8H4	−5 521	−2 248	−4 202	−4 962	−4 794	−5 615
Total Europe	GZCO	−5 849	−2 538	−4 440	−5 195	−5 090	−5 939
USA	GZCS	−239	−158	−198	−232	−305	−299
CURRENT BALANCE							
EMU members	LTLW	−553	1 922	−5 027	−15 448	−21 664	−22 920
EU members	AA4J	−6 299	−719	−10 247	−21 970	−27 822	−32 240
Total Europe	LERC	−18 656	−13 846	−19 141	−27 440	−34 899	−46 128
USA	BFVD	9 628	6 267	11 460	14 844	15 726	19 513

1 EMU Members: Austria, Belgium, Finland, France, Germany, Greece, Irish Republic, Italy, Luxembourg, Netherlands, Portugal, Spain.
2 EU and Europe include transactions with European Union institutions.

9.9 UK offical transactions with institutions of the EU

£ million

		1995	1996	1997	1998	1999	2000	2001	2002	2003	2004	2005
Credits												
Exports of services												
UK charge for collecting duties and levies(net)[1]	QWUE	251	235	240	212	208	217	525	487	489	544	561
Current transfers													
Other sectors													
Agricultural Guarantee Fund[2]	EBGL	2 392	3 931	3 063	2 935	2 781	2 571	2 336	2 381	2 691	2 909	3 216
European Social Fund	HDIZ	755	804	615	783	434	659	370	412	427	433	900
European Coal & Steel Community Grant	FJKP	39	29	5	1	–	–	1	–	–	2	–
Central government													
Fontainebleau abatement	FKKL	1 208	2 411	1 733	1 377	3 171	2 084	4 560	3 099	3 560	3 592	3 655
Other EU receipts	GCSD	25	13	6	7	5	–	8	13	10	12	–
Capital transfers													
Other sectors													
Agricultural Guidance Fund	FJXL	48	30	57	56	47	82	26	–	2	49	80
European Regional Development Fund	HBZA	437	620	812	357	285	989	543	296	622	1 062	1 393
Other capital transfers from EU Institutions[2]	EBGO	–	524	178	43	–	–	322	–	–	–	–
Total credits	GCSL	5 155	8 597	6 709	5 771	6 931	6 602	8 691	6 688	7 801	8 603	9 805
Debits												
Current transfers													
Other sectors													
Customs duties and agricultural levies [3]	FJWD	2 458	2 318	2 291	2 076	2 024	2 086	2 069	1 919	1 937	2 145	2 220
Sugar levies [3]	GTBA	55	26	91	42	46	44	31	25	18	25	24
European Coal & Steel Community production levy [3]	GTBB	–	–	–	–	–	–	–	–	–	–	–
VAT based contribution [4]	HCML	4 635	4 441	3 646	3 758	3 920	4 104	3 624	2 720	2 775	1 764	1 980
VAT adjustment [4]	FSVL	210	30	–249	470	–109	100	–49	88	–35	25	19
Central government													
GNP fourth resource[5]	HCSO	1 639	2 488	2 655	3 516	4 403	4 243	3 859	5 259	6 622	7 565	8 597
GNP adjustments[5]	HCSM	187	–34	–197	404	229	136	–1	76	150	–16	135
Total GNP based fourth own resource contribution	NMFH	*1 826*	*2 454*	*2 458*	*3 920*	*4 632*	*4 379*	*3 858*	*5 335*	*6 772*	*7 549*	*8 732*
Inter-government agreements	HCBW	–	–	–	–	–	–	–	–	–	–	–
EU non-budget (miscellaneous)	HRTM	–	–	–	–	–	–	–	–	–	–	–
Other current transfers to EU institutions	GVEG	8	8	31	–1	11	6	24	10	18	–3	106
Total debits	GCSM	9 192	9 277	8 268	10 265	10 524	10 719	9 557	10 097	11 485	11 505	13 081
Balance (UK net contribution to the EU)	BLZS	–4 037	–680	–1 559	–4 494	–3 593	–4 117	–866	–3 409	–3 684	–2 902	–3 276

1 Before 1989 this is netted off the VAT contribution but cannot be identfified separately.
2 Other capital transfers from EU institutions are included indistiguishably with Agricultural Guarantee Fund receipts before 1996.
3 EU traditional own resource.
4 Third own resource contribution.
5 Fourth own resource contribution.

9.10 Trade in services
By type of service 2004

£ million

	Transportation	Travel	Communications	Construction	Insurance	Financial	Computer and information	Royalties and license fees	Other business services	Personal, cultural and recreational	Government	Total services
Exports												
European Union (EU25) total	6 655	6 211	1 847	80	934	8 496	3 810	1 942	12 243	1 007	772	43 997
Belgium and Luxembourg	419	274	46	..	44	896	489	67	1 355	24	..	3 623
Denmark	517	194	17	–	40	182	48	87	382	92	4	1 563
France	1 019	900	127	7	161	1 743	402	217	1 547	..	..	6 327
Germany	1 278	982	440	2	164	1 582	458	288	2 199	212	30	7 635
Ireland	682	804	418	51	107	544	930	504	1 418	32	11	5 501
Italy	448	569	232	6	146	672	164	108	780	..	..	3 288
Netherlands	697	479	195	4	87	1 326	741	220	2 563	87	21	6 420
Russia	68	158	11	24	28	300	12	14	230	7	14	866
Spain	713	618	150	3	50	548	116	73	523	30	13	2 837
Switzerland	144	297	39	–	95	644	592	398	1 673	48	3	3 933
Turkey	75	72	7	..	22	70	..	6	73	8	8	360
Argentina	22	73	..	..	12	11	..	2	17	..	..	143
Brazil	102	73	3	..	20	23	5	15	47	..	6	307
Canada	321	496	21	..	285	232	19	64	277	9	..	1 736
Chile	32	–	..	–	14	5	..	2	..	..	3	84
Mexico	54	106	2	..	42	40	..	33	..	..	5	349
United States of America	3 697	2 651	437	13	2 121	4 379	1 362	1 918	7 143	555	455	24 731
China	432	395	8	–	18	74	11	59	258	..	..	1 269
Hong Kong	217	221	26	..	44	263	..	15	237	20	3	1 058
India	210	315	18	1	28	84	34	18	..	..	67	981
Japan	465	330	159	–	142	992	32	343	1 382	50	4	3 899
Malaysia	58	159	7	–	20	50	18	30	190	7	3	542
Philippines	53	20	2	..	14	..	..	3	14	..	1	128
Saudi Arabia	131	108	10	1	17	60	30	7	177	..	..	546
Singapore	294	97	10	..	34	152	21	..	209	..	1	2 109
South Korea	178	395	5	..	34	124	20	42	96	..	..	903
Taiwan	75	115	3	–	19	163	10	21	68	..	..	495
Thailand	65	120	5	..	13	35	..	18	38	..	5	317
Australia	570	592	58	..	188	234	50	139	302	118	..	2 261
South Africa	272	192	22	..	88	145	37	58	217	20	..	1 064
Other	3 065	3 609	..	..	868	..	..	..	..	..	..	22 542
Global total	16 373	15 414	2 933	303	4 965	20 281	6 373	6 704	30 305	2 145	2 021	107 817
Imports												
European Union (EU25) total	10 480	18 161	1 303	35	441	1 806	1 083	1 387	7 057	367	1 629	43 749
Belgium and Luxembourg	429	471	49	..	43	198	42	151	521	26	..	1 956
Denmark	401	118	23	..	11	26	23	53	125	19	..	809
France	1 699	3 707	253	8	67	397	195	510	1 414	57	100	8 407
Germany	1 555	683	185	3	106	360	288	190	1 471	39	1 018	5 898
Ireland	603	1 037	180	9	42	142	250	55	553	13	27	2 911
Italy	1 066	1 434	78	3	35	123	21	41	575	..	..	3 495
Netherlands	764	603	124	2	52	251	84	282	966	70	22	3 220
Russia	96	108	20	..	10	41	2	2	..	..	4	503
Spain	2 194	5 955	191	–	26	104	27	15	389	13	158	9 072
Switzerland	317	398	40	1	11	168	105	261	404	51	3	1 759
Turkey	213	530	23	..	9	9	..	1	23	..	..	818
Argentina	14	41	..	..	–	..	..	1	5	..	..	69
Brazil	82	84	5	..	4	4	..	2	23	..	6	219
Canada	231	479	36	–	14	35	10	63	126	..	..	1 036
Chile	46	16	..	–	..	..	..	..	..	–	1	72
Mexico	73	254	..	..	1	14	..	1	15	..	3	367
United States of America	2 589	3 567	309	1	112	956	569	2 827	2 794	409	619	14 752
China	263	185	15	..	30	7	..	1	61	..	..	567
Hong Kong	246	122	23	–	17	123	..	..	142	..	4	702
India	274	517	58	..	7	10	112	5	105	..	..	1 095
Japan	254	112	22	–	23	260	9	281	604	6	8	1 579
Malaysia	71	100	..	..	6	7	..	..	15	..	2	210
Philippines	28	54	8	..	..	..	..	..	..	..	–	110
Saudi Arabia	49	25	8	..	3	7	7	–	..	..	1	217
Singapore	116	55	7	..	10	43	21	4	343	..	..	612
South Korea	80	44	..	–	9	55	..	2	52	..	..	248
Taiwan	51	30	..	–	7	42	..	–	..	..	–	195
Thailand	126	462	12	..	5	6	..	..	28	..	..	644
Australia	233	895	34	..	6	46	17	26	118	8	..	1 388
South Africa	297	449	26	..	9	35	31	5	64	..	..	924
Other	4 211	8 338	..	..	..	..	..	..	..	..	..	18 045
Global total	18 671	30 873	2 372	142	830	3 982	2 012	5 007	14 547	904	2 559	81 899

Symbols used in this table:
 .. Indicates that data might be disclosive and have therefore been omitted
 - Indicates that the data is nil or less than £500,000

9.11 Trade in services
By type of service 2005

£ million

	Trans-port-ation	Travel	Commun-ications	Cons-truction	Insu-rance	Finan-cial	Computer and infor-mation	Royal-ties and license fees	Other busi-ness services	Personal, cultural and recrea-tional	Govern-ment	Total services
Exports												
European Union (EU25) total	6 664	7 629	2 077	257	604	9 280	3 431	2 224	11 707	721	790	45 384
Belgium and Luxembourg	431	288	56	..	31	996	..	35	1 102	13	8	3 336
Denmark	512	227	17	2	26	180	61	90	326	83	5	1 529
France	1 064	1 003	148	14	96	1 403	396	196	1 509	95	25	5 949
Germany	890	1 120	364	9	107	2 067	485	287	2 275	125	31	7 760
Ireland	772	964	470	187	68	613	992	473	1 170	45	14	5 768
Italy	597	644	240	..	95	841	..	115	633	..	16	3 442
Netherlands	819	488	311	15	60	1 335	443	385	2 674	41	18	6 589
Russia	96	149	11	..	18	395	..	20	283	60	15	1 081
Spain	904	828	190	5	33	747	148	69	540	27	15	3 506
Switzerland	183	430	34	..	63	797	..	362	1 871	40	4	4 248
Turkey	117	121	5	..	15	88	..	7	96	..	9	479
Argentina	26	25	2	..	5	17	..	5	..	..	4	96
Brazil	138	99	..	..	10	29	4	29	31	..	7	353
Canada	192	452	30	..	178	240	21	67	315	5	..	1 519
Chile	39	9	2	−	8	..	..	3	..	..	3	90
Mexico	69	48	2	..	28	49	..	22	..	..	5	298
United States of America	4 043	2 522	325	9	−254	5 029	1 327	1 844	7 033	516	430	22 824
China	410	521	10	..	10	113	14	60	176	12	..	1 339
Hong Kong	357	228	29	..	23	272	14	19	220	..	3	1 209
India	281	320	18	−	21	104	20	28	..	..	72	1 102
Japan	616	297	125	..	85	1 322	27	387	1 146	40	..	4 050
Malaysia	75	146	10	−	11	54	8	19	100	4	4	431
Philippines	84	19	2	..	7	5	..	2	16	..	1	150
Saudi Arabia	150	136	10	..	10	68	..	10	305	..	4	705
Singapore	343	110	10	..	26	160	17	..	167	..	1	2 479
South Korea	197	90	3	1	26	162	..	51	154	..	5	712
Taiwan	93	86	..	2	13	179	7	25	..	..	20	544
Thailand	65	57	5	..	9	51	4	20	26	..	..	244
Australia	855	663	56	1	134	229	..	141	325	98	..	2 584
South Africa	300	265	17	..	54	110	35	65	199	25	..	1 083
Other	3 256	4 513	..	..	562	..	..	..	..	..	..	25 624
Global total	17 974	16 868	3 036	522	1 578	23 260	5 832	7 313	30 738	1 966	2 036	111 123
Imports												
European Union (EU25) total	11 155	19 027	1 584	282	464	2 221	1 209	1 303	8 036	288	1 667	47 236
Belgium and Luxembourg	405	457	62	3	46	178	58	..	550	11	..	1 902
Denmark	438	107	23	9	14	27	28	50	111	16	8	831
France	1 730	3 761	279	9	73	347	158	421	1 604	39	97	8 518
Germany	1 501	757	360	81	109	533	437	221	1 786	19	1 062	6 866
Ireland	611	1 092	210	119	34	178	248	63	573	13	24	3 165
Italy	1 143	1 729	119	..	34	196	20	32	727	..	53	4 119
Netherlands	800	577	117	12	62	419	64	301	939	62	22	3 375
Russia	116	98	20	..	14	61	..	1	215	..	3	542
Spain	2 539	5 911	185	29	32	142	23	28	471	12	135	9 507
Switzerland	328	454	47	6	10	214	114	336	603	49	3	2 164
Turkey	251	677	23	..	9	16	..	1	29	..	..	1 013
Argentina	16	35	..	−	..	6	..	−	11	..	..	76
Brazil	95	85	8	..	5	..	..	2	28	..	6	263
Canada	229	529	40	..	12	49	9	77	135	..	1	1 128
Chile	47	18	..	−	..	..	..	−	..	..	..	77
Mexico	82	204	6	..	1	7	..	..	12	..	..	317
United States of America	2 666	3 803	287	4	106	1 126	524	2 729	2 827	365	397	14 834
China	282	220	15	..	35	11	..	1	94	..	4	668
Hong Kong	276	148	18	..	21	104	..	3	138	..	..	716
India	289	626	74	..	7	13	122	4	107	..	..	1 247
Japan	328	161	22	−	23	374	10	370	822	..	2	2 121
Malaysia	63	111	..	..	5	7	..	..	24	..	−	220
Philippines	34	79	8	..	..	..	..	..	..	..	..	141
Saudi Arabia	62	63	..	..	5	8	..	..	68	..	1	214
Singapore	138	64	15	..	11	56	10	..	368	..	..	687
South Korea	80	37	3	..	..	50	..	−	72	..	2	254
Taiwan	46	29	..	..	6	54	..	..	72	..	−	223
Thailand	127	425	11	−	5	..	..	1	19	..	2	613
Australia	438	999	38	..	8	56	33	62	152	..	4	1 808
South Africa	318	489	24	..	11	34	12	19	79	10	..	1 002
Other	4 623	9 061	..	..	..	..	..	..	..	..	..	19 456
Global total	20 101	32 806	2 664	455	880	4 866	2 110	4 986	15 973	788	2 438	88 067

Symbols used in this table:
.. Indicates that data might be disclosive and have therefore been omitted
- Indicates that the data is nil or less than £500,000

9.12 Trade in goods and services
Top fifty UK trading partners

Exports				Imports			
Goods (£211bn in 2005)		Services (£111bn in 2005)		Goods (£278bn in 2005)		Services (£88bn in 2005)	
	05 04		05 04		05 04		05 04
USA	1 1	USA	1 1	Germany	1 1	USA	1 1
Germany	2 2	Germany	2 2	France	2 3	Spain	2 2
France	3 3	Netherlands	3 3	USA	3 2	France	3 3
Ireland	4 4	France	4 4	Netherlands	4 4	Germany	4 4
Netherlands	5 5	Ireland	5 5	Belgium	5 5	Italy	5 5
Belgium	6 6	Switzerland	6 6	China	6 7	Netherlands	6 6
Spain	7 7	Japan	7 7	Italy	7 6	Ireland	7 7
Italy	8 8	Spain	8 9	Norway	8 10	Switzerland	8 9
United Arab Emirates[1]	9 13	Italy	9 8	Spain	9 9	Japan	9 11
Switzerland	10 12	Belgium	10 9	Ireland	10 8	Greece	10 8
Sweden	11 9	Australia	11 11	Japan	11 11	Australia	11 13
Japan	12 10	Singapore	12 12	Hong Kong	12 12	Belgium	12 10
Canada	13 11	The Channel Islands	13 16	Sweden	13 13	Portugal	13 12
Hong Kong	14 14	Kuwait	14 14	Russia	14 15	Sweden	14 14
China	15 16	Sweden	15 15	Denmark	15 18	India	15 15
India	16 17	Norway	16 18	Canada	16 14	Cyprus	16 16
Australia	17 15	Denmark	17 17	South Africa	17 19	Canada	17 17
Denmark	18 18	Canada	18 13	Switzerland	18 16	Turkey	18 19
Norway	19 19	Cayman Islands	19 20	Singapore	19 17	South Africa	19 18
Turkey	20 20	China	20 19	Turkey	20 20	Norway	20 21
Singapore	21 22	Hong Kong	21 22	South Korea	21 21	Austria	21 22
South Africa	22 21	India	22 23	India	22 25	Denmark	22 20
Russia	23 26	South Africa	23 21	Finland	23 24	Hong Kong	23 23
Portugal	24 24	Russia	24 26	Austria	24 22	Singapore	24 28
South Korea	25 25	United Arab Emirates[1]	25 27	Poland	25 29	China	25 29
Poland	26 28	Finland	26 24	Taiwan	26 23	Barbados	26 24
Saudi Arabia	27 23	Bermuda	27 29	Australia	27 28	Kuwait	27 27
Finland	28 30	Poland	28 36	Portugal	28 27	Thailand	28 25
Greece	29 27	Greece	29 30	Hungary	29 31	Russia	29 30
Israel	30 29	Luxembourg	30 28	Malaysia	30 26	Poland	30 39
Austria	31 31	South Korea	31 25	Brazil	31 32	Egypt	31 36
Malaysia	32 32	Saudi Arabia	32 34	Thailand	32 30	New Zealand	32 32
Czech Republic	33 33	Nigeria	33 32	Saudi Arabia	33 34	Pakistan	33 33
Taiwan	34 34	Kazakhstan	34 33	Czech Republic	34 33	United Arab Emirates[1]	34 31
Brazil	35 36	Portugal	35 31	United Arab Emirates[1]	35 36	Czech Republic	35 34
Hungary	36 35	Hungary	36 42	Botswana	36 35	Malta	36 37
Nigeria	37 37	Taiwan	37 37	Israel	37 38	The Channel Islands	37 26
Romania	38 41	Austria	38 38	Luxembourg	38 37	Finland	38 41
Mexico	39 40	Turkey	39 43	Indonesia	39 39	Mexico	39 35
Thailand	40 39	British Virgin Islands	40 41	Romania	40 40	Kazakhstan	40 37
Egypt	41 38	Malaysia	41 35	Latvia	41 42	Brazil	41 44
Pakistan	42 46	Pakistan	42 –	Philippines	42 43	South Korea	42 42
Iran	43 42	New Zealand	43 40	Vietnam	43 41	Bulgaria	43 –
Kuwait	44 45	Israel	44 39	Greece	44 44	Luxembourg	44 40
New Zealand	45 43	Qatar	45 47	Bangladesh	45 45	Hungary	45 –
Oman	46 48	Bangladesh	46 –	New Zealand	46 46	Israel	46 43
Indonesia	47 44	Egypt	47 49	Costa Rica	47 47	Taiwan	47 48
Cyprus	48 50	Czech Republic	48 45	Pakistan	48 48	Malaysia	48 46
Qatar	49 49	Azerbaijan	49 –	Chile	49 –	Saudi Arabia	49 45
Philippines	50 –	Mexico	50 44	Sri Lanka	50 –	Jamaica	50 47

1 United Arab Emirates includes Abu Dhabi, Dubai, Sharjah, Ajman, Umm al Qaiwain, Ras al Khaimah and Fujairah

9.13 World total and G7 countries trade in services[1]

| | US$ million | | | | | | | | | % | |
| | | | | | | | | | | The UK as a percentage of: | |
	World	US	Canada	Japan	France	Germany	Italy	UK[2]	G7 total	World	G7
Exports											
1994	1 048 743	199 030	23 958	58 297	75 521	66 248	53 681	69 887	546 622	6.7	12.8
1995	1 230 521	217 460	26 128	65 274	84 090	81 838	61 619	79 816	616 225	6.5	13.0
1996	1 315 397	236 890	29 243	67 712	83 529	85 408	65 660	90 531	658 973	6.9	13.7
1997	1 364 176	254 317	31 596	69 303	80 790	82 735	66 991	101 701	687 433	7.5	14.8
1998	1 390 623	261 127	33 836	62 412	84 958	84 496	67 549	112 640	707 018	8.1	15.9
1999	1 437 298	280 169	36 117	60 998	82 085	83 920	58 788	119 133	721 210	8.3	16.5
2000	1 522 736	296 852	40 230	69 238	80 917	83 150	56 556	120 790	747 733	7.9	16.2
2001	1 525 758	285 296	38 804	64 516	82 298	88 714	57 676	121 028	738 332	7.9	16.4
2002	1 639 236	291 344	39 759	65 712	86 130	103 008	60 439	135 214	781 606	8.2	17.3
2003	1 884 561	305 857	42 624	77 621	98 759	123 476	71 767	158 682	878 786	8.4	18.1
2004	2 234 795	340 418	47 534	97 611	110 313	141 852	83 706	197 521	1 018 955	8.8	19.4
Imports											
1994	1 069 179	131 920	32 530	106 356	57 675	105 942	48 238	59 914	542 575	5.6	11.0
1995	1 255 631	141 500	33 473	122 626	66 117	127 200	55 050	65 680	611 646	5.2	10.7
1996	1 335 402	150 629	35 906	129 988	67 275	135 977	57 605	73 031	650 411	5.5	11.2
1997	1 371 654	166 478	38 013	123 454	64 164	129 647	59 227	78 598	659 581	5.7	11.9
1998	1 391 035	181 390	38 156	111 833	67 728	135 120	63 379	88 328	685 934	6.3	12.9
1999	1 450 265	199 871	40 573	115 158	63 524	141 001	57 707	97 129	714 963	6.7	13.6
2000	1 540 811	225 340	44 118	116 864	61 044	137 254	55 601	100 147	740 368	6.5	13.5
2001	1 551 747	223 940	43 843	108 249	56 861	141 916	57 753	100 259	732 821	6.5	13.7
2002	1 655 452	233 717	44 653	107 940	68 907	144 814	63 166	109 926	773 123	6.6	14.2
2003	1 887 207	256 627	50 732	111 529	82 863	172 084	74 332	127 360	875 527	6.7	14.5
2004	2 222 335	296 071	57 303	135 514	97 523	193 110	81 987	150 039	1 011 547	6.8	14.8
Balances											
1994		67 110	−8 572	−48 059	17 846	−39 694	5 443	9 973	4 047		
1995		75 960	−7 345	−57 352	17 973	−45 362	6 569	14 136	4 579		
1996		86 261	−6 663	−62 276	16 254	−50 569	8 055	17 500	8 562		
1997		87 839	−6 417	−54 151	16 626	−46 912	7 764	23 103	27 852		
1998		79 737	−4 320	−49 421	17 230	−50 624	4 170	24 312	21 084		
1999		80 298	−4 456	−54 160	18 561	−57 081	1 081	22 004	6 247		
2000		71 512	−3 888	−47 626	19 873	−54 104	955	20 643	7 365		
2001		61 356	−5 039	−43 733	25 437	−53 202	−77	20 769	5 511		
2002		57 627	−4 894	−42 228	17 223	−41 806	−2 727	25 288	8 483		
2003		49 230	−8 108	−33 908	15 896	−48 608	−2 565	31 322	3 259		
2004		44 347	−9 769	−37 903	12 790	−51 258	1 719	47 482	7 408		

1 G7 country data is not yet available for 2005
2 The analysis of UK data is based on the all accounts totals shown in table 3.1

Sources: G7 and world data provided by IMF;
UK data provided by ONS

Chapter 10

Geographical breakdown of International Investment Position

Summary

The latest available geographical breakdown of the UK's International Investment Position (IIP) is for the end of 2004. The geographical breakdown of the IIP lags that of the Current Account as much of the data is sourced from annual inquiries, which are not available until 12 months after the reference year.

Foreign direct investment geographical breakdown levels are derived from annual inquiries to outward and inward direct investors in the UK. Portfolio investment consists of equity and debt securities holdings, in the form of bonds and notes and money market instruments. Information on the geographical breakdown of UK holdings of portfolio investment assets are broadly based on the UK contribution to the IMF's Co-ordinated Portfolio Investment Survey (CPIS).

Geographical breakdowns of UK banks' deposits abroad and loans made abroad are derived from banking data supplied by the Bank of England. This information is also used to apportion securities dealers' deposits abroad. Country breakdowns of UK private sector (excluding banks and securities dealers) deposits with banks abroad are derived from the banking statistics of countries in the Bank for International Settlements (BIS) reporting area. Geographical breakdowns of foreign deposits with UK banks are derived from banking data, with foreign loans made to securities dealers apportioned in the same way. Country breakdowns of UK private sector (excluding banks and securities dealers) loans from abroad are derived from the banking statistics of countries in the BIS reporting area.

Geographical International Investment Position

At the end of 2004, the UK's net IIP was -£110.9 billion with reported assets totalling £3,960.7 billion and reported liabilities totalling £4,071.6 billion. These are respectively equal to 337 per cent and 346 per cent of GDP (GDP at current market prices, as published in *UK National Accounts: the Blue Book 2006*). In 2004, the UK had a net asset position with Euro Area countries, EU25 and Australasia but net liability positions with the Americas, Asia and Africa.

Geographical breakdown of assets

Of the assets held by UK residents at the end of 2004, 54 per cent were issued in Europe. In total 46 per cent were held in EU25 countries and 43 per cent in Euro Area countries.

The European destinations most popular with UK investors were Germany, the Netherlands and France, which accounted for £339.0 billion, £313.7 billion and £274.8 billion of UK assets respectively at the end of 2004.

Investments in the Americas amounted to £1,141.3 billion, 29 per cent of UK investment holdings abroad. The majority of these, £869.1 billion, were in the United States of America (USA), equal to 22 per cent of total UK holdings.

An additional 12.1 per cent of UK assets were investments in Asia. Japan, the UK's main investment partner in this region, accounted for 6.6 per cent of total UK assets, at £262.0 billion. Investments in Australasia, Africa and International Organisations accounted for only 3.1 per cent of total assets.

Geographical breakdown of liabilities

The distribution of liabilities by region largely mirrored that of assets at the end of 2004. Investments in the UK from Europe amounted to 56 per cent of total investments. EU25 countries accounted for £1,708.7 billion, or 42 per cent of total investment, and Euro Area countries for £1,577.0 billion, or 39 per cent.

Investment in the UK from the Americas was 29 per cent of total UK liabilities at the end of 2004, while the USA itself accounted for 23 per cent of total investment into the UK. Investments in the UK from Asia totalled 12.4 per cent of UK liabilities. Australasia, Africa and International Organisations combined accounted for just 3.0 per cent of total investments in the UK overall.

Geographical breakdown of direct investment

UK direct investments abroad contributed 17.4 per cent to the total stock of UK assets at the end of 2004, at £689.0 billion. Of these investments, 53 per cent were in holdings issued by countries in the EU25, and 22 per cent in the USA. The country in the EU accounting for the highest level of UK assets was the Netherlands, with total investments in that country amounting to £134.4 billion.

Direct investment in the UK equalled 9.4 per cent of the total level of UK liabilities in 2004. The USA accounted for 33 per cent of direct investment into the UK and the EU for a further 45 per cent. The country in the EU with the most significant direct investment into the UK was Germany with total investments of £50.2 billion. This was closely followed by the

Netherlands at £48.1 billion worth of UK liabilities and France at £41.4 billion. Investment into the UK by the USA, however, was much higher than for any of these European countries.

Geographical breakdown of portfolio investment

UK portfolio investments assets at the end of 2004 stood at £1092.3 billion, 28 per cent of total UK assets. The geographical breakdown of UK portfolio investment assets is compiled using the UK's contribution to the CPIS. The largest issuer of UK holdings in the UK's 2004 CPIS return was the USA, contributing to over 25 per cent of total investments. The UK sectors investing most heavily in the USA were insurance companies and pension funds investing around 43 per cent while 38 per cent of investment was from banks. Of the assets held in the USA, 56 per cent were long term debt issues.

Residents of the EU were the issuers of 40 per cent of UK portfolio investment holdings at the end of 2004. £79.3 billion of these were issued by Germany and £77.5 billion were issued by France; short term debt issues accounted for £10.2 billion of the latter. Broken down by sector, 45 per cent of German issues were attributed to banks. UK banks also invested heavily in Italy, at £40.9 billion, and the Cayman Islands, at £35.7 billion, while UK insurance companies and pension funds invested £25.3 billion in France. Just over half of portfolio investment in Asia was in Japanese issued holdings, 72 per cent of which were equities.

Twenty-five countries accounted for nearly 90 per cent of investments held by UK residents. In 2004 there were 92 countries in which the UK invested less than £1 billion each. The sum of these investments accounted for only 1.2 per cent of total portfolio investment assets.

Portfolio investment liabilities are derived from the CPIS returns of other countries reporting assets held in the UK. The country holding most portfolio investments in the UK is the United States, at £379.1 billion. Of the USA's total holdings in the UK at the end of 2004, 62 per cent were equity holdings. Ireland also reported high levels of investment in the UK, at £108.8 billion. Short-term debt holdings account for 40 per cent of these.

Geographical breakdown of other investment

The UK's other investment assets totalled £2,156.2 billion at the end of 2004. £1,000.3 billion was invested in EU countries; 25 per cent of this was in Germany, 16 per cent in France and 11.5 per cent in the Netherlands. Investment by UK residents in the USA was £442.7 billion, or 21 per cent of total other investment assets. A significant proportion of UK assets were also held in Japan, at £189.6 billion, or 8.8 per cent, at the end of 2004.

Other investment liabilities totalled £2,509.4 billion in 2004. The USA accounted for 16.7 per cent of total other investment liabilities, at £418.5 billion. A total of 76 per cent of German holdings in the UK were other investments, at £341.6 billion. Other investment also made up a significant proportion of Switzerland's investment in the UK, at £213.2 billion, which was 85 per cent of total UK liabilities to Switzerland.

Figure 10.1

International Investment Position

10 largest holders of UK assets 2004

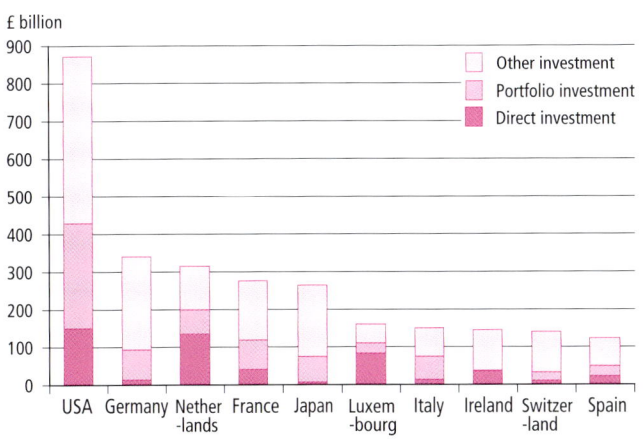

Figure 10.2

International Investment Position

10 largest issuers of UK liabilities 2004

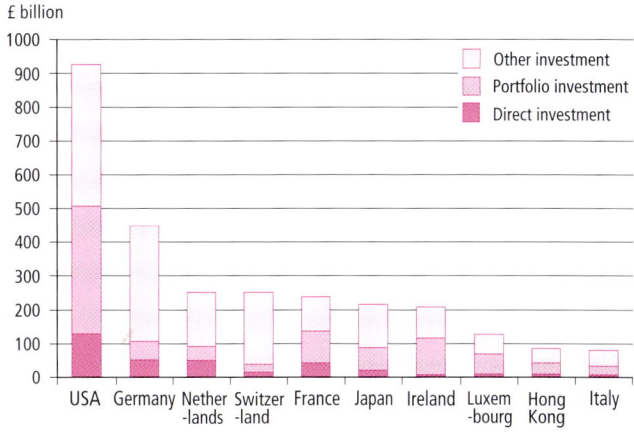

Time series: comparisons

Following a slight fall between 2001 and 2002, there was an increase in total UK IIP assets, from £3,142.4 billion in 2002 to £3,960.7 billion in 2004. Investment in Europe increased from £1,942.7 billion in 2003 to £2,127.2 billion in 2004. Investment in the USA rose from £783.3 billion in 2003 to £869.1 billion in 2004. Most countries showed a fall in levels between 2001 and 2002 followed by a rise in the level of investment to 2003 and 2004; the USA was the largest mover in each of the years. UK assets in Switzerland, however, fell between 2003 and 2004.

Figure 10.3
International Investment Position
Total assets by region

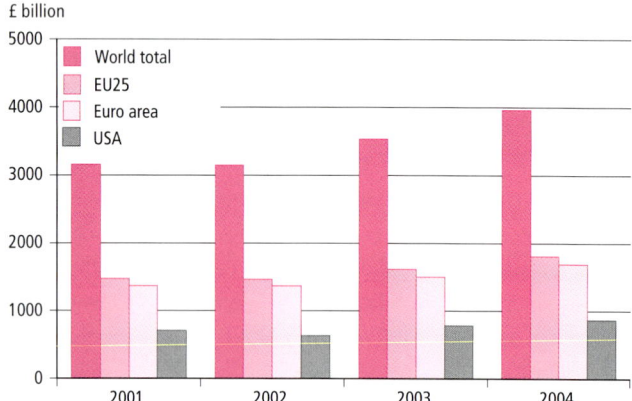

Figure 10.4
International Investment Position
Total liabilities by region

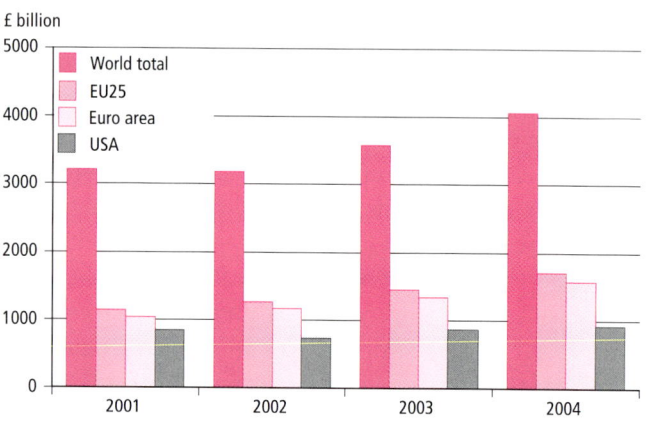

IIP liabilities also increased overall in 2004, having dipped slightly in 2002. There were increases in investments in the UK from the EU, rising from £1,447.8 billion in 2003 to £1,708.7 billion in 2004. Investment by the USA in the UK also rose; from £865.8 billion in 2003 to £925.0 billion in 2004. Switzerland was one of the few countries to show a fall in UK liabilities in 2004, by £15.0 billion.

The increase in both assets and liabilities between 2003 and 2004 reflected movements in global equity prices. The FTSE all share index, an indicator of movements in liabilities, rose 9.3 per cent between the end of 2003 and the end of 2004, following a rise of 16.6 per cent between 2002 and 2003. The global stock market index series (excluding the UK), an indicator of movements in assets, rose 13.1 per cent between 2003 and 2004, following a 32.4 per cent increase between 2002 and 2003. Both the FTSE and global indices fell between 2001 and 2002.

Rates of return

Regional rates of return are calculated by dividing income earned and paid on investments by the total value of the investment. Taking the EU as an example, the UK earned £57.8 billion from its average investments of £1,712.3 billion in 2004, equivalent to an annual rate of return of 3.4 per cent. (The average investment in 2004 is calculated by taking the mean of the end 2003 and end 2004 levels.) In 2004 the UK earned a 3.8 per cent rate of return on its total external assets and paid out a 3.0 per cent rate of return on external liabilities. In all years since 2002 the UK has earned a higher rate of return with its main partners and globally on its external assets than it pays out on its liabilities.

Regional rates of return

	External assets (per cent)				External liabilities (per cent)		
	2002	2003	2004		2002	2003	2004
EU 25	3.9	3.6	3.4	**EU 25**	3.2	3.2	3.2
Total Europe	3.8	3.6	3.5	**Total Europe**	3.1	3.0	3.0
USA	4.0	4.0	3.7	**USA**	3.4	3.4	3.5
Total Asia	3.6	3.5	3.8	**Total Asia**	3.0	2.0	2.2
Rest of the World	4.8	4.3	5.0	**Rest of the World**	2.9	2.8	2.6
World total	3.9	3.7	3.8	**World total**	3.1	3.0	3.0

10.1 International Investment Position: UK assets by type of investment
Balance sheets valued at end of year
2004

£ billion

	Assets				Liabilities			
	Type of Investment				Type of Investment			
	Direct	Portfolio	Other	Total	Direct	Portfolio	Other	Total
Belgium	8.4	11.7	78.1	98.2	3.8	11.9	79.5	95.2
France	39.0	77.5	158.4	274.8	41.4	93.6	102.5	237.5
Germany	11.9	79.3	247.8	339.0	50.2	55.2	341.6	447.0
Ireland	34.0	1.2	109.0	144.2	6.0	108.8	92.3	207.1
Italy	12.1	60.9	76.3	149.3	4.8	27.2	46.4	78.4
Luxembourg	80.7	28.1	50.7	159.5	6.6	61.3	57.1	125.0
Netherlands	134.4	64.7	114.7	313.7	48.1	42.6	159.4	250.1
Spain	20.9	27.5	73.9	122.2	5.3	27.9	33.5	66.7
Total EU25	367.8	442.0	1 000.3	1 810.1	172.5	499.7	1 036.6	1 708.7
Norway	4.9	7.0	16.3	28.2	0.2	12.6	14.5	27.3
Switzerland	8.7	22.3	108.1	139.1	13.6	22.9	213.2	249.7
Total EFTA	14.0	30.5	125.3	169.9	14.6	36.0	230.0	280.7
Total Europe	401.7	504.4	1 221.1	2 127.2	191.7	598.1	1 475.3	2 265.1
Canada	8.1	7.7	25.1	40.9	12.6	20.4	15.6	48.6
USA	149.8	276.5	442.7	869.1	127.3	379.1	418.5	925.0
Total America	194.2	346.7	600.4	1 141.3	144.8	413.1	621.3	1 179.3
Hong Kong	21.1	15.1	19.6	55.8	6.8	33.7	43.3	83.8
Japan	5.8	66.6	189.6	262.0	18.2	68.5	126.8	213.5
Singapore	6.2	5.5	31.0	42.7	0.9	15.8	60.7	77.5
Total Asia	50.6	126.6	303.1	480.3	30.5	122.3	352.0	504.8
Australia	21.2	27.8	13.6	62.6	16.6	10.6	16.0	43.2
Total Australasia and Oceania	23.1	30.9	15.9	69.8	16.8	11.7	17.4	45.9
South Africa	11.0	5.0	4.4	20.3	0.3	18.4	11.5	30.1
Total Africa	19.4	5.9	11.6	36.8	0.5	18.5	37.3	56.3
International Organisations	–	12.5	4.1	16.6	–	14.2	6.0	20.2
World Total	689.0	1 092.3	2 156.2	3 960.7	384.4	1 177.8	2 509.4	4 071.6

10.2 Geographical breakdown of International Investment Position: UK assets
Balance sheets valued at end of year

£ million

		2001	2002	2003	2004
Europe					
European Union (EU)					
Austria	HCZI	15 685	14 859	17 881	20 855
Belgium	A54T	63 340	74 404	81 755	98 176
Cyprus	A3OB	2 207	1 646	1 692	1 840
Czech Republic	HDFF	2 641	2 049	2 114	2 093
Denmark	HDRV	20 249	22 351	26 205	35 510
Estonia	HDSA	265	195	229	314
Finland	HDSJ	19 555	16 857	15 598	15 340
France	HDSL	226 750	217 446	233 182	274 837
Germany	HDQJ	294 389	293 789	318 211	339 049
Greece	HDSM	17 846	19 940	22 220	25 293
Hungary	HDXA	4 193	4 757	4 425	5 625
Ireland	HDZG	97 269	97 730	119 976	144 152
Italy	HEGT	134 672	131 275	149 468	149 312
Latvia	HFHZ	102	120	62	110
Lithuania	HFHY	156	166	116	150
Luxembourg	A5BW	114 593	109 465	138 993	159 482
Malta	A3OC	588	718	1 188	–386
Netherlands	HFID	300 595	311 793	286 019	313 707
Poland	HFII	5 334	6 222	6 993	6 956
Portugal	HFIJ	20 511	21 504	26 795	23 431
Slovak Republic	HFJI	436	591	787	374
Slovenia	HFJH	495	309	414	659
Spain	HDSC	60 981	57 583	93 576	122 222
Sweden	HFJA	51 319	39 111	44 699	51 650
European Central Bank	HBKQ	–	–	–	–
EU Institutions	HBKP	17 654	20 981	21 842	19 356
Total EU25	A2K3	1 471 825	1 465 861	1 614 440	1 810 107
European Free Trade Association (EFTA)					
Iceland	HDZZ	1 128	908	1 487	2 028
Liechtenstein	HFCE	840	819	957	516
Norway	HFIE	18 038	19 755	23 338	28 241
Switzerland	HCZZ	154 324	149 140	173 892	139 103
Total EFTA	HBKW	174 330	170 622	199 674	169 888
Other Europe					
Albania	HBLA	–	–8	1	20
Belarus	HCZX	1	1	1	1
Bulgaria	HCZU	1 102	364	428	642
Croatia	HDWZ	635	1 052	1 168	1 131
Romania	HFIX	558	713	1 009	1 578
Russia	HFIY	7 613	8 390	11 987	16 367
Turkey	HFJK	7 191	8 375	6 742	8 703
Ukraine	HFJM	106	166	235	815
Serbia and Montenegro	HFJQ	49	51	37	44
Other	HFIP	79 161	90 661	106 977	117 935
Total Europe	HDRW	1 742 571	1 746 248	1 942 699	2 127 231
Americas					
Argentina	HCPD	5 196	3 108	2 971	3 247
Brazil	HCZW	11 776	9 210	8 989	9 205
Canada	HCZY	46 108	50 187	42 217	40 933
Chile	HDER	3 360	3 397	3 254	3 481
Colombia	HDEZ	2 113	2 655	2 719	2 412
Mexico	HFIB	10 162	7 788	8 078	10 372
United States of America	HFJN	704 905	634 061	783 348	869 100
Uruguay	HFJO	301	323	234	144
Venezuela	HFJP	1 807	1 732	1 091	1 598
Other Central American Countries	HFIR	107 238	122 743	147 488	193 781
Other	HFIV	912	1 426	1 960	7 025
Total Americas	HDRZ	893 878	836 630	1 002 349	1 141 298
Asia					
China	HDES	5 941	6 482	6 587	8 985
Hong Kong	HDSN	39 853	31 387	45 284	55 819
India	HDZX	5 613	5 120	6 953	8 565
Indonesia	HDZD	3 037	2 591	2 846	3 118
Iran	HDZY	1 114	1 590	2 487	3 405
Israel	HDZK	1 671	2 301	2 446	2 597
Japan	HEIC	206 061	222 620	218 242	262 039
Malaysia	HFIC	4 940	5 233	5 555	6 782
Pakistan	HFIH	1 228	847	901	1 077
Philippines	HFIG	2 220	2 491	3 256	2 435
Saudi Arabia	HFIZ	6 021	5 824	5 869	5 141
Singapore	HFJG	49 996	41 470	44 445	42 728
South Korea	HEJH	11 498	14 883	17 294	19 606
Taiwan	HFJL	7 667	5 463	11 483	20 514
Thailand	HFJJ	3 053	3 730	4 307	4 521
Residual Gulf Arabian Countries	HFIS	21 126	21 980	22 037	25 894
Other Near & Middle Eastern Countries	HDSG	1 122	1 092	576	1 435
Other	HFIT	5 770	3 834	5 170	5 667
Total Asia	HDSF	377 931	378 938	405 738	480 328
Australasia & Oceania					
Australia	HCZT	44 938	48 256	51 319	62 568
New Zealand	HFIF	5 134	5 969	5 318	6 717
Other	HFIU	368	407	396	556
Total Australasia & Oceania	HDSH	50 440	54 632	57 033	69 841
Africa					
Egypt	HDSB	2 253	1 915	1 706	1 956
Morocco	HFIA	573	323	472	527
South Africa	HFJR	11 352	16 594	17 759	20 349
Other North Africa	HFIW	601	1 427	977	1 031
Other	HFIQ	8 257	7 792	10 718	12 980
Total Africa	HDRY	23 036	28 051	31 632	36 843
International Organisations	HBKR	13 419	15 641	15 629	16 613
Unallocated	D39B	26 184	56 796	56 959	65 317
Reserve Assets	LTEB	25 649	25 469	23 794	23 250
World total	HBQA	3 153 108	3 142 405	3 535 833	3 960 721

10.3 Geographical breakdown of International Investment Position: UK liabilities
Balance sheets valued at end of year

£ million

		2001	2002	2003	2004
Europe					
European Union (EU)					
Austria	HFMW	16 871	17 310	18 181	21 026
Belgium	A56Z	64 220	79 652	81 592	95 198
Cyprus	A3SB	3 299	3 157	3 039	3 471
Czech Republic	HFNG	3 337	2 122	1 989	1 993
Denmark	HFNJ	21 426	20 938	28 354	35 351
Estonia	HFNO	122	302	227	335
Finland	HFOA	5 844	6 716	11 658	12 823
France	HFOB	160 766	183 913	194 285	237 478
Germany	HFNI	264 943	305 292	358 666	446 986
Greece	HFOD	13 638	17 378	18 425	23 496
Hungary	HFOG	722	1 226	908	1 301
Ireland	HFOI	112 932	136 448	165 139	207 110
Italy	HFOO	69 051	75 387	78 774	78 423
Latvia	HFOX	186	315	249	327
Lithuania	HFOW	101	176	130	283
Luxembourg	A5E4	81 455	83 377	114 920	124 994
Malta	A3SC	1 140	1 347	1 510	1 044
Netherlands	HFPD	208 841	207 506	226 381	250 055
Poland	HFPN	4 390	3 077	3 050	3 833
Portugal	HFPO	6 733	9 581	10 999	12 644
Slovak Republic	HFQH	396	151	548	1 310
Slovenia	HFQG	983	914	746	247
Spain	HFNR	36 202	46 037	56 006	66 726
Sweden	HFQE	29 358	26 200	31 833	39 207
European Central Bank	HFMP	–	–	–	–
EU Institutions	HFMO	38 200	38 873	40 239	43 087
Total EU25	A2M3	1 145 156	1 267 395	1 447 848	1 708 748
European Free Trade Association (EFTA)					
Iceland	HFOM	461	282	801	1 377
Liechtenstein	HFOU	1 980	2 176	2 166	2 299
Norway	HFPG	16 026	19 926	26 530	27 325
Switzerland	HFNC	232 418	240 875	264 628	249 670
Total EFTA	HFMT	250 885	263 259	294 125	280 671
Other Europe					
Albania	HFMU	86	52	58	89
Belarus	HFNA	69	70	136	224
Bulgaria	HFMY	509	499	495	628
Croatia	HFOF	828	1 349	1 442	1 498
Romania	HFQB	520	412	601	769
Russia	HFQC	9 337	13 605	16 960	23 146
Turkey	HFQJ	3 206	2 521	2 566	3 679
Ukraine	HFQL	208	1 228	1 461	2 625
Serbia and Montenegro	HFQP	213	457	351	447
Other	HFPT	194 730	203 333	204 674	242 576
Total Europe	HFNK	1 605 747	1 754 179	1 970 717	2 265 100
Americas					
Argentina	HFMV	462	247	225	242
Brazil	HFMZ	3 213	2 683	4 213	5 404
Canada	HFNB	40 650	34 782	42 287	48 563
Chile	HFND	961	775	1 042	1 136
Colombia	HFNF	1 313	573	799	743
Mexico	HFPB	5 658	2 575	2 776	2 902
United States of America	HFQM	847 829	735 305	865 846	924 974
Uruguay	HFQN	171	113	788	228
Venezuela	HFQO	1 030	1 012	1 105	1 091
Other Central American Countries	HFPV	118 696	106 751	153 731	191 161
Other	HFPZ	3 985	2 503	2 859	2 809
Total Americas	HFNN	1 023 968	887 319	1 075 671	1 179 253
Asia					
China	HFNE	8 106	9 184	7 090	10 050
Hong Kong	HFOE	82 698	71 165	78 868	83 821
India	HFOK	8 935	9 121	12 073	18 274
Indonesia	HFOH	3 067	2 791	2 681	2 176
Iran	HFOL	5 992	3 610	3 350	3 057
Israel	HFOJ	6 752	6 292	5 939	6 081
Japan	HFOP	221 130	201 905	175 063	213 462
Malaysia	HFPC	6 039	4 257	4 337	7 734
Pakistan	HFPM	2 207	3 411	3 710	2 619
Philippines	HFPJ	1 181	1 034	582	880
Saudi Arabia	HFQD	16 089	15 641	13 747	16 226
Singapore	HFQF	68 251	68 937	67 629	77 468
South Korea	HFOS	4 215	4 030	5 464	5 162
Taiwan	HFQK	4 533	3 483	7 477	10 367
Thailand	HFQI	4 035	1 314	3 443	2 995
Residual Gulf Arabian Countries	HFPW	39 734	32 652	27 839	33 358
Other Near & Middle Eastern Countries	HFNX	5 173	5 655	6 331	6 378
Other	HFPX	3 803	3 667	4 167	4 720
Total Asia	HFNT	491 940	448 149	429 790	504 828
Australasia & Oceania					
Australia	HFMX	26 064	26 645	32 006	43 183
New Zealand	HFPH	2 546	2 484	2 217	2 505
Other	HFPY	178	177	173	197
Total Australasia & Oceania	HFNZ	28 788	29 306	34 396	45 885
Africa					
Egypt	HFNQ	6 611	6 055	6 309	6 152
Morocco	HFOZ	876	734	813	841
South Africa	HFQQ	18 485	20 460	27 645	30 092
Other North Africa	HFQA	4 471	3 851	4 266	5 350
Other	HFPU	14 347	13 892	13 333	13 905
Total Africa	HFNM	44 790	44 992	52 366	56 340
International Organisations	HFMQ	16 313	14 772	16 997	20 183
Unallocated	D4BO	–	–	–	–
World total	HBQB	3 211 546	3 178 717	3 579 937	4 071 589

10.4 Geographical breakdown of International Investment Position: Net
Balance sheets valued at end of year

£ million

		2001	2002	2003	2004
Europe					
European Union (EU)					
Austria	IDBU	−1 186	−2 451	−300	−171
Belgium	A58F	−880	−5 248	163	2 978
Cyprus	A3V5	−1 092	−1 511	−1 347	−1 631
Czech Republic	IDCE	−696	−73	125	100
Denmark	IDCH	−1 177	1 413	−2 149	159
Estonia	IDCM	143	−107	2	−21
Finland	IDCS	13 711	10 141	3 940	2 517
France	IDCT	65 984	33 533	38 897	37 359
Germany	IDCG	29 446	−11 503	−40 455	−107 937
Greece	IDCU	4 208	2 562	3 795	1 797
Hungary	IDCX	3 471	3 531	3 517	4 324
Ireland	IDCZ	−15 663	−38 718	−45 163	−62 958
Italy	IDDE	65 621	55 888	70 694	70 889
Latvia	IDDJ	−84	−195	−187	−217
Lithuania	IDDI	55	−10	−14	−133
Luxembourg	A5FI	33 138	26 088	24 073	34 488
Malta	A3V6	−552	−629	−322	−1 430
Netherlands	IDDN	91 754	104 287	59 638	63 652
Poland	IDDS	944	3 145	3 943	3 123
Portugal	IDDT	13 778	11 923	15 796	10 787
Slovak Republic	IDEI	40	440	239	−936
Slovenia	IDEH	−488	−605	−332	412
Spain	IDCO	24 779	11 546	37 570	55 496
Sweden	IDEF	21 961	12 911	12 866	12 443
European Central Bank	IDBN	–	–	–	–
EU Institutions	IDBM	−20 546	−17 892	−18 397	−23 731
Total EU25	A2NH	326 669	198 466	166 592	101 359
European Free Trade Association (EFTA)					
Iceland	IDDD	667	626	686	651
Liechtenstein	IDDH	−1 140	−1 357	−1 209	−1 783
Norway	IDDO	2 012	−171	−3 192	916
Switzerland	IDCA	−78 094	−91 735	−90 736	−110 567
Total EFTA	IDBR	−76 555	−92 637	−94 451	−110 783
Other Europe					
Albania	IDBS	−86	−60	−57	−69
Belarus	IDBY	−68	−69	−135	−223
Bulgaria	IDBW	593	−135	−67	14
Croatia	IDCW	−193	−297	−274	−367
Romania	IDEC	38	301	408	809
Russia	IDED	−1 724	−5 215	−4 973	−6 779
Turkey	IDEK	3 985	5 854	4 176	5 024
Ukraine	IDEM	−102	−1 062	−1 226	−1 810
Serbia and Montenegro	IDEQ	−164	−406	−314	−403
Other	IDDU	−115 569	−112 672	−97 697	−124 641
Total Europe	IDCI	136 824	−7 931	−28 018	−137 869
Americas					
Argentina	IDBT	4 734	2 861	2 746	3 005
Brazil	IDBX	8 563	6 527	4 776	3 801
Canada	IDBZ	5 458	15 405	−70	−7 630
Chile	IDCB	2 399	2 622	2 212	2 345
Colombia	IDCD	800	2 082	1 920	1 669
Mexico	IDDL	4 504	5 213	5 302	7 470
United States of America	IDEN	−142 924	−101 244	−82 498	−55 874
Uruguay	IDEO	130	210	−554	−84
Venezuela	IDEP	777	720	−14	507
Other Central American Countries	IDDW	−11 458	15 992	−6 243	2 620
Other	IDEA	−3 073	−1 077	−899	4 216
Total Americas	IDCL	−130 090	−50 689	−73 322	−37 955
Asia					
China	IDCC	−2 165	−2 702	−503	−1 005
Hong Kong	IDCV	−42 845	−39 778	−33 584	−28 002
India	IDDB	−3 322	−4 001	−5 120	−9 709
Indonesia	IDCY	−30	−200	165	942
Iran	IDDC	−4 878	−2 020	−863	348
Israel	IDDA	−5 081	−3 991	−3 493	−3 484
Japan	IDDF	−15 069	20 715	43 179	48 577
Malaysia	IDDM	−1 099	976	1 218	−952
Pakistan	IDDR	−979	−2 564	−2 809	−1 542
Philippines	IDDQ	1 039	1 457	2 674	1 555
Saudi Arabia	IDEE	−10 068	−9 817	−7 878	−11 085
Singapore	IDEG	−18 255	−27 467	−23 184	−34 740
South Korea	IDDG	7 283	10 853	11 830	14 444
Taiwan	IDEL	3 134	1 980	4 006	10 147
Thailand	IDEJ	−982	2 416	864	1 526
Residual Gulf Arabian Countries	IDDX	−18 608	−10 672	−5 802	−7 464
Other Near & Middle Eastern Countries	IDCQ	−4 051	−4 563	−5 755	−4 943
Other	IDDY	1 967	167	1 003	947
Total Asia	IDCP	−114 009	−69 211	−24 052	−24 500
Australasia & Oceania					
Australia	IDBV	18 874	21 611	19 313	19 385
New Zealand	IDDP	2 588	3 485	3 101	4 212
Other	IDDZ	190	230	223	359
Total Australasia & Oceania	IDCR	21 652	25 326	22 637	23 956
Africa					
Egypt	IDCN	−4 358	−4 140	−4 603	−4 196
Morocco	IDDK	−303	−411	−341	−314
South Africa	IDER	−7 133	−3 866	−9 886	−9 743
Other North Africa	IDEB	−3 870	−2 424	−3 289	−4 319
Other	IDDV	−6 090	−6 100	−2 615	−925
Total Africa	IDCK	−21 754	−16 941	−20 734	−19 497
International Organisations	IDBO	−2 894	869	−1 368	−3 570
Unallocated	D4BP	39 813	72 522	56 959	65 317
Reserve Assets	LTEB	25 649	25 469	23 794	23 250
World total	IDBP	−58 438	−36 312	−44 104	−110 868

Part 4

Supplementary information

Balance of payments and the relationship to national accounts

This section is intended to help users of the Pink Book gain a better understanding of how the data fit within the broader economic accounts framework. It can be read as a stand-alone, although it makes several references to Blue Book tables and so readers are advised to have access to these if possible.

Introduction

Conceptually, the balance of payments, including the international investment position, form part of the broader system of the UK national accounts. The national accounts provide a comprehensive and systematic set of statistics for the UK economy, with information on economic transactions, other changes in the levels of assets and liabilities, and the levels of assets and liabilities themselves. The UK national accounts have generally been compiled according to the *European System of Accounts (ESA95)*. Linkages between the UK balance of payments and national accounts are reinforced by the fact that the UK balance of payments are compiled at the same time as the national accounts, as a component of the sector accounts and using many common data sources.

The national accounts are a closed system in which both ends of every transaction involving a resident economic entity are recorded. A set of accounts is introduced to capture transactions that involve economic relationships with non-resident entities. These accounts are known as the *rest of the world accounts* and are presented from the perspective of non-residents rather than residents. Consequently, entries in the balance of payments (which show transactions from the perspective of residents) are reversed in the presentation of the rest of the world accounts. The accounts for resident entities, which consist of the production, income and accumulation accounts, are described in more detail below.

Two important accounting differences occur when one compares the balance of payments and the national accounts. First, each transaction is recorded twice in the balance of payments (double entry) and four times in the national accounts (quadruple entry). This is because in the balance of payments the activity of only one transactor is recorded, that of the resident entity (with a non-resident entity), whereas in the national accounts the activity of both transactors is recorded (i.e. the activity of either two residents or a resident and a non-resident). Second, in the balance of payments, transactions are shown from the perspective of the resident entity; whereas in the national accounts, transactions are shown from the perspective of the resident in the production, income and accumulation accounts, and from the perspective of the non-resident in the rest of the world account.

Relationship between national accounts and balance of payments concepts and classifications

Because the balance of payments, including the international investment position, forms an integral part of the national accounts, there is complete concordance between them in concept and classification, although the extent of cross-classifications may differ between the two systems.

The balance of payments and national accounts identify resident producers and consumers identically, and both invoke the same concepts of economic territory and centre of economic interest. Both use market prices as the primary concept of valuation of transactions and they adopt identical concepts of accrual accounting. The systems use identical conversion procedures to convert transactions which take place in foreign currency, to UK currency.

While for some purposes it would be convenient if classifications used in the rest of the world accounts and the balance of payments accounts were identical, differences between the two are justifiable because on occasion they serve different purposes. For example, in the balance of payments financial account, precedence is given to classification of transactions by type of investment (i.e. direct, portfolio, reserve assets, other), whereas in the rest of the world financial account the instrument of investment is the primary classification. More important is the fact that concepts, definitions and classifications are consistent between the two systems.

The production, income and capital accounts of the national accounts

The national accounts tables reflect the basic aspects of economic life (production, income, consumption, accumulation and wealth). For many analysts, *Gross Domestic Product (GDP)* is the key economic aggregate as it measures the total value added for the UK economy in any period. GDP may be measured as:

- the total value of output less the cost of goods and services used in the production process (intermediate consumption). This is referred to as the *output (or production) approach;*

- the value of income accruing from the production process to each of the factors of production (plus net taxes on production and imports). This is referred to as the *income approach;* or

- total final expenditure on goods and services during the period referred to as the *expenditure approach.*

Conceptually these measures are equal, but because different and imperfect data sources are used to measure each approach the measures may differ in practice. This difference is reflected in the statistical discrepancy item. The national accounts are regularly benchmarked to balanced annual supply and use (input-output) tables. This ensures that, except for the latest year, the three measures of GDP are equal on an annual basis, though there will still be a statistical discrepancy between the quarterly estimates based on the three approaches.

Blue Book table 1.2 presents the *Gross Domestic Product Account* for the whole economy, the derivation of GDP using the expenditure approach and the income approach. Table 1.7.1, the *Production Account,* shows the derivation of GDP using the production approach.

- The expenditure based measure of GDP is derived as final consumption expenditure by government and households, plus investment in fixed capital formation and changes in inventories, plus exports minus imports of goods and services, plus (or minus) the statistical discrepancy. Exports and imports are the same as the balance of payments components, exports and imports of goods and services.

- The income based measure of GDP shows the components of factor income, namely compensation of employees, gross operating surplus and mixed incomes, plus taxes less subsidies on production and imports.

- The production based measure of GDP is shown as total gross output at purchasers' prices less intermediate consumption.

For the purpose of discussion here, all values are in current prices.

Blue Book table 1.7.3 presents the *National Income and Use of Income Account,* showing the derivation of gross national income, gross disposable income and use of gross disposable income. Gross national income is equivalent to GDP plus primary income receivable from non-residents, less primary income payable to non-residents. These primary income items are the same as the balance of payments income components which are used in the derivation of gross saving (gross disposable income less consumption) and net saving (gross saving less consumption of fixed capital). Table 1.7.3 illustrates how the various balance of payments income and current transfers components affect the nation's saving. To derive gross disposable income, net secondary income receivable from non-residents is added to gross national income; secondary income items are equivalent to the net current transfer components in the balance of payments. The segment of table 1.7.3 dealing with use of gross disposable income shows the derivation of gross saving (gross disposable income less consumption) and net saving (gross saving less consumption of fixed capital).

Blue Book table 1.7.7, the *National Capital Account,* shows the link between gross saving and net lending/borrowing (to/from the rest of the world). The latter is derived as gross saving plus net capital transfers from non-residents less investment in fixed capital and inventories and the net acquisitions of non-produced, non-financial assets from non-residents. The items net capital transactions and net acquisitions of non-produced non-financial assets are both sourced from the balance of payments capital account. The capital account was introduced into the balance of payments to emphasise this clear relationship between the balance of payments and the national accounts.

The financial account and balance sheet of the national accounts

Net lending/borrowing is also the balance shown in *Blue Book* table 1.7.8, the *Financial Account.* The financial account shows how the net lending/borrowing is financed through a combination of transactions in financial assets and liabilities. As table 1.7.8 is a summary account for the economy, transactions between resident sectors are offset and eliminated. Therefore table 1.7.8 is also equivalent to the balance of payments financial account. However, there are some important differences in classification emphasis between table 1.7.8 and the balance of payments financial account. In table 1.7.8 the emphasis is on instrument of investment (currency and deposits, securities, loans, equity, etc.), while in the balance of payments financial account, the emphasis is on type of investment (direct investment, portfolio investment, and other investment). Both presentations give emphasis to the asset and liability classification.

It is worth noting that, if table 1.7.8 were expanded to include the financial transactions taking place between the various resident sectors, it would show the full financial account for the economy (which is published monthly in *Financial Statistics* and quarterly in *UK Economic Accounts*).

Blue Book table 1.7.9, the *National Balance Sheet,* shows the UK's non-financial assets (fixed assets, inventories, tangible and intangible non-produced assets such as land, copyright, etc.), financial assets, and liabilities and net worth at the end of the period. As table 1.7.9 is a summary account for the economy, financial assets and liabilities only measure financial claims by residents on non-residents and liabilities by residents to non-residents. In other words, in this table the financial assets and liabilities components are the international investment position statement for the UK. Claims and liabilities between resident sectors have been offset and eliminated. Again, there are some important classification differences between table 1.7.9 and the international investment position statement. In table 1.7.9 the emphasis is on instrument of investment, while in the international investment position statement the emphasis is on type of investment. Both presentations give emphasis to the asset and liability classification.

Rest of the world accounts of the national accounts

There are five accounts for the rest of the world in the national accounts shown in the *Blue Book*. These are:

(i) table 7.1.0, *the External account of goods and services*;

(ii) table 7.1.2, *the External account of primary incomes and current transfers*;

(iii) table 7.1.7, *the External capital account*;

(iv) table 7.1.8, *the External Financial Account*; and

(v) table 7.1.9, *the External Balance Sheet Accounts*.

The *External Financial Account* is published quarterly in *UK Economic Accounts*. As mentioned earlier, these accounts are required to close the system of national accounts and, while essentially the same as the balance of payments accounts and international investment position statement, they are compiled from the perspective of the non-resident transactor. Table 7.1.2 is essentially the current account of the balance of payments, table 7.1.7 the capital account, table 7.1.8 the financial account, and table 7.1.9 the international investment position. The reader should be able to readily identify the counterpart entries in all of these tables.

Transactions with the EU

Blue Book table 12.2 shows UK official transactions with institutions of the EU from a UK national accounts perspective. It has been re-created in the *Pink Book* as table 9.9 using balance of payments terminology.

Methodological notes

Trade in goods (chapter 2)

Introduction

The IMF Balance of Payments Manual, 5th edition (BPM5) defines trade in goods as covering general merchandise, goods for processing, repairs on goods, goods procured in ports by carriers, and non-monetary gold.

General merchandise (with some exceptions) refers to moveable goods for which real or imputed changes of ownership occur between UK residents and the rest of the world.

Goods for processing: this covers goods that are exported or imported for processing and that comprise two transactions: the export of a good and the re-importation of the good on the basis of a contract and for a fee OR the import of a good and the re-exportation of the good on the basis of a contract and for a fee. The inclusion of these transactions on a gross basis is an exception to the change of ownership principle. The value of the good before and after processing is recorded. This is included in total trade in goods but cannot be separately identified.

Repairs on goods: this covers repairs that involve work performed by residents on movable goods owned by non-residents (or vice versa). Examples of such goods are ships, aircraft and other transport equipment. The value recorded is the value of the repairs (fee paid or received) rather than the value of the goods before and after repair.

Goods procured in ports: this covers goods such as fuels, provisions, stores and supplies procured by UK resident carriers abroad or by non-resident carriers in the UK.

Non-monetary gold: this is defined as all gold not held as reserve assets (monetary gold) by the authorities. Non-monetary gold can be subdivided into gold held as a store of value and other (industrial) gold – for further information see page 167.

Coverage and other adjustments

The balance of payments statistics of trade in goods compiled by the Office for National Statistics (ONS) are derived principally from data provided by HM Revenue & Customs (HMRC) on the physical goods exported from and imported to the UK. However, this information is on a different basis to that required for balance of payments statistics. Accordingly in order to conform to the IMF definitions the ONS has to make various adjustments to include certain transactions which are not reported to HMRC and to exclude certain transactions which are reported to them but where there is no change of ownership. In addition, the value required for balance of payments purposes is the value of goods at the point of export (i.e. the Customs border of the exporting country) rather than the value of goods as they arrive in the UK. Therefore, the freight and insurance costs of transporting the goods to the UK needs to be deducted from the values recorded by HMRC. Table 2.4 summarises this transition onto a balance of payments for each of the last 11 years.

Overseas trade statistics compiled by HM Revenue & Customs (HMRC)

Statistics of the UK's overseas trade in goods have been collected for over 300 years by HMRC, formerly HM Customs and Excise (HMCE). Since 1993 these data comprise statistics of UK imports from and exports to countries outside the EU and statistics on trade with other EU Member States. Data are compiled from declarations made to HMRC by importers, exporters or their agents AND statistics of UK arrivals (imports) from and dispatches (exports) to other member states of the EU compiled from the Intrastat returns submitted by traders or their agents to HMRC.

Prior to 1993 statistics of UK imports from and exports to all countries in the world were compiled from declarations made to HMRC by importers, exporters or their agents.

Information on trade with EU countries

The Intrastat system has applied since 1993, with minor variations, in all EU member states. In the UK all VAT registered businesses are required to complete two additional boxes on their VAT returns, which are normally submitted quarterly. These show the total value of exports of goods to customers in other member states (dispatches) and the total value of imports of goods from suppliers in other member states (arrivals).

Traders whose annual value of arrivals or dispatches exceed given thresholds are required to provide an Intrastat declaration each month, showing full details of their arrivals and dispatches during the month. These thresholds are reviewed annually. For the calendar year 2005, the thresholds were £221,000, both for arrivals and for dispatches, the same as in 2004. These detailed Intrastat declarations cover approximately 97 per cent of the value of trade.

Link with VAT

The information on the VAT returns serves three purposes: (i) to establish a register of traders and to determine which exceed the thresholds; (ii) to provide a cross-check with the Intrastat declarations; and (iii) to provide figures on the total value of trade carried out by traders below the Intrastat thresholds.

Traders not registered for VAT and private individuals who move goods within the EU have no obligations under the Intrastat system and their trade is therefore not included in the statistics. Examples of commodities where this trade can be significant are works of art and racehorses.

Below threshold trade

The total values of arrivals and dispatches by traders below the Intrastat thresholds are available from their VAT returns. The figures are included in the month in which the VAT return is received by HMRC, although the VAT return itself may relate to a period of more than one month. Detailed information on below threshold trade is not available. However, it has been established that the pattern of that trade before the Intrastat system was introduced in 1 January 1993 was similar to that of traders just above the thresholds. Thus estimates enabling detailed allocations of below threshold trade can be made on this basis by HMRC.

Late response

Traders who have a legal responsibility to provide Intrastat declarations are required to do so by the end of the calendar month following the month to which the declaration relates. However, where traders have failed to provide returns to Intrastat by the due date, estimates of the total value of such trade are included. These are based on the trade reported by these traders in a previous period, and the growth rate since that period experienced by comparable traders who have provided returns for the current month.

Late declarations of trade with EU countries are subsequently incorporated into the month's figures to which they relate with a corresponding reassessment of the initial estimates for late response.

The methodology used to collect EU (Intrastat) data on Natural Gas and Electricity has been amended by the Commission of the European Communities (EC regulation no 1982/2004). As a result, from January 2005, HM Revenue & Customs (HMRC) has changed to collecting information relating to the trade in Natural Gas and Electricity directly from the pipeline and grid operators. This has removed the need for individual companies to submit Intrastat (EU) import and export declarations for these goods. The new methodology records the physical flow of Natural Gas and Electricity between the UK and its EU trading partners. Value data are estimated using the relevant market prices for gas and electricity.

Information on trade with non-EU countries

In general the figures for trade with non-EU countries show the trade as declared by importers and exporters or their agents and for which documentation has been received and processed by HMRC during the month.

Methodological notes

Importers are usually required to present a Customs declaration before they can obtain Customs clearance and remove the goods. The great majority of imports are cleared immediately by a computerised system. Furthermore the import statistics include documents received by HMRC up to the third working day after the end of the month. Therefore the import figures correspond fairly closely to goods actually imported during the calendar month. Generally speaking about 90 per cent by value and 85 per cent by number of all entries relate to the calendar month with the bulk of the remainder relating to the immediately preceding month.

Under the procedures for the control of exports, the principle is the same – namely that goods cannot be cleared for export until a Customs declaration has been made. Traders can, if they wish, submit a simplified declaration so that the goods can be exported, which has to be followed within 14 days after date of shipment with a complete export declaration. Moreover the processing of these complete export documents begins three working days before the end of the calendar month (two working days for December). Thus the export statistics compiled for a month (which are based on the date of receipt of the complete export documents) do not correspond with goods actually shipped in the calendar month. Generally both in terms of the value and the number of documents, 75 per cent relates to the calendar month with the bulk of the remaining 25 per cent relating to the immediately preceding month.

HMRC's New Export System (NES), which replaces manual (paper) Customs declarations with electronic submissions, requires electronic messages from the trade once the goods have been exported in order to provide the departure date. The new system has led to greater efficiency; improving processing and thereby speeding up the flow of information. This means that, in terms of the value of trade, the proportion allocated to the correct month has increased from September 2003 onwards.

Basis of valuation

For statistical purposes the UK adopts the valuation bases recommended in the *International Trade Statistics Concepts & Definitions* published by the United Nations.

The valuation of exports (dispatches) is on a free on board (f.o.b.) basis, i.e. the cost of goods to the purchaser abroad, including:

- packaging;
- inland and coastal transport in the UK;
- dock dues;
- loading charges; and
- all other costs such as profits, charges and expenses (e.g. insurance) accruing up to the point where the goods are deposited on board the exporting vessel or aircraft or at the land boundary of Northern Ireland.

The valuation of imports (arrivals) is on a *cost, insurance and freight* (c.i.f.) basis including:

- the cost of the goods;
- charges for freight and insurance;
- all other related expenses in moving the goods to the point of entry into the UK (but excluding any duty or tax chargeable in the UK).

When goods are re-imported after process or repair abroad the value includes the cost of the process or repair as well as the value of the goods when exported.

Arrivals from and dispatches to EU countries

As part of the simplification procedure to reduce the burden on business, in the UK most traders are permitted to provide a valuation for trade in goods with EU countries based on the invoice value. Large traders, currently those who have more than £14 million of trade in the year, are required to supply information on their delivery terms. Regular sample surveys to all traders are conducted by HMRC to establish conversion factors to adjust the invoice values to produce the valuation basis required for statistical purposes. Separate factors are imputed for a range of different delivery terms and for trade with each member state.

The value recorded for arrivals and dispatches includes any duties or levies that have been applied to goods originating in non-EU countries but which have since cleared EU Customs procedures in one EU country prior to moving onto other EU countries.

Imports from non-EU countries

The statistical value of imports of goods subject to duty is the same as the value for Customs purposes. This value is arrived at by the use of specific methods of valuation in the following order of preference:

(i) the transaction value of the imported goods (i.e. the price paid or payable on the goods);

(ii) the transaction value of identical goods;

(iii) the 'deductive method' – value derived from the selling price in the country of importation;

(iv) computed value based on the built-up cost of the imported goods.

Imported goods are valued at the point where the goods are introduced into the Customs territory of the EU. This means that costs for delivery of the imported goods to that point have to be included in the Customs value.

For all other goods (i.e. goods free or exempted from duty and goods subject to a specific duty) the statistical value is determined in relation to the point at which the goods enter the UK.

An amount expressed in foreign currency is converted to sterling by the importer using a system of "period rates of exchange" published by HMRC. These rates are normally operative for a four weekly period unless there is a significant movement in the exchange rate.

Treatment of Taxes

As described above, the value of all goods moving into and out of the UK is based on the transaction value recorded for Customs purposes or, in the case of trade in goods with EU countries, the invoice or contract value. In line with this principle, the values recorded exclude VAT. For trade in goods with non-EU countries, all other taxes such as duties and levies applied to goods after arrival in the UK are excluded. For trade in goods with EU countries, the value recorded for imports and exports includes any duties or levies that have been applied to goods originating in non-EU countries but which have since cleared Customs procedures prior to moving onto other EU countries. However excise duties are excluded from the value recorded for trade.

Balance of payments statistics for trade compiled by ONS

Table 2.4 summarises the transition from trade in goods statistics on an Overseas Trade Statistics basis (compiled and published by HMRC) to those on a Balance of Payments basis (compiled by ONS).

Valuation adjustments

Freight: the cost of freight services for the sea legs of dry cargo imports is estimated by applying freight rates (derived from the rates for a large sample of individual commodities imported from various countries) to tonnages of goods arriving by sea. For the land legs, estimates of freight rates per tonne-kilometre for different commodities and estimated distances are used. Estimates of rail freight through the Channel Tunnel are estimated from data provided by Le Shuttle and freight operators. The cost of freight on imports arriving by air is derived from information on the earnings of UK airlines on UK imports and the respective tonnages landed by UK and foreign airlines at UK airports. Pending investigations of an alternative methodology the cost of freight and insurance on oil and gas imports is projected from data formerly supplied by the Department of Trade and Industry.

Sources: tonnages from HMRC; information on freight rates from Chamber of Shipping, Civil Aviation Authority and road hauliers; information from Eurotunnel.

Insurance: the cost of insurance premiums on non-oil imports is estimated as a fixed percentage of the value of imports.

Source: ONS estimate.

Coverage adjustments

Second hand ships: to include purchases and sales of second-hand ships which are excluded from the Overseas Trade Statistics as the transactions are not notified to HMRC.

Source: inquiries to UK ship owners conducted by the Department for Transport until late 2005. Estimates are now provided by HMRC.

New ships delivered abroad: to include deliveries of new ships built abroad for UK owners while the vessel is still in a foreign port as the transactions are not notified to HMRC.

Source: inquiries to UK ship owners conducted by the Department for Transport until late 2005. Estimates are now provided by HMRC.

North Sea installations: to include goods (including drilling rigs) directly exported from and imported to the UK production sites in the North Sea. This adjustment is also used when there is a redistribution of the resources of fields which lie in both UK and non-UK territorial waters (e.g. the Frigg, Murchison and Statfjord). In these circumstances the contribution to (or reimbursement of) a proportion of the development costs has been treated as a purchase (or sale) of fixed assets at the date of the re-determination and appears as an adjustment to imports (exports) of goods.

Source: ONS inquiries to the petroleum and natural gas industry.

NAAFI: to exclude goods exported by the Navy, Army and Air Force Institute for the use of UK forces abroad since these are regarded as sales to UK residents.

Source: quarterly returns from NAAFI.

Goods not changing ownership: the Overseas Trade Statistics exclude temporary trade (i.e. goods that are to be returned to the original country within two years and there is no change of ownership). However, goods may well have originally been recorded as 'genuine' trade but which are subsequently returned to the original country. Examples of these 'returned goods' are goods traded on a 'sale or return' basis; goods damaged in transit and returned for replacement or repair; and contractor's plant. The same amount is deducted from both imports and exports for the month in which the return movement is declared to Customs.

Source: HMRC (non EU trade in goods identified by reference to Customs Procedure codes (CPCs) and by 'Nature of Transaction Code' on Intrastat submissions).

Gold: trade in gold (i.e. gold bullion, gold coin, unwrought or semi-manufactured gold and scrap) is reported to HMRC but it is excluded from the statistics of total exports and imports published in the Overseas Trade Statistics. However, trade in ores and concentrates and finished manufactures of gold (e.g. jewellery) are included in total exports and imports.

For Balance of Payments purposes, all trade in non-monetary gold should be included under trade in goods. Non-monetary gold is defined as all gold not held as reserve assets (monetary gold) by the authorities. Non-monetary gold can be subdivided into gold held as a store of value and other (industrial) gold. The UK currently makes adjustments to include industrial gold. In exports, the adjustment reflects the value added in refining gold and producing proof coins. In imports, the adjustment reflects the value of gold used in finished manufactures (such as jewellery and dentistry).

Within the transactions of the London Bullion Market, the UK cannot currently distinguish between monetary gold and non-monetary gold held as a store of value. Accordingly, the UK has obtained an exemption from adopting IMF recommendations, as specified in BPM5 and for the time being these transactions are included in the Financial Account.

The treatment of non-monetary gold is being reviewed as part of the worldwide process to revise the IMF Balance of Payments manual. Current proposals can be found on the IMF website www.imf.org/external/np/sta/bop/iss.htm. The main proposal is that the concept of non-monetary gold would be replaced by two categories – allocated gold (a commodity) and unallocated gold (a financial instrument). UK BoP will continue current practice until the treatments defined in the revised manual are implemented.

Source: ONS estimate.

Letter post: to include exports by letter post which are not included in the Overseas Trade Statistics.

Sources: books – ONS estimate based on historic information from publishers and booksellers; other items – ONS estimate based on historic sample inquiry made by the former Post Office.

Additions and alterations to ships: to include work carried out abroad on UK owned ships and work carried out in UK yards on foreign owned ships.

Sources: Inquiries to UK ship owners conducted by the Department for Transport, (imports) until late 2005, then HMRC, and ONS estimates (exports).

Repairs to aircraft: to include the value of repairs carried out in the UK on foreign owned aircraft.

Source: ONS estimate.

Goods procured in ports: to include fuels, provisions, stores and supplies purchased for commercial use in ships, aircraft and vehicles. (Estimates of goods dispatched are recorded by HMRC.)

Sources: Chamber of Shipping and Civil Aviation Authority for goods procured in foreign ports by UK transport companies (imports); UK oil companies, Civil Aviation Authority, BAA, municipal airports and port authorities for goods procured in UK ports by overseas transport companies (exports).

Smuggling of alcohol and tobacco: Customs provide volume figures for smuggled goods entering the UK based on published estimates of revenue loss and revenue evasion through smuggling. This information is supplemented by information on the average prices for alcohol and tobacco goods in France and Belgium from the published sources of the statistical and banking institutions in those countries in order to estimate the value of smuggled alcohol and tobacco entering the UK.

Sources: HMRC, INSEE and National Bank of Belgium

Territorial coverage adjustment: for the purposes of the Overseas Trade Statistics, "UK" is defined as Great Britain, Northern Ireland, the Isle of Man, the Channel Islands and the Continental Shelf (UK part). Therefore the Overseas Trade Statistics exclude trade between these different parts of the UK but include their trade with other countries.

For balance of payments purposes the Channel Islands and the Isle of Man are not considered part of the UK economic territory. Adjustments are made to exports to include UK exports to those islands and to exclude their exports to other countries; and to imports to include UK imports from those islands and to exclude their imports from other countries.

Source: ONS estimate.

Other adjustments

Diamonds: much of the World's trade in rough (uncut) diamonds is controlled from London by the Diamond Trading Company, part of De Beers. Prior to 2001, in order not to distort the trade statistics, all imports into and exports from the UK of uncut diamonds which remain in the ownership of foreign principles are excluded from the Overseas Trade Statistics by HMRC.

In addition the value of diamonds imported into the UK can be reassessed after the diamonds have been cleared by Customs. Prior to 2001, this adjustment reflects these changes in valuation. From 2001 the procedure for recording movements of diamonds was changed so that all trade was included in the Overseas Trade Statistics by HMRC. From 2001, this adjustment removes movements of diamonds where no change of ownership has taken place.

Source: Diamond Trading Company.

Adjustments to imports for the impact of VAT Missing Trader Intra-Community (MTIC) fraud: VAT intra-Community missing trader fraud is a systematic, criminal attack on the VAT system, which has been detected in many EU Member States. In essence, fraudsters obtain VAT registration to acquire goods VAT free from other Member States. They then sell on the goods at VAT inclusive prices and disappear without paying over the VAT from their customers to the tax authorities. The fraud is often carried out very quickly, with the fraudsters disappearing by the time the tax authorities follow up the registration with their regular assurance activities.

Acquisition fraud is where the goods are imported from the EU into the UK by a trader who then goes missing without completing a VAT return or Intrastat declaration. The 'missing trader' therefore has a VAT free supply of goods, as they make no payment of the VAT monies due on the goods. He sells the goods to a buyer in the UK and the goods are available on the home market for consumption.

Carousel fraud is similar to acquisition fraud in the early stages, but the goods are not sold for consumption on the home market. Rather, they are sold through a series of companies in the UK and then re-exported to another Member State. Goods may be imported and exported several times, hence the goods moving in a circular pattern or 'carousel'.

The VAT system (and therefore the Intrastat collection of trade statistics) picks up the exports of any 'carouselled' goods, but does not pick up the associated import at the time the carouselled goods entered the UK. As a consequence, UK import statistics have been under reported.

Originally, most carousel chains only involved EU member states. More recently (from the beginning of 2004), there has been an increase in carousel chains that involve non-EU countries, e.g. Dubai and Switzerland. However, the MTIC trade adjustments are added to the EU import estimates derived from Intrastat returns as it is this part of the trading chain that is not recorded. Changes to the pattern of trading associated with MTIC fraud can therefore make it difficult to analyse trade by commodity group and by country. In particular, adjustments affect trade in capital goods and intermediate goods – these categories include mobile phones and computer components.

ONS and HMRC have agreed a methodology to estimate for the impact of MTIC on the trade statistics. The method used relies heavily on information uncovered during HMRC's operational activity. As such it cannot be detailed for risk of prejudicing current activity, including criminal investigations and prosecutions and more generally undermining HMRC's ability to tackle the fraud effectively. The method specifically excludes adjustments for the acquisition variant of the fraud which cannot be quantified at present. HMRC set up a project in 2005 to review the methodology for producing the estimates of the impact on the trade statistics. Estimates may change as the analysis of the fraud continues. The UK is the first member state to make adjustments in their trade statistics for this type of fraud.

Source: HMRC estimate

Adjustment for under-recording and for currency and other valuation errors: these adjustments compensate for the following types of error:

- failure on the part of traders or their agents to submit details of shipments;
- incorrect valuations recorded;
- declarations wrongly given in foreign currency instead of sterling.

Regular reviews show the adjustments for non-EU trade remaining broadly constant over time. Those for EU trade have reduced since the early days of the Intrastat system. The adjustments, expressed as percentages of total trade excluding oil and erratics, are shown in Table 1.

Adjustments to estimates for late response:
A review of the introduction of the Intrastat system carried out in 1994 identified a number of difficulties in the initial monthly estimates of trade with EU countries provided by HMRC. The following describes the adjustments made by ONS to cope with these difficulties.

The HMRC method of estimation for late response relies on linking the values of trade reported by traders in the current period with previous periods. Problems can arise when traders change their VAT registration (perhaps as a result of an internal reorganisation, mergers or sales). Similarly problems can arise when a trader starts submitting returns for the first time. If the trader then becomes a late responder there may be no history of previous trade upon which to base an estimate. The current HMRC adjustments make an allowance for this, but recent changes in the overall trader profile, with an increasing proportion of smaller traders, means the current methodology needs to be revised. In the meantime, ONS and HMRC have agreed that ONS makes an initial adjustment of +£30 million to both exports and imports (reducing to zero over the following two months).

Furthermore, some traders may submit first declarations for a month that do not include all their trade in that month. Later declarations are then received for the rest of their trade. The pattern of receipt at HMRC of these partial returns is analysed to enable ONS to make initial adjustments to both exports and imports to anticipate these later declarations. These initial adjustments are progressively reduced in subsequent months as late declarations are processed.

Currently the profile of these adjustments is as shown in Table 2.

When Intrastat was introduced it was envisaged that all declarations in respect of any particular month would be made within 6 months of the end of that month. As a consequence HMRC computer programs were designed to recalculate its initial estimates for late response for six months after those estimates first appear in the Overseas Trade Statistics. However the reality is that some declarations are still being received and processed after that six month period. These are being included as additions to the value of reported trade with no corresponding reduction in the value of estimated trade. Accordingly, in order to eliminate this element of double counting the ONS makes a negative adjustment to the value of estimated trade equal to the value of these late amendments. Note where the value of late amendments exceeds the value of estimated trade the level of estimated trade is set to zero.

Source: ONS estimate.

Price and volume indices

Any difference between time periods in the total value of trade reflects changes in prices as well as changes in the levels of the underlying economic activity (e.g. the physical amounts of goods exported or imported). Separation of these changes greatly enhances the interpretation of the data and, for this reason, the ONS compiles separate data measuring changes in price and changes in volume. These data are presented in index number form.

References

Aggregate estimates of trade in goods, seasonally adjusted and on a balance of payments basis, are published monthly by National Statistics in a First Release. More detailed figures are available from the Time Series Data Service and are also contained in the *Monthly Review of External Trade Statistics (Business Monitor MM24)* which is available, free of charge, in electronic format as a PDF on the National Statistics website.

The latest *Trade in goods First Release* can be found at: www.statistics.gov.uk/StatBase/Product.asp?vlnk=1119

The *Monthly Review of External Trade Statistics*, previously published as *MM24*, can be found at: www.statistics.gov.uk/StatBase/Product.asp?vlnk=613&Pos=&ColRank=2&Rank=256

An article entitled 'UK visible trade statistics – the Intrastat system' was published in *Economic Trends*, August 1994.

An article describing MTIC fraud and its effect on BoP and the UK National Accounts was published in *Economic Trends No. 597*, August 2003. A copy can be found at: www.statistics.gov.uk/cci/article.asp?id=402. A follow-up report was published on 17 February 2005 which summarises the work carried out since July 2003 to review the estimates of the impact on the trade figures; a copy can be found at www.statistics.gov.uk/cci/article.asp?id=1066

A fuller version of these methodological notes appears in *Statistics on Trade in Goods (Government Statistical Service Methodological Series 10)*. It also describes the methodology employed to derive volume and price indices and is available on the National Statistics website at: www.statistics.gov.uk/StatBase/Product.asp?vlnk=3134.

Trade in services (chapter 3)

Introduction

Trade in services covers the provision of services by UK residents to non-residents and vice versa. Trade in services are disaggregated into eleven broad categories of services, as follows:

(a) Transportation (Sea, Air and Other) – Passenger, freight and other

(b) Travel (Business and Personal)

(c) Communications services

(d) Construction services

(e) Insurance services

(f) Financial services

(g) Computer and information services

(h) Royalties and licence fees

(i) Other business services (Merchanting and other trade-related services; operational leasing services; miscellaneous business, professional and technical services)

(j) Personal, cultural and recreational services (Audio-visual and related services; other cultural and recreational services)

(k) Government services

Separate tables appear at Chapter 3 of this publication for each of the above categories except construction services, which are shown in the trade in services summary table 3.1.

The change from an industry to product based presentation on implementation of BPM5 in 1998 meant that trade in services data at the individual product level could not always be constructed back in time. Preparation to collect trade in services by product commenced in 1996, with the introduction of the International Trade in Services (ITIS) survey. A full product based dataset

Table 1

	Exports to:		Imports from:	
	EU	non-EU	EU	non-EU
Under recording	+¼%	+1½%	+¼%	0
Currency errors	0	–½%	0	0
Other valuation errors	0	–¼%	0	0

* (+1% for 1997 & +1½% 1993–1996)

Source: Sample surveys made by HMCE

is available from this date. Account totals, and some additional product estimates have been constructed back to 1991 or 1992, based on the relationship between the new ITIS data and the previous industry based data. It was not valid to project this relationship further back in time. For the transport, travel, royalties and government services accounts, there were only small changes from the industry based data, and it was possible to construct longer time series.

Construction services (shown within Table 3.1)

Construction services cover work done on construction projects and installations by employees of an enterprise in locations outside their resident economic territory. The source of information is the International Trade in Services (ITIS) survey. For construction services, where a permanent base is established which is intended to operate for over a year, the enterprise becomes part of the host economy and its *transactions are excluded from the trade in services account.* Transactions where a permanent base is established are recorded under direct investment, within investment income.

Transportation services (Table 3.2)

The transportation account covers sea, air and other (i.e. rail, land, and pipeline) transport. It includes the movement of passengers and freight, and other related transport services, including chartering of ships or aircraft with crew, cargo handling, storage and warehousing, towing, pilotage and navigation, maintenance and cleaning, and commission and agents' fees associated with passenger/freight transportation.

Freight and the valuation of UK trade in goods

The trade in goods estimates included in the balance of payments value *imports* as they arrive in the UK f.o.b. (free on board) at the frontiers of the exporting country. This is net of the *cost of freight* to the UK border and any loss and damage incurred in transit to the UK. For UK importers who purchase goods f.o.b. and arrange transport themselves, their payment for the goods at the exporting countries' frontiers comprises:

(i) the value included in the trade in goods estimates (which is net of subsequent loss and damage);

(ii) the value of loss and damage incurred in transit.

In addition, such importers bear the costs of:

(iii) freight services outside the exporting countries;

(iv) insurance services (the excess of insurance premiums paid for the journeys over claims made).

Where importers purchase goods c.i.f. (cost, insurance and freight) on arrival in the UK, items (ii) to (iv) are paid by the foreign exporters in the first instance. The c.i.f. prices are set accordingly, however, and the UK importers are regarded as bearing the costs of items (i) to (iv).

Therefore, irrespective of the payment basis, items (ii) to (iv) represent costs to UK importers additional to the trade in goods entries (item i). Item (ii), the value of loss and damage, is part of the price paid to the foreign exporter and so always represents a debit entry in the balance of payments accounts. Items (iii) and (iv), freight and insurance services, also represent debit entries when provided by non-residents; where such services are provided by UK residents there is no balance of payments entry. The debit entries above relating to freight are included in imports of transportation services.

The estimates of trade in goods cover exports valued f.o.b. The valuation of exports at the UK frontier must, by definition, include any subsequent loss or damage en route to the importer. Therefore, unlike imports, there is no need to make an explicit adjustment for loss and damage to exports. However, foreign importers must additionally bear the costs of freight and insurance services for the journeys outside the UK and where such services are provided by UK residents this gives rise to credit entries in the services accounts.

The f.o.b. value for UK imports includes the cost of transport within the exporting country. Where this service is provided by a UK operator then the trade valuation of imports overstates the balance of payments effect and an offsetting credit entry is therefore included under "Road transport". Similarly, an offsetting debit entry is included for foreign operators' carriage of UK exports within the UK.

Sea transport

Exports by UK operators consist of freight services on UK exports (but not imports – see "Freight and the valuation of UK trade in goods", above) and on cross-trades, the carriage of non-resident passengers and the provision to them of services, and the chartering of ships to non-residents. Exports also include port charges and other services purchased in the UK by non-resident operators. Conversely, imports comprise services purchased abroad by UK operators, their chartering of ships from non-residents, and the carriage by non-resident operators of UK imports (but not exports) and goods on UK coastal routes and UK passengers.

Statistics relating to UK operators are provided by the Chamber of Shipping (CoS), which conducts inquiries into its members' participation in foreign trade. Until 1995, inquiries covering all CoS members were made every four years, with sample surveys for intervening years. Since 1995, the CoS has surveyed all its members annually.

Exports

Passenger revenue: the value of services provided to non-resident passengers comprises fares and passengers' expenditure on board. Since UK operators are not able to distinguish between fares received from UK residents and non-residents, fares collected abroad are assumed to represent fares received from non-residents (passenger revenue collected abroad from UK residents is thought to be small and is likely to be counter-balanced by that collected in the UK from foreign residents). An estimate of passengers' expenditure on board is added, taking the non-residents' proportionate share of the total to be the same as for fares.

Freight: earnings consist of freight services on UK exports and are based on data supplied to the Chamber of Shipping. Time charter receipts include receipts for charters with crew. Time charters without crew are included within the operational leasing component of Other Business Services (Table 3.9).

Disbursements: estimates of disbursements in the UK by foreign operators are formed from a variety of sources. UK income from port charges, towage, handling costs and other port related services was collected in 1996 from a survey of port authorities. Crews' expenditure is estimated from information on numbers of visiting seamen, supplied by the Home Office. Regular returns are received on light dues from Trinity House. Estimates of expenditure on ships stores and on bunkers are now included within the trade in goods data. Time charter payments made to UK residents are included under "Ships owned or chartered-in by UK residents".

Imports

Passenger revenue: estimates of passenger fares paid to non-resident operators are derived mainly from the results of the International Passenger Survey which is described in the notes below on "Travel". A further allowance is made for on board sales of goods and services. Passenger fares paid to non-resident operators for fly-cruises, however, together with other expenditure by UK passengers on board non-resident shipping, is included, but not separately identified, in "Travel" imports.

Freight: estimates of freight services on UK imports provided by non-resident operators are compiled as follows; the estimates of total freight services (provided by ALL operators) on the sea legs of UK imports of goods are taken as the starting point, as described in chapter 9. Chamber of Shipping estimates of the element provided by

Table 2 £ million

	Exports	Imports
First published estimates	+600	+650
Second estimates	+250	+250
Third estimates	+120	+150
Fourth estimates	+50	+70
Fifth estimates	+10	+20
All subsequent estimates	0	0

Source: HMCE

UK operated ships are then, deducted to obtain the non-resident operators element which is then used in the transportation account. Charter payments cover payments for charters with crew.

Disbursements: disbursements abroad include payments for canal dues, the maintenance of shore establishments, port charges, agency fees, handling charges, crews' expenditure, pilotage and towage, light dues and other miscellaneous port expenditure abroad. Payments for bunkers, ships stores and other goods purchased are now included within the trade in goods data.

Air transport

The exports of UK airlines comprise the carriage of non-resident passengers to, from or outside the UK, the carriage of UK exports of goods (but not imports – see "Freight and the valuation of UK trade in goods", above) and cross-trades and the chartering of aircraft to non-residents. Exports also include airport charges and services purchased in the UK by foreign airlines. Purchases of fuel and other goods are included within trade in goods.

Imports include expenditure abroad by UK airlines on airport charges, crews' expenses, charter payments, etc. They also include payments to foreign airlines for the carriage etc. of UK imports of goods (but not exports) and of UK mail; and for the carriage of UK passengers on flights covered by tickets for journeys to or from the UK (the carriage of UK passengers on other non-resident flights is included under "Travel").

The transactions of UK airlines are derived from returns supplied by the airlines to the Civil Aviation Authority.

Exports

Passenger revenue: this relates to all tickets sold outside the UK and used on UK aircraft, together with receipts from carrying passengers' excess baggage. An exercise by British Airways plc demonstrated that the value of tickets sold abroad to UK residents is roughly counter-balanced by sales in the UK to non residents.

Freight: this consists of freight services on UK exports and the carriage of non-resident airmails, and is based on data supplied to the Civil Aviation Authority.

Disbursements and other revenue: these comprise expenditure in the UK by non-resident airlines on landing fees, other airport charges, handling charges, crews' expenses, office rentals and expenses, salaries and wages of staff at UK offices, commissions to agents and advertising. The estimates are based on returns from the Civil Aviation Authority, BAA plc and municipal airports on their receipts from non-resident airlines for air traffic control, landing fees and other airport charges; and survey information collected from large non-resident airlines operating in the UK on their other UK expenses. Purchases of fuel and other goods are now included within trade in goods.

Also included are receipts from the charter or hire of aircraft, and gross receipts of sums due from non-resident airlines under pooling arrangements and for services such as consultancy and engine overhaul.

Imports

Passenger: the information on fares paid by UK passengers to non-resident airlines is derived from the International Passenger Survey; see notes on "Travel" below.

Freight: estimates of non-resident airlines' freight on UK imports are derived by subtracting from the estimates of total freight on imports of goods arriving by air (see chapter 9) the element provided by UK airlines, the residual being the freight services supplied by non-resident airlines. Other imports comprise payments to non-resident airlines for carrying UK airmails as reported by the Royal Mail Group to the Civil Aviation Authority.

Disbursements and other payments: disbursements abroad include airport landing fees, other airport charges, charter payments, crews' expenses, the operating costs of overseas offices, agents' commissions, advertising, settlements with non-resident airlines under pooling arrangements, and miscellaneous expenditure abroad. Purchases of fuel and other goods are now included within trade in goods.

Other Transport

This covers the movement of passengers and freight, and other related transport services, by rail, road and pipeline.

Rail: this consists primarily of expenditure on fares and rail freight through the Channel tunnel. Passenger revenue estimates are based on numbers of passengers through the tunnel and average fare information. Estimates of rail freight through the tunnel are based on data provided by Le Shuttle and freight operators.

As the tunnel operators are a joint UK/French enterprise, half of passenger and freight transactions are taken to accrue to the UK part of the business. All tickets sold in France are assumed to be sold to non-UK residents (likewise, all tickets sold in the UK are assumed sold to UK residents). Of these, 50 per cent are assumed to accrue to the UK as they represent exports of rail transport services.

Road: exports comprise the earnings of UK road hauliers for the carriage outside the UK of UK exports of goods and the carriage within the exporting countries of UK imports (although such earnings from lorries leaving the UK via the Northern Ireland land boundary are only included from 2002). Estimates of numbers of journeys to various countries are derived from the International Road Haulage Survey, and rates for each journey are estimated from trade and other sources.

Imports include payments to all non-resident land transport operators for the carriage of UK imports of goods between the frontiers of the exporting countries and the foreign sea ports. Estimates are made by subtracting from the estimate of total freight on imports for land legs (as described in chapter 9) an estimate of the element earned by UK operators (derived as for exports). Imports also include the earnings of non-resident road hauliers for carrying UK exports and imports within the UK, although estimates of the trade with the Republic of Ireland are only included from 1996. These are estimated from the statistics of ferry movements of foreign registered lorries, average loads, and average lengths of haul within the UK and estimated freight rates. The disbursements abroad by UK road hauliers, and in the UK by non-resident road hauliers, are included within "Travel".

Pipeline: this covers the cost of transport of oil freight via undersea pipelines. Data are derived from a survey of North Sea Oil and Gas companies.

Travel (Table 3.3)

Travel covers goods and services provided to UK residents during trips of less than one year abroad (and provided to non-residents during similar trips in the UK). Transport to and from the UK is excluded and shown as passenger services under transportation (see above). Internal transport within the country being visited is included within travel.

A traveller is defined as an individual staying, for less than one year, in an economy of which he/she is not a resident. The exceptions are those military and diplomatic personnel, whose expenditure is recorded under government services. The one year rule does not apply to students and medical patients, who remain residents of their country of origin, even if the length of stay in another economy is more than a year.

The estimates are based primarily on the International Passenger Survey, which seeks information on expenditure from samples of non-resident visitors leaving the UK and of UK residents returning from abroad. For package tourists, estimates of the transport elements are deducted from the reported total package costs. Estimates of the expenditure of UK residents visiting the Republic of Ireland and of Irish residents visiting the UK have been covered by the survey since the second quarter of 1999. Prior to this, data were derived from statistics published by the Irish Central Statistics Office.

Business travel

Business travel is divided into expenditure by seasonal and border workers (individuals who work some or all of the time in economic territories that differ from their resident households) and other business travel. Estimates are based on the International Passenger Survey.

Personal travel

Personal travel covers holidays, visits to friends and relatives, the expenditures of people visiting for education and health reasons and miscellaneous purposes. Visits for more than one purpose, where none is distinguished as the main purpose, are classified as other.

Education related travel exports covers the tuition fees and other expenditure of students who are funded from abroad and studying in the UK (imports covers the expenditure of UK students studying abroad). The figures also include the fees and other expenditure of pupils in UK private schools and students at other colleges and language schools. Income received direct from abroad by examining bodies and correspondence course colleges is included within personal, cultural and recreational services.

Fees and other expenditure paid by non-resident students for higher education are collected via a special International Passenger Survey (IPS) trailer which commenced in 1997.

Health related travel covers the cost of medical and other expenses of those travelling abroad for medical treatment. Estimates are based on information supplied to the IPS.

Communication services (Table 3.4)

Communication services covers two main categories of international transactions: telecommunications (telephone, telex, fax, email, satellite, cable and business network services) and postal and courier services. Information is obtained through the ONS International Trade in Services survey (ITIS) and direct from the Royal Mail Group.

Insurance services (Table 3.5)

Insurance services cover the provision of various types of insurance to non-residents by resident insurance enterprises and vice versa. Insurance services include freight insurance on goods being imported or exported, direct insurance (life, accident, fire, marine, aviation etc.) and reinsurance. The amounts recorded in the accounts reflect the service charge earned on the provision of insurance services. This is equal to net premiums from abroad (premiums less claims), plus property income attributed to policy holders, less the change in the reserves for foreign business, less foreign expenses. The figures for insurance companies' and brokers' underwriting activities are derived from annual inquiries conducted by the ONS. Lloyd's of London underwriting activity is based on data supplied by the Corporation of Lloyd's; they also include receipts for management services provided to overseas members of Lloyd's syndicates.

Life insurance and pension funds

Life insurance covers underwriting services associated with long term policies. Data are collected in the ONS inquiry into insurance companies. Pension fund services include service charges relating to occupational and other pension schemes, but not compulsory social security services.

Freight

Treatment of freight insurance is consistent with the f.o.b. valuation of trade in goods (see "freight and the valuation of trade in goods" above). That is, non-resident importers pay for freight and insurance on journeys outside the UK. Where such services are provided by UK residents, this gives rise to a credit entry.

Other direct insurance

Other direct insurance covers accident and health insurance; marine, aviation and other transport insurance; fire and property insurance; pecuniary loss insurance; general liability insurance, and other (such as travel insurance and insurance related to loans and credit cards).

Reinsurance

Reinsurance represents subcontracting parts of risks, often to specialised operators, in return for a proportionate share of the premium income. Reinsurance may relate to packages which mix several types of risks. Exports of services are estimated as the balance of flows between resident reinsurers and non-resident insurers. Imports are estimated as the balance of flows between resident insurers and non-resident reinsurers.

Auxiliary insurance services

This covers insurance broking and agency services, insurance and pension consultancy services, evaluation and adjustment services, actuarial services, salvage administration services, regulatory and monitoring services on indemnities and recovery services. These are measured by net brokerage earnings on business written in foreign currencies, and sterling business known to relate to non-residents. The main source of information on auxiliary insurance services is the ITIS survey.

Financial services (Table 3.6)

Financial services cover financial intermediary and auxiliary services other than those of insurance companies and pension funds. They include intermediary service fees associated with letters of credit, bankers' acceptances, lines of credit, financial leasing and foreign exchange transactions. Also included are commissions and other fees related to transactions in securities; e.g. brokerage, underwriting, arrangements of swaps, options and other hedging instruments etc.; commissions of commodity futures traders; and services related to asset management, financial market operational and regulatory services, security custody services etc. Estimates are based on returns from the Bank of England (for banks), ITIS, and directly from other sources including the Baltic Exchange.

From the 2001 edition of the *Pink Book,* the service earnings of financial institutions are presented on a gross exports and imports basis. This treatment is consistent with the BPM5 edition of the accounts. Trade in services transactions covered by type of financial institution are detailed below:

Monetary financial institutions (banks and building societies)

This covers UK banking services giving rise to:

(i) commissions for credit and bill transactions such as advising, opening and confirming documentary credits, collection of bills, etc.;

(ii) spread earnings (net service earnings through spreads on market making) including those on transactions in foreign exchange, securities and derivatives;

(iii) fees and commissions on foreign exchange dealing;

(iv) commission on new issues of securities, investment management and securities transactions;

(v) commission on derivatives transactions; and

(vi) banking charges, income arising from lending activities, fees and commissions in respect of current account operations, overdraft facilities, executor and trustee services, guarantees, securities transactions and similar services.

Estimates are based on inquiries carried out annually from 1986 to 1990 and for some earlier years. A quarterly survey was run in 1991. A new survey was introduced in 1992 to collect data on UK banks' current account transactions including services. A further new survey was introduced in 2004, which enabled the collection of spread earnings on foreign exchange, securities and derivatives transactions – the data prior to 2004 is estimated by the Bank of England largely on the basis of information on the volumes of transactions and movements in spreads. The survey is completed quarterly by a selected sample of banks and annually by the full UK banking population.

Fund management companies

From 2001, information on investment management fees and fees generated from advisory and other related functions has been collected via the ITIS survey. Earlier estimates were derived from a survey of companies whose main activity is fund management. Earnings are net of any foreign expenses by the institutions concerned. They exclude earnings of insurance companies, which are covered by separate returns made to the Office for National Statistics (see above, under "Insurance Services").

Securities Dealers

The earnings of securities dealers are derived from a survey run by ONS. From the 1998 edition of the *Pink Book*, security dealers' spread earnings (service earnings through market making activities) are included as part of securities dealers' overseas earnings. This treatment is consistent with the domestic accounts as described in the European System of Accounts (1995). Estimates of these spread earnings are based on information on acquisitions and realisations of various classes of securities derived from ONS inquiries, together with the bid and offer prices for certain international bonds.

Baltic Exchange

This covers the brokerage and other service earnings of members of the Exchange for chartering, sales and purchases of ships and aircraft and other associated activities. Estimates are based on a survey of Exchange members.

Other

This includes commissions etc. received from abroad by UK residents (other than MFIs and oil companies, whose earnings are included elsewhere) for dealings in physical goods and in futures and options contracts. From 1990 to 2004 ONS carried out an annual survey of dealers in physical commodities. This data is now collected via the ITIS survey. The foreign earnings of financial futures and options dealers are assumed to have moved in line with the corresponding total earnings of such dealers reported in statutory returns to supervisory bodies.

This component also includes those financial services not included elsewhere, including financial service transactions (exports and imports) picked up from the ITIS survey, service charges on purchases of International Monetary Fund resources and estimates of imports of net spread earnings, which are based on the UK's share of world turnover data for cross-border foreign exchange and derivatives transactions and the UK's share of global imports of financial services.

Computer and information services (Table 3.7)

Computer and information services cover computer data and news related service transactions including databases, such as development, storage and on-line time series; data processing; hardware consultancy; software implementation; maintenance and repair of computers and peripheral equipment; news agency services; and direct, non-bulk subscriptions to newspapers and periodicals. Information is obtained from the ITIS survey.

Royalties and license fees (Table 3.8)

Royalties and licence fees cover the exchange of payments and receipts for the authorised use of intangible, non-produced, non-financial assets and proprietary rights (such as patents, copyrights, trademarks, industrial processes, franchises etc.) and with the use, through licensing agreements, of produced originals or prototypes (such as manuscripts and films).

The heading includes royalties, licenses to use patents, trade marks, designs, copyrights, etc.; manufacturing rights and the use of technical "know-how"; amounts payable or receivable in respect of mineral royalties; and royalties on printed matter, sound recordings and performing rights. Data are obtained through the ITIS survey. Film royalties from the ONS Films and TV inquiry are also included. Royalties incorporated in the contract prices of UK exports and imports of goods are recorded under "Trade in Goods". The outright sale of a copyright is treated as a sale of a non-produced, non-financial asset and is recorded within the Capital Account (Table 6.1).

Other business services (Table 3.9)

Other business services cover a range of services including merchanting and other trade-related services, operational leasing (rental) without operators and miscellaneous business, professional and technical services.

Merchanting and other trade related services

Merchanting is defined as the purchase of a good by a resident from a non-resident and the subsequent resale of the good to another non-resident, without the good entering the compiling economy. The difference between the purchase and sale price is recorded as the value of merchanting services provided.

Estimates of the net profits of UK firms from third country trade in goods are derived from ONS surveys. From 1990 to 2004 ONS carried out a specific sample survey of export houses, but information from these institutions is now collected via the ITIS survey, which has always collected information from other institutions on merchanting and trade related services. This component also covers fees charged for ship classifications and other related services, including information supplied by Lloyd's Register of Shipping.

Operational leasing

Operational leasing covers leasing (other than financial leasing) and charters of ships, aircraft and other transportation equipment without crews. Operational leasing data are derived from the ITIS survey and from the Chamber of Shipping.

Miscellaneous business, professional and technical services

Miscellaneous services include legal, accounting, management consulting, recruitment and training and public relations; advertising and market research and development; architectural, engineering and other technical services; agricultural, mining and on-site processing services associated with agricultural crops (protection against disease or insects), forestry, mining (analysis of ores) etc.; and other services such as placement of personnel, security and investigative services, translation, photographic etc. This item includes data from a number of different data sources, the most important of which is the ITIS survey.

Estimates of the earnings of solicitors are based on surveys held in respect of 1980 and annually since 1986 by the Law Society (in which amounts forwarded to barristers are included). From the 2000 edition of Pink Book, earnings of solicitors are collected as part of the ITIS survey. Other legal services also included estimates of the overseas earnings of UK barristers as supplied by the Commercial Bar Association.

Estimates of banks' and securities dealers' management services appear in the other business services account.

The North Sea oil and gas exports data mainly consists of work done abroad by UK owned drilling rigs and offshore supply boats and by UK seismic survey contractors, services provided by UK residents to the owners of foreign drilling rigs, the treatment of Norwegian oil and gas at the Seal Sands and St. Fergus terminals and the transporting of Norwegian gas to the latter terminal and receipts of the UK company operating the Murchison field from the Norwegian partners in respect of their share of the operating costs of the field. The imports item comprises services such as the hire of drilling rigs and marine support vessels, consultancy, diving and insurance (premiums less claims). The estimates are based on returns to the ITIS survey by companies classified to the industry (Class 11.20 of the Standard Industrial Classification, 1992).

Personal, cultural and recreational services (Table 3.10)

Personal, cultural and recreational services are divided into audio-visual and related services and other. The first category covers services and associated fees relating to the production of motion pictures (on film or video tape), radio and television programmes (live or on tape), and musical recordings. It includes rentals, fees received by actors, directors, producers etc. The second category covers all other personal, cultural and recreational services including those associated with museums, libraries, archives, provision of correspondence courses by teachers or doctors etc. Income received direct from abroad by examining bodies and correspondence course colleges is also included. Most of the information is obtained from the ITIS survey but there is a special ONS inquiry for the film and television industry.

Government services (Table 3.11)

Government services include all transactions by embassies, consulates, military units and defence agencies with residents of staff, military personnel etc. in the economies in which they are located. Other services included are transactions by other official entities such as aid missions and services, government tourist information and promotion offices, and the provision of joint military arrangements and peacekeeping forces (e.g. United Nations). Information comes directly from government departments (including the Ministry of Defence and the Foreign and Commonwealth Office), foreign embassies and United States Air Force bases in the UK.

Exports

Expenditure by foreign embassies/consulates in the UK: this comprises the cost of operating and maintaining Commonwealth High Commission offices, foreign embassies and consulates in the UK, including the personal expenditure of diplomatic staff, but excluding the salaries of locally engaged staff which are included within income; and similar expenditure by the UK offices of non-territorial organisations. In 1993 ONS conducted an inquiry to all high commission offices, embassies, consulates and international organisations in the UK. This figure has been updated for subsequent years using information obtained from several key high commissions and embassies and information on the number of diplomats in the UK.

Military units and agencies: this includes expenditure by the United States Air Force (USAF) in the UK (excluding the pay of locally engaged staff which is included within compensation of employees), together with receipts for services provided by UK military units in the UK and elsewhere to non-residents, such as military training schemes, which is sourced from the Defence Analytical Services Agency (DASA).

European Union institutions exports: these are services of the UK government in collecting the UK contributions to the EU Budget, and services provided at the site of the EU's Joint European Torus project in Oxfordshire.

Other: this comprises goods and services which the government provides to non-residents under its economic aid programmes (these are offset under "Bilateral aid" transfer debits) and miscellaneous goods and services supplied by the UK government to foreign countries, including the reimbursement from other member states of the EU for treatment given by the National Health Service to their nationals.

Imports

Expenditure abroad by UK embassies and consulates: goods and services provided by local residents to UK embassies, High Commission offices, Consulates and the British Council account for most of this heading. It also includes the goods and services provided by local residents to UK diplomatic and other non-military personnel stationed abroad, excluding the salaries of locally engaged staff. The source for this information is the Foreign and Commonwealth Office.

Expenditure abroad by UK military units and agencies: this includes expenditure on food, equipment, fuel and services purchased locally. These items are recorded partly on a net basis – that is, after deducting receipts arising locally. The source for this information is DASA.

Other: this includes goods and services provided by local residents to the UK Government, excluding military and diplomatic expenditure. It covers expenditure abroad of the British Council and the reimbursement to other member states of the EU for medical treatment given to UK nationals.

References

United Kingdom Trade in Services, UKA1

UKA1 has been discontinued as a separate publication. All of the tables that were in Section A of *UKA1* are now included in the *Pink Book*. New tables that were formerly in *UKA1* but not in the *Pink Book* have been added to Chapter 9. The tables that were formerly in Sections B and C of *UKA1* are now in a web-only publication which focuses on the results of the ITIS survey.

Old editions of *UKA1* can be found at the following web address: www.statistics.gov.uk/StatBase/Product.asp?vlnk=3343&Pos=&ColRank=1&Rank=256

The publication containing the ITIS survey results can be found at the following web address: www.statistics.gov.uk/statbase/Product.asp?vlnk=14407

Sea transport

An annual analysis describing the international activities of the UK shipping industry is published by the Department for Transport, in *Transport Statistics Great Britain* (The Stationery Office).

Transport Statistics Great Britain, 2005 edition can be found at: www.dft.gov.uk/intradoc-cgi/nph-idc_cgi?qckQuery=transport+statistics&IdcService=GET_SEARCH_RESULTS&SrchType=q&qckSection=&x=27&y=14

Air transport

Information relating to passenger expenditure is published by the Civil Aviation Authority in CAA Monthly and Annual Statistics.

CAA statistics are available at: www.caa.co.uk/default.aspx?categoryid=80

Travel

Details are published regularly in National Statistics monthly *First Releases* and quarterly *Business Monitors (MQ6)*, both titled *Overseas Travel and Tourism,* and in the annual publication *Travel Trends.*

Overseas Travel and Tourism First Releases can be found at: www.statistics.gov.uk/StatBase/Product.asp?vlnk=8168&Pos=&ColRank=1&Rank=272

MQ6 can be found at: www.statistics.gov.uk/StatBase/Product.asp?vlnk=1905&Pos=1&ColRank=1&Rank=256

Travel Trends can be found at: www.statistics.gov.uk/StatBase/Product.asp?vlnk=1391&Pos=&ColRank=1&Rank=272

Income (chapter 4)

Introduction

The income account covers compensation of employees and investment income. For compensation of employees, estimates for total credits, debits and the balance appear at Table 4.1 but no detailed breakdown of the account is available. Investment income is broken down into four main categories; direct investment, portfolio investment, other investment and reserve assets.

Compensation of employees

Compensation of employees comprises wages, salaries, and other benefits, in cash or in kind, earned by individuals in economies other than those in which they are residents, for work paid for by residents of those economies. Employees in this context, include seasonal or other short term workers (less than one year), and border workers who have centres of economic interest in their own economies. Compensation of employees also includes pay received by local (host country) staff of embassies, consulates and military bases as such entities are considered non-resident of the host economy.

Personal expenditure made by non-resident seasonal and border workers in the economies in which they are employed is recorded under travel within trade in services. Wages and salaries are recorded gross, with taxes paid recorded under current transfers.

Credits

There are three components:

(i) wages, salaries and other benefits earned by UK seasonal and border workers, together with employers' contributions. The International Passenger Survey has been amended to collect this information alongside expenditure of non-resident seasonal and border workers from 1998. Estimates for earlier years are based on the growth of travel and average earnings data;

(ii) wages and salaries earned by UK employees in US military bases in the UK. Information has been supplied to ONS by US military bases;

(iii) wages and salaries earned by UK employees of foreign embassies in the UK. In 1993, ONS conducted an inquiry to all high commission offices, embassies, consulates and international organisations in the UK, asking for information on expenditure – including that on locally employed staff. This figure has been updated for subsequent years using information from a small sample of key embassies.

Debits

There are two components:

(i) wages, salaries and other benefits earned by non-resident workers employed in the UK for less than one year. The International Passenger Survey has been amended to collect this information alongside expenditure of non-resident seasonal and border workers from 1998. Estimates for earlier years are based on the growth of travel and average earnings data;

(ii) wages, salaries and other benefits earned by foreign workers working in UK embassies and military bases abroad. Information on *pay of locally engaged staff* in UK embassies and military bases abroad is obtained from HM Treasury's Combined Online Information System (COINS) and the Ministry of Defence (MOD).

Investment income (Table 4.1 and 4.2)

The investment income account covers earnings (e.g., profits, dividends and interest payments and receipts) arising from foreign investment in financial assets and liabilities. Credits are the earnings of UK residents from their investments abroad and other foreign assets. Debits are the earnings of foreign residents from their investments and funds held in the UK and other UK liabilities. The flow of investment is recorded separately from the earnings in the *Financial account,* although reinvested earnings of companies with foreign affiliates are a component of both – see *Earnings on direct investment* below. The total value of UK assets and liabilities held at any time is also recorded separately under the *International Investment Position.* The presentation of these three sections is almost identical, although there are small differences in coverage in some cases, mainly because full information is not available for all items.

Earnings on the credit side of the account cover such items as interest on UK residents' deposits with banks abroad, profits earned by UK companies from their foreign affiliates, and dividends and interest received by UK investors on their portfolio investments in foreign companies' securities, etc. Similarly, debits cover earnings by foreign investors on deposits held with UK banks, profits of foreign companies from their investments in their affiliates in the UK, and dividends and interest paid to foreign investors on their holdings of UK bonds and shares, including British government stocks, etc.

Earnings on assets and liabilities are defined to include all profits earned and interest and dividends paid to UK residents from non-residents or to non-residents by UK residents. They are, where possible, measured net of income or corporation taxes payable without penalty during the recording period by the enterprise to the economy in which that enterprise operates and, in the case of profits, after allowing for depreciation. Dividends are recorded when they are paid (on a cash basis), whereas interest is recorded on an accruals basis.

Profits and dividends include the (credit) earnings from foreign affiliates of UK registered companies and the (debit) earnings of profits and dividends by UK based affiliates of foreign based companies. Conceptually, stock appreciation and other unrealised capital gains and losses should be excluded from the flows entered in the balance of payments accounts, because they represent only valuation changes. However, data on these are included in banking sector statistics provided by the Bank of England. Profits retained abroad by foreign affiliates or retained in the UK by affiliates of foreign companies are included in the flows of earnings and offset in the financial account. All interest flows between UK residents and non-residents are in principle included.

Earnings on direct investment (Table 4.3 and 4.4)

A direct investment relationship exists if the investor has an equity holding in an enterprise, resident in another country, of 10 per cent or more of the ordinary shares or voting stock. The direct investment relationship extends to branches, subsidiaries and to other businesses where the enterprise has significant shareholding.

Credits

Direct investment earnings include interest on loan capital, profits from branches or other unincorporated enterprises abroad and the direct investor's share of the profits of subsidiary and associate companies. It includes the direct investor's portion of reinvested earnings, which is also treated as a new investment flow out of the parent's country into the affiliate's and appears in the financial account (Table 7.3) as an offsetting entry to the earnings one.

Estimates of profits are made after providing for depreciation, the companies' own estimates of depreciation being used. Although depreciation is estimated at replacement cost in the national accounts, there is little doubt that the estimates in the balance of payments are, in the main, measured at historic cost (different treatments of depreciation result in different entries in the current and financial accounts, but the sum of the two entries will always be the same). Refunds of tax made retrospectively under double-taxation agreements are included in the period when they were made rather than the earlier periods in which they could be deemed to have accrued. Dividend receipts and payments include subsidiaries payments of withholding tax. Estimates for reinvested earnings are not collected separately but are derived by deducting dividends paid from total subsidiaries' profits.

Monetary financial institutions (banks): information on the direct investment earnings of UK registered banks, from their foreign branches, subsidiaries and associates are collected by the Bank of England from a selection of banks quarterly and from all banks which are, or have, a direct investment enterprise annually.

Insurance companies and other financial intermediaries: an annual inquiry forms the basis for estimates of direct investment earnings by UK insurance companies and other financial intermediaries; these results are supplemented by a quarterly survey. Earnings from foreign property by financial companies are also included here. They are estimated from the levels of such assets held by financial companies and information on their total income from abroad. In line with international standards, the earnings of other financial intermediaries include those of all holding companies.

Private non-financial and public corporations: earnings, both credits and debits, of all private and public non-financial corporations are estimated from the results of ONS's annual direct investment inquiry. This inquiry covers a sample of UK companies that either have foreign affiliates or are affiliated to a foreign parent. Returns are imputed for companies which are not approached in the inquiry but which are known to have direct investment links. Results of the annual inquiry are available about twelve months after the end of the year and are published in a National Statistics *First Release* and in *Business Monitor MA4*. The estimates for the latest year are based on a quarterly inquiry.

Copies of the FDI First Release can be found at: www.statistics.gov.uk/StatBase/Product.asp?vlnk=728&Pos=1&ColRank=1&Rank=224

Copies of the FDI Business Monitor can be found at: www.statistics.gov.uk/StatBase/Product.asp?vlnk=9614&Pos=1&ColRank=1&Rank=224

Earnings on foreign assets by the household sector: this comprises household sector investment in property abroad. Investment in property includes the ownership of 'second homes' located outside the UK. Estimates of property ownership have been reassessed this year to take on new information from the Department for Communities and Local Government (DCLG)'s Survey of English Housing (SEH). The SEH collects information from English households on the number of properties owned outside the UK. These estimates have been grossed to include all UK households. Average dwelling prices are applied as well as an estimate of property rental. These methodological changes have resulted in revisions to property flows, levels and income. For more information see the *Economic Trends* article: www.statistics.gov.uk/CCI/article.asp?ID=1176&Pos=6&ColRank=2&Rank=224

Debits

Estimates for income earned from direct investment in the UK are based on the same inquiries to banks, financial institutions and private non-financial corporations as credits.

Earnings on portfolio investment (Table 4.5 and 4.6)

Credits

A large part of the total earnings of UK residents on equity securities and bonds and notes are earned on investments that are not considered to have led to the acquisition of a foreign affiliate (i.e. less than 10 per cent ownership), and so are classified as portfolio rather than direct investment.

Earnings of UK residents on portfolio investment abroad are sub-divided into earnings on equity securities and earnings on debt securities; earnings on debt securities are further sub-divided into earnings on bonds and notes and earnings on money market instruments.

Earnings on equity securities: earnings on equity securities consist of dividends received by UK residents on their holdings of shares of foreign registered companies.

Earnings on debt securities: earnings on bonds and notes consist of interest received by UK residents on their holdings of foreign government and municipal loan stock and bonds of foreign registered companies; earnings on money market instruments consist of earnings of UK residents on holdings of foreign issued commercial paper, certificates of deposit etc.

Estimates of earnings by monetary financial institutions (banks and building societies) are derived from statutory inquiries conducted by the Bank of England.

Estimates of earnings by insurance companies and pensions funds, and securities' dealers, are largely derived from ONS inquiries.

Estimates of earnings by other financial intermediaries and private non-financial corporations are derived from survey-based asset levels to which rates of return on comparable assets shown by financial institutions are applied.

Estimates of the household sector largely consist of earnings by members of Lloyd's of London which are supplied annually by Lloyd's. They include portfolio investment income on funds which are held abroad to support business underwritten in those countries. This income, which is generally reinvested in these foreign funds (see *Portfolio Investment*), is net of earnings distributed to Lloyd's foreign members. Also included are estimates of income from holdings of foreign equities acquired by UK households in exchange for their holdings of UK equities following an acquisition by a foreign direct investor. Typically, such acquisitions are funded by the issuance of shares by the investing company, rather than a cash payment. Significant levels of household ownership are most likely to exist when the UK company is a demutualised building society or privatised public utility.

Debits

Foreign earnings on portfolio investment in the UK are sub-divided into earnings on equity securities and earnings on debt securities; earnings on debt securities are further sub-divided into earnings on bonds and notes and earnings on money market instruments.

Earnings on equity securities: Estimates of foreign earnings from UK equity securities consist of dividends paid to foreign holders of UK company ordinary shares. These estimates are calculated from Stock Exchange data on dividend payments, which are applied pro-rata to levels of non-resident holdings of UK shares derived from ONS's share ownership surveys.

Earnings on debt securities:

(i) Earnings on bonds and notes: Interest on UK foreign currency bonds and notes issued by central government relates to bonds issued by HM Government (the latest of which is the $3 billion 5-year eurobond issued in 2003). Data are estimated from the liability level and known interest rates.

Foreign earnings on British government stocks (gilts) are estimated from information on the levels outstanding and appropriate rates of interest. These earnings are calculated gross of UK income tax. Most gilts are issued by the UK government at a discount to the redemption value. This is recorded as interest accruing over the lifetime of the gilt.

(ii) *Earnings on money market instruments:* Foreign earnings on UK money market instruments consist of earnings on foreign holdings of UK treasury bills, certificates of deposit and commercial paper. Estimates of interest paid to foreign holders of treasury bills are calculated on the basis of levels outstanding and appropriate interest rates. Estimates of foreign earnings on holdings of UK certificates of deposit and commercial paper are derived from statistical inquiries conducted by ONS and the Bank of England, and from information supplied by the UK's Debt Management Office.

Earnings on other investment (Table 4.7 and 4.8)

Credits

Earnings of UK residents on other investment abroad are sub-divided into earnings on trade credit, loans, deposits and other assets.

Trade credit: only a minimal amount of data is available within trade credit. See Financial account notes for detail.

Earnings on loans: earnings on loans are sub-divided into earnings on long-term loans and earnings on short-term loans; short-term loans are those which are repaid in full within one year.

It is not possible to separate out UK monetary financial institutions' (MFIs) earnings on lending abroad from their earnings on deposits abroad. Estimates for earnings on such loans are therefore included indistinguishably within earnings on deposits (see below).

On long-term loans, earnings which are separately identifiable consist of earnings on loans by UK banks guaranteed by the Export Credit Guarantee Department (ECGD), earnings on loans by the ECGD, and earnings on loans by the Commonwealth Development Corporation (CDC). Data on earnings from these loans are derived from information supplied by the Bank of England, the ECGD and the CDC.

On short-term loans, the earnings which are separately identifiable mainly consist of earnings on loans by non-governmental sectors other than MFIs. Earnings on such loans are derived from banking statistics.

Earnings on deposits: estimates of earnings on deposits relate to private sector earnings.

Estimates for MFIs' earnings abroad are sub-divided into earnings on sterling deposits abroad and earnings on foreign currency deposits abroad.

Included under the heading of MFIs' earnings on deposits are earnings on MFIs' foreign lending as it is not possible to separate out UK banks' earnings on their lending abroad from earnings on their deposits abroad. Earnings from lending consist of the interest received by UK banks on overdrafts and loans to non-residents. In this context UK banks means all banks in the UK, including (with effect from 1 April 1998) the Banking Department of the Bank of England. The figures are based on returns made by banks to the Bank of England.

Estimates of securities dealers' earnings on deposits abroad are derived from an ONS statistical inquiry.

Estimates of earnings on deposits abroad for the UK private sector other than banks and securities dealers are largely estimated from levels of such assets (mainly those reported in banking statistics of countries in the BIS reporting area) and appropriate rates of interest. Adjustments are made to remove as far as possible the effects of incomplete coverage and breaks in the reported assets series.

Earnings on other assets: until 2001 earnings from trusts and annuities were estimated from Inland Revenue data on all reported interest and dividend receipts from abroad. From 2001 Inland Revenue have ceased to collect this data, and from this point the data should be regarded as being of lower quality.

Debits

Foreign earnings on other investment in the UK are subdivided into earnings on trade credit, loans, deposits and other liabilities.

Trade credit: Only a minimal amount of data is available within trade credit. See Financial account notes for detail.

Earnings on loans: this covers interest on loans raised from commercial banks abroad and the European Investment Bank (EIB).

It is not possible to separate out earnings on foreign loans to UK banks from earnings on foreign deposits with UK banks. The estimates for foreign earnings on UK banks' loans from abroad are therefore included indistinguishably within earnings on deposits.

Interest paid on central government long-term fixed-interest loans such as Lend-Lease and the Lines of Credit is reported by HM Treasury. Interest on the Very Short-term Financing Facility (VSTFF) taken out during 1992 and repaid in 1993 is also included here. Estimates of interest on local authorities' and public corporations' borrowing from abroad are made by the Bank of England on the basis of levels outstanding and appropriate discount rates.

Estimates of foreign earnings on securities dealers' loans from abroad are derived from an ONS statistical inquiry.

For estimates of foreign earnings on loans to the UK private sector (excluding monetary financial institutions and securities dealers) most interest payments are estimated from levels of liabilities to banks abroad (as published in the BIS international banking statistics) and appropriate interest rates. Information on interest paid by the UK non-bank private sector to the EIB is supplied by the EIB.

Earnings on deposits: foreign earnings on deposits with UK MFIs are sub-divided into earnings on deposits with banks, and earnings on deposits with building societies.

It is not possible to separate out foreign earnings on deposits with UK banks from foreign earnings on loans to UK banks. The estimates for foreign earnings on loans to UK banks are therefore included indistinguishably within earnings on deposits. Foreign earnings on deposits with UK banks consist of interest on foreign residents' deposits in sterling and foreign currencies. They include the interest paid on deposits which are the counterpart to foreign currency loans made to HM Government and, under the public sector Exchange Cover Scheme, to local authorities and other public bodies. Estimates are made from banking statistics.

Estimates of interest paid abroad on deposits with UK building societies are estimated by applying appropriate interest rates to levels outstanding.

Earnings on other liabilities: imputed income to foreign households from UK insurance companies' technical reserves is recorded in the balance of payments because households are regarded as owning the net equity of pension funds and life assurance reserves; i.e., the funds set aside for the purpose of satisfying the claims and benefits foreseen. The estimates are derived from data collected on ONS statistical inquiries.

Earnings on reserve assets (Table 4.1)

Interest received on the official foreign exchange reserves and on the UK's holdings of Special Drawing Rights with the IMF and other remuneration received from the IMF (related to its holdings of sterling), is recorded within the Exchange Equalisation Account by the Bank of England.

Current transfers (chapter 5)

Introduction

Most entries in the balance of payments accounts represent resources provided (goods and services exported or imported or the use of investments) or changes in financial assets and liabilities. Most transactions between UK residents and non-residents give rise to two such entries, which are theoretically recorded in the accounts with opposite signs. For some transactions however, only one such entry appears. Examples are a gift of goods sent abroad (which appears as a positive entry under "Trade in Goods") and a transfer to abroad of financial assets (which appears as a positive entry in the financial account). Some of the entries in this section represent the counterpart to such entries (the value of the gift of goods or of the assets transferred, with a negative sign in both the examples).

Transfers are separately identified as either current or capital. Capital transfers relate to the transfer of ownership of a fixed asset, or the forgiveness of a liability by a creditor, when no counterpart is received in return. Counterparts to the financial account entries resulting from money being brought to, or taken from, the UK by migrants are included within the *Capital account*.

Current transfers are sub-divided into those of central government and other sectors. UK's contributions to and receipts from the European Union budget are recorded on a gross basis.

Central government current transfers

Central government transfers include receipts, contributions and subscriptions from or to European Union (EU) institutions and other international bodies, bilateral aid and military grants. Information mainly comes from government departments (HM Treasury, Foreign & Commonwealth Office and Department for International Development).

Credits

These mainly comprise receipts of the UK central government from EU institutions, taxes on income, and social contributions paid by non-resident workers.

Current taxes on income and wealth: these are the receipts of the UK government from taxes on the incomes of non-resident seasonal and border workers working in the UK (the incomes themselves are recorded as compensation of employees) and withholding taxes paid abroad by UK direct investment corporations. The former are estimated on the basis of the compensation of employees information derived from the International Passenger Survey and the latter from the ONS inquiries into foreign direct investment.

Social contributions: these represent social contributions paid to the UK National Insurance Fund by non-residents.

EU institutions: these receipts comprise the VAT Abatement and other smaller, miscellaneous EU receipts. From the 1998 edition of the Pink Book, the VAT Abatement is treated as a credit entry to the UK balance of payments, rather than simply netted off VAT based contributions.

Debits

These comprise payments by the UK central government to international organisations and other non-residents.

Social Security benefits: these mainly consist of National Insurance Fund retirement and war pensions paid abroad.

European Union institutions: these payments are mainly the Central Government part of the UK contribution to the EU budget.

Other international organisations: this includes contributions to the military budget of NATO, contributions to the European Regional Development Fund and agencies of the United Nations to provide economic assistance to developing countries, and subscriptions to cover the administrative expenses of various other international bodies.

Bilateral Aid: this covers technical co-operation and non-project grants (project grants are included within capital transfers as they fund capital projects). Technical co-operation covers the provision of technical "know-how" to developing and transitional countries either as qualified manpower or as facilities for the training of nationals of these countries. It is wholly-funded by the UK Government and is included as a credit in Trade in Services. Non-project grants are cash grants to developing countries for use in financing imports and budgetary support, together with the value of goods and services provided by the UK government as food aid or disaster relief.

Military Grants: these consist of cash grants for military purposes and the value of goods and services of a military nature provided without charge to foreign countries and international organisations by the UK government.

Other sectors' transfers

Other sectors' transfers cover current taxes paid, receipts and payments to EU institutions, net non-life insurance premiums and claims, and other payments and receipts of households, including workers remittances.

Credits

Private social contributions: this consists of the actual social contributions paid by non-residents, plus the imputed contribution supplement, less the service charge. Data is sourced from ONS surveys to pension funds.

Receipts from EU institutions: comprise those in respect of the EU's Agricultural Guarantee Fund and Social Fund. They are treated as non-government transfers within the national accounts and balance of payments, as the UK government acts as an agent for the ultimate beneficiary of the transfer.

Net non-life insurance premiums: comprise the actual premiums received from non-residents plus the imputed premium supplement, less the insurance service charge. The sources for these data are the ONS surveys of insurance corporations, which collect premiums by type of insurance product, and Lloyd's of London.

Net non-life insurance claims: these are based on information supplied to the International Trade in Services survey on insurance claims received from non-resident insurance companies.

Other receipts of households: consists of three main components:

(i) Workers remittances, estimated as the savings from work of UK nationals temporarily resident in Middle East oil exporting countries, estimated from the number of UK passport holders resident in these countries, and assumed average savings per worker. These data are supplemented by information in the global transfer debits of the countries concerned.

(ii) Pension payments and other transfers (excluding immigrants assets) from OECD countries, estimated mainly from information supplied by these countries on their payments to the UK.

(iii) Similar transfers from other countries. These are estimated from published current transfer debits figures, supplemented by bilateral information on payments to the UK, supplied directly to ONS. Also included are UK receipts from voluntary aid agencies or non-profit institutions serving households (NPISH's).

Debits

Current taxes on income: these are taxes on the incomes of UK seasonal and border workers (recorded as Compensation of employees) working abroad and withholding taxes paid abroad by UK direct investment corporations. The former are estimated on the basis of the compensation of employees information derived from the International Passenger Survey and the latter from the ONS inquiries into foreign direct investment.

Private social benefits: comprise private pensions paid abroad, plus the change in net equity in pension fund reserves of non-residents. The data source is the ONS survey of pension funds.

Payments to EU institutions: these comprise agricultural and sugar levies, customs duties and VAT based contributions.

Net non-life insurance premiums: this covers premiums paid by UK companies to non-resident insurance companies collected via the International Trade in Services survey.

Net non-life insurance claims: this covers settlement of claims by UK insurance companies to non-resident claimants, which are regarded as a transfer debit. The total of claims equals the total of net premiums (service charges having been deducted), as the essential function of non-life insurance is to redistribute resources. The sources for these data are the ONS surveys of insurance corporations, and Lloyd's of London.

Other payments of households: these include a number of separate components:

(i) Cash transfers from UK households to non-residents. Data were obtained from exchange control records until 1979. Estimates for later years are based on counterpart information supplied by a number of countries on their receipts from the UK. These data are used in conjunction with historical information collected in the Family Expenditure Survey and, for recent years, the trend in UK personal disposable income.

(ii) Payments abroad by voluntary aid agencies or non-profit institutions serving households (NPISHs). These estimates are based on data supplied by the Institutions.

(iii) The estimated value of gifts sent abroad by parcel post.

Capital account (chapter 6)

The capital account comprises two components: capital transfers and the acquisition/disposal of non-produced, non-financial assets.

Capital Transfers

Capital transfers are those involving transfers of ownership of fixed assets, transfers of funds associated with the acquisition or disposal of fixed assets, and cancellation of liabilities by creditors without any counterparts being received in return. As with current transfers, they can be sub-divided into central government transfers and other sectors transfers. The main sources of information are government departments (Department for International Development and HM Treasury) and the Bank of England. Compensation payments from the EU related to the destruction of animals to combat BSE and foot and mouth disease are also included here.

Central government capital transfers

These consist of debt forgiveness and project grants (there are no receipts in recent years).

Debits

Debt forgiveness is defined as the voluntary cancellation of debt between a creditor, in this case the UK government, and a debtor in another country. Data are supplied by the Department for International Development. Project grants are cash grants to developing countries for the establishment of production and infrastructure facilities. Such transfers are distinguished from current transfers as they are conditional on the acquisition of fixed assets. Data are supplied by the Department for International Development.

Other sectors capital transfers

These include migrant's transfers, debt forgiveness and capital transfers from European Union Institutions.

Credits

Migrants' Transfers: these are recorded as being equal to the net worth of the migrants, as they arrive in the UK. Estimates are based on information on number of migrants and average assets being transferred as supplied to the International Passenger Survey. These data are supplemented by information on migrants to and from Ireland and asylum seekers, which are not covered by the IPS.

Transfers from EU Institutions: regional development fund and agricultural guidance fund receipts from the EU are considered to be capital rather than current transfers as they relate to infrastructure projects. Data are supplied by HM Treasury. Other capital transfers include agricultural compensation scheme payments relating to the destruction of animals to combat BSE and Foot and Mouth Disease.

Debits

Migrants' transfers: these represent the net worth of emigrants as they leave the UK. Estimates are based on information on the number of migrants and average assets being transferred as supplied to the International Passenger Survey. These data are supplemented by information on migrants to and from Ireland, which are not covered by the IPS.

Debt forgiveness: this consists of non-government debt forgiveness by monetary financial institutions and public corporations. Data on monetary financial institutions is supplied by the Bank of England and data on public corporations is supplied by the Export Credit Guarantee Department.

Sales/Purchases of non-produced, non-financial assets

This heading covers intangibles such as patents, copyrights, franchises, leases and other transferable contracts, goodwill etc. and transactions involving tangible assets that may be used or needed for the production of goods and services but have not themselves been produced, such as land and sub-soil assets. The use of such assets is recorded under trade in services as royalties and license fees; only the outright purchase or sale of such assets is recorded in the capital account.

The International Trade in Services (ITIS) survey has collected information on the sale and purchase of copyrights, patents and transferable contracts from 1996. Such transactions are indistinguishable from other areas of the current account for years before 1996.

Financial account (chapter 7)

Introduction

The financial account covers transactions which result in a change of ownership of financial assets and liabilities between UK residents and non-residents. The financial account is broken down into five main categories: direct investment, portfolio investment, financial derivatives, other investment and reserve assets.

In the balance of payments accounts, the term "investment" has a wide coverage. It does not refer only to the creation of physical assets but also, for example, to the purchase (or sale) of paper assets, such as shares, bonds and other securities. Investment also covers the financing of trade movements and other financial transactions between related companies in the UK and abroad. These "other financial transactions" consist mainly of borrowing and lending by banks, both transactions by UK banks with non-residents and transactions of banks abroad with UK residents. Such borrowing and lending may be associated with UK trade in goods. For example, a non-resident may borrow from a UK bank to pay a UK exporter; alternatively he may use money already on deposit with the bank. Such borrowing or use of deposits will be included in the appropriate item in the financial account offsetting the entry under trade in goods.

Banking transactions may also arise from the financing of other financial transactions. For example, a UK company may borrow from a foreign bank in order to finance investment ("direct investment") in one of its subsidiary companies abroad. In this case, both the bank borrowing and the investment would be recorded in this section of the accounts and the two entries would offset each other; the investment would increase UK assets abroad while the borrowing would increase UK liabilities to foreign residents.

The total value of assets and liabilities held at the end of each year is recorded separately under the International Investment Position (see Chapter 8) and the income earned from them is recorded under investment income within the income account (see Chapter 4). The presentations of these sections are almost identical although there are small differences in coverage in some cases, mainly because full information is not available for all items. The financial account tables appearing at Chapter 7 show net debits (UK assets) above net credits (UK liabilities), in order to allow easier read across with the investment income and international investment position tables which appear at chapters 4 and 8.

Direct investment (Table 7.3 and 7.4)

The term "direct investment" defines a group of transactions between enterprises, usually companies, that are financially and organisationally related and are situated in different countries. Such related enterprises – "affiliates" – comprise subsidiaries, associates and branches. Further details are given in the Glossary. Direct investment refers to investment that is made to add to, deduct from, or acquire, a lasting interest in an enterprise operating in an economy other than that of the investor and which gives the investor an effective voice in the management of the enterprise. Equity investment in which the investor does not have an effective voice in the management of the enterprise (i.e., the investor has less than 10 per cent of the voting shares) are regarded as portfolio investments. The estimates of direct investment include the investor's share of the reinvested earnings of the subsidiary or associated company, the net acquisition of equity capital, changes in inter-company accounts and changes in branch/head office indebtedness.

Investment abroad by UK residents

Direct investment abroad by UK residents comprises net investment by UK companies in their foreign branches, subsidiaries or associated companies. The figures of outward investment also cover the transactions of a number of concerns which were previously classified as public corporations. Transactions of central government are excluded from direct investment.

Direct investment abroad includes property transactions by both institutional investors and households. Investment in property includes the ownership of 'second homes' located outside the UK. Estimates of property ownership have been reassessed this year to take on new information from the DCLG's Survey of English Housing (SEH). The SEH collects information from English Households on the number of properties owned outside the UK. These estimates have been grossed to include all UK households.

For further information on property investment, see *Earnings on Direct Investment* under the *Investment Income* section.

Investment in the UK by foreign residents

Direct investment in the UK by foreign residents predominantly includes net investment by foreign companies in branches, subsidiaries or associated companies in the UK.

Methodological notes

The Pink Book: 2006 edition

Estimates of direct investment are mainly derived from quarterly and annual inquiries by the ONS and the Bank of England, the combined results of which are published periodically in National Statistics *First Releases* and *Business Monitor MA4*; the latter provides geographical analyses.

Copies of the *First Release* can be found at: www.statistics.gov.uk/StatBase/Product.asp?vlnk=728&Pos=1&ColRank=1&Rank=224

Copies of *Business Monitor MA4* can be found at: www.statistics.gov.uk/StatBase/Product.asp?vlnk=9614&Pos=1&ColRank=1&Rank=224

Limited information on property transactions is obtained by the Inland Revenue, the Office for National Statistics and the Bank of England. Some of this is published in *Financial Statistics*, and that relating to transactions by insurance companies and pension funds, in *Business Monitor MQ5*. From 2003 Inland Revenue ceased to collect this data, and from this point the data should be regarded as being of lower quality.

Copies of *Business Monitor MQ5* can be found at: www.statistics.gov.uk/StatBase/Product.asp?vlnk=502&Pos=&ColRank=1&Rank=256

Portfolio investment (Table 7.5 and 7.6)

Portfolio investment is sub-divided into investment in equity securities and investment in debt securities; investment in debt securities is further sub-divided into investment in bonds and notes and investment in money market instruments.

Investment abroad by UK residents

Transactions in equity securities: these represent net transactions by UK residents in shares of foreign registered companies.

Transactions in bonds and notes: transactions in bonds and notes, within debt securities, consists of net transactions in foreign government and municipal loan stock, and bonds of foreign registered companies.

Investment abroad by Lloyd's of London, including in members' premiums trust funds and overseas regulatory deposits, is included under household sector transactions in bonds and notes.

Estimates of portfolio investment transactions by UK MFIs, insurance companies and pension funds, and other financial intermediaries, are obtained from inquiries. Estimates for securities dealers' foreign investment, within other financial intermediaries, are based on integrated financial returns, with transactions aligned with changes in balance sheets. Adjustments are made to the reported data for insurance companies to remove the commission charges and other local costs included in the gross acquisitions and sales figures which are not appropriate to the financial account.

Estimates of portfolio investment transactions of private non-financial corporations are derived from asset levels at each year-end, measured in the ONS's Financial Assets and Liabilities inquiry.

Transactions in money market instruments: these consist of transactions in foreign issued commercial paper and certificates of deposit. Estimates are derived from statistical surveys undertaken by the ONS and the Bank of England.

Investment in the UK by foreign residents

Transactions in equity securities: the main source for estimates of transactions in ordinary shares is the portfolio investment inquiry run by the Bank of England. Other data are collected by ONS inquiries. Data are adjusted to take account of total levels of foreign investment in shares as indicated by the results of ONS's annual Share Register Survey.

Transactions in debt securities:

(i) *Transactions in bonds and notes:* this includes foreign net acquisitions and disposals of bonds and notes.

Foreign transactions in bonds and notes issued by HM Government are subdivided into transactions in UK foreign currency bonds and notes and transactions in other central government bonds.

Foreign transactions in British government stocks consists of net transactions by central banks, international organisations and private foreign residents in government and government guaranteed stocks. It is measured from banking statistics and other Bank of England sources. Most gilts are issued by the UK government at a discount to the redemption value.

Foreign transactions in bonds issued by local authorities and public corporations have been zero in recent years. They are measured from official records.

Total foreign transactions in bonds and notes issued by non-governmental sectors are obtained by assuming that any net transactions in UK securities not attributable to the domestic sectors of the UK (using all available data sources) are attributable to foreign residents. The further breakdown of these data by sector and instrument is derived from data provided by the Bank of England supplemented by best estimates.

(ii) *Transactions in money market instruments:* these consist of net acquisitions of UK treasury bills, certificates of deposit and commercial paper. Foreign residents' net transactions in Treasury bills exclude any bills held by the Bank of England as the sterling counterpart of foreign currency deposits arising from central bank assistance. Estimates of foreign transactions in UK certificates of deposit and commercial paper are derived from statistical inquiries conducted by ONS and the Bank of England, and from information supplied by the UK's Debt Management Office.

Financial derivatives (7.1)

Financial derivatives include options (on currencies, interest rates, commodities, indices, etc.), traded financial futures, warrants and currency and interest rate swaps. Estimates for financial derivatives are currently unavailable except for settlement receipts/payments on UK banks' interest rate swaps and forward rate agreements, which are supplied by the Bank of England. An article examining the use of derivatives in the UK accounts was published in the May 2005 edition of *Economic Trends*. It can be found at: www.statistics.gov.uk/downloads/theme_economy/ET618.pdf

Other investment (Table 7.7 and 7.8)

Other investment is sub-divided into trade credit, loans, currency and deposits, and transactions in other assets.

Investment abroad by UK residents

Trade credit: represents the extent to which the flows of payments for imports and exports follow or precede the flows of goods or services in the current account. Lending activity to facilitate trade, including those loans underwritten by the Export Credit Guarantee Department (ECGD), is treated as loans and not trade credit within the accounts *(see loans)*. Trade credit between related firms (i.e. credit received or extended between a UK business and a foreign affiliate or parent company) is treated as an investment in the affiliate or parent company, and is therefore recorded under direct investment.

At present only a minimal amount of data is recorded within trade credit. Some data previously recorded in this area has been reclassified as bank lending (see above), and is now within the loans data in other investment abroad. Other data are no longer suitable for inclusion and have been removed from the accounts, generally back to 1999.

Loans: these are sub-divided into long-term and short-term loans; short term loans are those which are repaid in full within one year. Long-term loans consist of inter-government loans by the UK central government, loans by the Commonwealth Development Corporation (CDC) (a public corporation), loans by UK banks guaranteed by the ECGD, and loans by the ECGD itself. Inter-government loans covers drawings on and repayments of loans between the UK government and foreign governments. Estimates for loans by the CDC are obtained directly from the Corporation, UK banks' loans data are supplied by the Bank of England, whilst information on loans by the ECGD is supplied direct by the Department.

Estimates for short-term loans mainly consist of loans by UK banks and other financial institutions (within "other sectors") and are derived from banking statistics.

Currency and deposits: estimates of UK residents' deposits abroad relate to private sector deposits.

Deposits abroad by UK MFIs are sub-divided into sterling and foreign currency deposits by UK banks. Some transactions in banks' foreign assets and liabilities taking place between two UK residents are also included, sometimes indistinguishably. However, these are matched by offsetting entries elsewhere in the accounts.

Estimates of MFIs' sterling deposits abroad are derived from banking statistics. Estimates for foreign currency deposits abroad have been calculated from the end-quarter balance sheets as reported by all UK banks and similar institutions to the Bank of England. Adjustments have been made to the reported changes in balance sheets to exclude revaluations resulting from changes in exchange rates.

Estimates of securities dealers' deposits abroad are derived from their asset levels reported to ONS.

Estimates of the UK private sector (excluding monetary financial institutions and securities dealers) are based on counterpart information obtained from the Bank for International Settlements (BIS). Due to limitations in the coverage of the BIS data, statistical adjustments have been applied from 1994 to improve the overall coherence of the sector financial accounts. The financial flows are estimated from changes in levels adjusted for exchange rate movements. They omit, as far as possible, the effects of any discontinuities in the levels series.

Estimates for transactions in foreign notes and coin by the UK private sector other than monetary financial institutions are based on tourists' expenditure. Transactions in non-monetary gold are included here. Net transactions in gold which is held as a financial asset by listed institutions in the London Bullion Market (LBM) are covered. These estimates are currently derived from data collected from banking statistics. The treatment of non-monetary gold is being reviewed as part of the worldwide process to revise the IMF Balance of Payments manual. Current proposals can be found on the IMF website www.imf.org/external/np/sta/bop/iss.htm. The main proposal is that the concept of non-monetary gold would be replaced by two categories – allocated gold (a commodity) and unallocated gold (a financial instrument). UK BoP will continue current practice until the treatments defined in the revised manual are implemented.

Other assets: this includes central government subscriptions to international organisations and covers capital subscriptions to international lending bodies other than the IMF, i.e. regional development banks, the International Finance Corporation and the International Fund for Agricultural Development. Some transactions are in the form of non interest-bearing promissory notes and are included in the accounts as the subscriptions fall due, irrespective of the time of encashment of the notes. The information is obtained from official records.

The entry for UK banks' and ECGD's debt forgiveness offsets the corresponding entry in the capital account. Other sectors' short-term assets largely relate to assets of UK insurance companies and pension funds and other financial intermediaries other than those classified under portfolio investment, estimates for which are obtained from ONS statistical inquiries.

Investment in the UK by foreign residents

Trade credit: only a minimal amount of data is recorded within trade credit. See outward investment notes for details.

Loans: these are sub-divided into long-term and short-term loans; the former are further sub-divided into drawings and repayments. It is not possible to separate out loans from abroad to UK banks from foreign deposits with UK banks; all such transactions are therefore assumed to be deposits.

Long-term loans consist of drawings and repayments by central government, local authorities and public corporations. Public corporations' borrowing directly from foreign residents under the exchange cover scheme is included. Repayments under the scheme by former public corporations that have since been privatised are included under repayments from central government, to whom their foreign debt was transferred following privatisation; such debt is known as novated debt. In recent years only local authorities have engaged in long-term borrowing from abroad; estimates are obtained from the Department for Communities and Local Government (DCLG). Estimates for other long-term loans are largely obtained from the Bank of England.

Estimates for central government short-term loans from abroad cover the Very Short-term Financing Facility (VSTFF), which was taken out during 1992 and repaid in 1993. Estimates for securities dealers' short-term loans from abroad are estimated from levels of liabilities reported in an ONS inquiry. Since 1995 statistical adjustments have been applied to the data for securities dealers' short-term loans in order to improve the overall coherence of the sector financial accounts.

Estimates of borrowing by UK residents other than banks are based on data reported to the Bank for International Settlements (BIS), and are generally confined to borrowing from commercial banks based within the BIS reporting area (see glossary). The data relate to levels of liabilities; flows have been estimated from changes in levels, adjusted to remove the effects of exchange rate movements and discontinuities in coverage. Due to limitations in coverage of the BIS data, statistical adjustments have been applied to the estimates since 1994 in order to improve the overall coherence of the sector financial accounts. Additional information on borrowing from the European Investment Bank (EIB) is obtained from the EIB.

Currency and deposits: these are sub-divided into transactions in sterling notes and coins, and deposits from abroad with UK monetary financial institutions including deposit liabilities of the UK central government.

Estimates of transactions in sterling notes and coin by private foreign residents (other than monetary financial institutions) are based on ONS statistics of tourists' expenditure. While sterling bank notes are issued by the Bank of England, which is classified to monetary financial institutions, coins are issued by the Royal Mint, which is classified to the central government sector. In the absence of any separate data for notes and coin, it is assumed that notes make up 90 per cent of total notes and coin.

Foreign deposits with UK monetary financial institutions are sub-divided into deposits with banks and deposits with building societies. It is not possible to separate out foreign deposits with UK banks from foreign loans to UK banks. The estimates for foreign loans to UK banks are therefore included indistinguishably within deposits.

Within deposits with UK monetary financial institutions, estimates for sterling deposits are derived from banking statistics and include both current and deposit accounts. Foreign currency deposits comprise all external borrowing denominated in foreign currencies by UK banks (sometimes described as Euro currency transactions). They consist of changes in deposits with, and other lending to, UK banks from abroad. These transactions may be a reflection of (i.e. the counterpart to) a variety of other foreign or domestic transactions by UK banks. These other transactions could be: foreign currency lending to UK residents (which are not balance of payments transactions); net purchases of foreign securities by the banks (which are included in direct or portfolio investment abroad as appropriate); any switching of banks' liabilities between foreign currencies (including gold) and sterling; or any change in the amount of foreign currency capital raised by banks.

Estimates for foreign currency deposits with UK monetary financial institutions have been calculated from the end-quarter balance sheets as reported by all UK banks and building societies to the Bank of England. Adjustments have been made to the reported changes in balance sheets to exclude revaluations resulting from changes in exchange rates.

Deposit liabilities of UK central government include short-term inter-government loans and transactions with non-residents under minor government accounts in the form of changes in balances not attributable elsewhere in the accounts. In recent years this has consisted entirely of balances held by the Paymaster General on the European Union (EU) account.

Other liabilities: these are sub-divided into long-term and short-term liabilities.

Long-term liabilities consist of net equity of foreign households in life assurance reserves and in pension funds and prepayments of premiums and reserves against outstanding claims which are recorded in the balance of payments because households are regarded as owning the net equity of pension funds and life assurance reserves; i.e., the funds set aside for the purpose of satisfying the claims and benefits foreseen. The estimates are derived from data collected on ONS statistical inquiries.

Short-term liabilities largely consists of additions to insurance companies' technical reserves, estimates for which are derived from ONS statistical inquiries, and non-interest bearing notes, estimates for which are obtained from the Bank of England. Non-interest-bearing notes are issued by HM Government and are held by international organisations.

Reserve assets (Table 7.9)

This item consists of the sterling equivalent, at current rates of exchange, of drawings on, and additions to the gold, convertible currencies and Special Drawing Rights (SDRs) held in the Exchange Equalisation Account; and of changes in the UK reserve position in the IMF. From July 1979 convertible currencies also include European Currency Units acquired from swaps with the European Monetary Co-operation Fund (until December 1993), the European Monetary Institute (until December 1997) and the European Central Bank (from 1998). The swap arrangement was terminated in December 1998.

International investment position (chapter 8)

Introduction

The international investment position brings together the available estimates of the levels of identified UK external assets (foreign assets owned by UK residents) and identified UK external liabilities (UK assets owned by foreign residents) at the end of each calendar year.

The presentation of the international investment position is almost identical to the presentation of investment income, within the income account (see Chapter 4) and the financial account (see Chapter 7) although there are small differences in coverage in some cases, mainly because full information is not available for all items.

Changes in balance sheet levels will reflect not only transactions in the corresponding assets and liabilities but also changes in valuation and certain other changes. Changes in valuation will occur in the following circumstances:

(i) where assets and liabilities are denominated in foreign currencies, their sterling value may change because of changes in foreign exchange rates;

(ii) where assets and liabilities are regularly bought and sold (e.g. British government stocks, UK and foreign company securities), the current market value may be different from the value at which they were acquired;

(iii) where the holders of assets and liabilities change their values in preparing their accounts to reflect what is thought to represent the current position (e.g. bad debts may be written off and direct investment assets may be written up or down in the books of the investing company).

In addition to changes in the valuation of identical underlying assets and liabilities, changes in recorded levels of external assets and liabilities will also reflect some changes in coverage which introduce discontinuities in the series.

Assessment of the international investment position

Because of the very varied data sources used to derive the estimates for the international investment position, there are some inconsistencies between the different figures in the tables, resulting particularly from different methods of valuation. Wherever possible, figures are at market values. However, for significant items such as direct investment, the figures are at book values and are subject to all the limitations of data taken from accounting balance sheets as a reflection of current market values. To the extent that the conventional valuation basis for direct investment is book values, or, in the case of banks, often historical cost values, an up-to-date valuation closer to market values is likely to be higher.

In addition, some assets and liabilities are measured very imperfectly (e.g. for a number of items levels of assets and liabilities are not directly reported but derived from cumulating recent identified transactions and allowing for estimated valuation changes). The balance between the estimates of identified external assets and liabilities has always been an imperfect measure of the UK's debtor/creditor position with the rest of the world.

To the extent that net errors and omissions reflect unrecorded or misrecorded financial transactions, the external balance sheet will tend to fail to capture the corresponding levels of assets and liabilities, although much will depend on the categories of assets and liabilities concerned:

(a) where both levels and transactions are reported (e.g. portfolio investment by most financial intermediaries), there may be similar deficiencies to estimates of both levels and transactions, although levels may tend to be more accurate to the extent they are derived from annual accounting data;

(b) where only levels are reported and transactions are derived from changes in levels, allowing as far as possible for valuation changes, (e.g. non-portfolio transactions of UK and foreign banks), there may be errors in the estimates of transactions (e.g. in allowing for valuation changes) with no corresponding error in levels;

(c) where only transactions are reported and levels are calculated by cumulating transactions and allowing for valuation changes, e.g. inward portfolio investment in UK company bonds, errors in recording transactions will lead to corresponding errors in levels. Thus if part of the net errors and omissions represents such missing portfolio investment inflows, the identified net assets figures will be overstated.

Allocation of Special Drawing Rights

These are issued to the UK by the IMF but are not regarded by them as a liability of the UK and do not form part of total external liabilities in this table.

Direct investment levels (Table 8.3 and 8.4)

Investment abroad by UK residents

Direct investment abroad by UK residents: this represents the stock of investment in foreign branches, subsidiaries and associates and in real estate abroad. Figures for insurance companies, other financial intermediaries and private non-financial corporations are based on ONS survey data. The annual Foreign Direct Investment Inquiry collects balance sheet information to produce estimates of the net book value of direct investment for the end of each year. The figures to 2004 are based on the annual inquiry data and the 2005 figures are a projection taking into account flows of direct investment, exchange rate changes and other projected revaluations.

The surveys relate to total net asset values attributable to investing companies, i.e. book values of fixed assets less accumulated depreciation provisions plus current assets less current liabilities. The book values of direct investments are likely to be less than the values at written down replacement cost and less than the market values. There are no official estimates of the market value of UK direct investment assets and liabilities. However, research by Cliff Pratten (Department of Applied Economics, University of Cambridge) indicated that, on certain assumptions, the market value of UK direct investments abroad at end-1989 might be about double their book value, while the market value of foreign direct investment in the UK might be just under double their book values at the same point of time. However there are considerable uncertainties in making such estimates.

The comparison between transactions in the balance of payments account and changes in total assets and liabilities is not affected by allowances for depreciation of fixed assets as charged to the profit and loss account; such allowances are deducted before arriving at the earnings included in the current account, and the provision for depreciation is regarded as maintaining the total book value of the existing assets. Similarly, the comparison is unaffected by the treatment of reinvested earnings from direct investments, since these appear both in the current account as earnings and in the financial account as a flow of capital adding to the stock of assets. However, the values are affected by the treatment applied in their consolidated accounts by UK companies to value newly acquired foreign companies. Under both merger and acquisition accounting the increase in the net book value can be less than the net investment to complete the acquisition. The difference represents goodwill and the other costs associated with the transaction that are written off directly against reserves.

Direct investment by insurance companies and the household sector include estimates of all property investments together with related foreign loans of non-bank financial institutions. For more information on household property investment, see the *Investment Income* section.

The figures for UK MFIs have been based on periodic censuses of foreign assets and liabilities carried out by the Bank of England, the latest data available is for end-2004; values for other years are estimated by similar methods to those used for other companies. From December 1998 a new annual report form was introduced for banks. The level of investment is defined as the sum of reporting institutions' investment in ordinary and preference shares, loan and working capital and other capital funds and reserves of their foreign affiliates; less certain funds raised by foreign affiliates through the issue of loan stocks and subsequently redeposited with their UK parents.

Investment in the UK by foreign residents

Direct investment in the UK by foreign residents: this represents the stock of investment by companies incorporated abroad in their UK branches, subsidiaries and associates. The estimates relate to book values and are measured in the same way as those for direct investment abroad. The latest year estimates are based on accumulated flows. Foreign direct investment in private non-financial corporations includes foreign residents' holdings of UK real estate not held through companies trading in the UK. It is estimated from the financial flows and appropriate indicators of market prices.

Portfolio investment levels (Table 8.5 and 8.6)

Portfolio investment abroad is sub-divided into equity securities and debt securities; debt securities are further sub-divided into bonds and notes and money market instruments.

Investment abroad by UK residents

Equity securities, and debt securities: bonds and notes: equity securities consists of UK residents' holdings of shares of foreign registered companies. Investment in bonds and notes consists of holdings by UK residents of foreign government and municipal loan stock and bonds of foreign registered companies.

The total is calculated using a combination of banking statistics, the results of the Bank of England's portfolio investment inquiry to banks on their customers' transactions, and information from ONS statistical inquiries to insurance companies and pension funds and other financial intermediaries including securities dealers.

Estimates for Lloyd's of London fall within household sector investment in bonds and notes; estimates are derived from data supplied by Lloyd's. Estimates of assets held by the household sector other than Lloyd's of London, together with estimates of assets held by private non-financial corporations, are derived from the quarterly Financial Assets and Liabilities survey.

Debt securities: money market instruments: this consists of holdings of foreign issued commercial paper and certificates of deposit. Estimates are derived from statistical surveys undertaken by ONS and the Bank of England.

Investment in the UK by foreign residents

Equity securities: the market value of inward portfolio investment in listed ordinary shares is based on the results of annual share ownership surveys. Adjustments are made to exclude holdings of a direct investment nature and to establish the beneficial ownership of nominee share holdings (the latest *Share Ownership* Report, covering end-2004, was published by ONS in June 2005).

Editions of the Share Ownership Report can be found at: www.statistics.gov.uk/StatBase/Product.asp?vlnk=930&Pos=1&ColRank=1&Rank=272

Debt securities:

(i) *Bonds and notes:* Levels of investment in UK foreign currency bonds and notes issued by HM Government are translated to sterling at end-year middle-market rates.

Levels of British government stocks held by foreign central banks, international organisations and private foreign residents are measured from banking statistics and other Bank of England sources including the Central Gilts Office.

Levels of inward investment in bonds and notes issued by UK MFIs and other sectors are estimated from information derived from Bank of England and London Stock Exchange records of UK company bond issues, accumulated financial transactions, and price and exchange rate movements.

(ii) *Money market instruments*: this consists of foreign holdings of UK treasury bills, commercial paper and certificates of deposit. Estimates are derived from statistical surveys undertaken by ONS and the Bank of England, and from information supplied by the UK's Debt Management Office.

Other investment levels (Table 8.7 and 8.8)

Investment abroad by UK residents

Other investment abroad by UK residents is sub-divided into trade credit, loans, currency and deposits, and other assets. For notes on trade credit, loans, and central government subscriptions to international organisations, see under "Other investment abroad by UK residents " under *Financial account* (Chapter 7).

Currency and deposits: Estimates of UK residents' holdings of foreign notes and coins and of deposits abroad relate to the private sector.

Estimates of foreign notes and coin covers estimated holdings (excluding gold coin) by UK residents. Data for MFIs are obtained from the Bank of England. Data for other sectors are derived from transactions with an allowance for exchange rate movements.

Deposits abroad by UK MFIs are derived from banking data collected by the Bank of England. Estimates of securities dealers' deposits abroad are derived from an ONS statistical inquiry.

Estimates of other UK private sector deposits with banks abroad are derived from the banking statistics of countries in the BIS reporting area (as defined in the Glossary) obtained from the Bank for International Settlements. They include the working balances of various UK companies. Due to the limitations in the coverage of the BIS data, statistical adjustments have been applied to the financial flows data since 1994 to improve the overall coherence of the sector financial accounts. In order to maintain consistency between financial flows and balance sheet levels corresponding coherence adjustments have been applied to the International Investment Position.

Other assets: Other sectors' long-term and short-term assets largely relate to assets other than bonds and shares of UK insurance companies, pension funds and other financial intermediaries. Estimates are obtained from ONS statistical inquiries.

Investment in the UK by foreign residents

Other investment in the UK by foreign residents is sub-divided into trade credit, loans, currency and deposits, and other liabilities. For notes on trade credit, and loans to central government, local authorities, public corporations and securities dealers, see under "Other investment in the UK by foreign residents" under *Financial account* (Chapter 7).

Short-term loans to the UK private sector other than MFIs and securities dealers: estimates for such loans are derived mainly from the banking statistics of countries in the BIS reporting area. Adjustments have been made to eliminate overlap with other items. The limitations in the BIS data has resulted in statistical adjustments being applied to the financial flows data from 1994 to improve the overall coherence of the sector financial accounts. In order to maintain consistency between financial flows and balance sheet levels corresponding coherence adjustments have been applied to the IIP. Borrowing from the European Investment Bank is also included.

Currency and deposits: levels of sterling notes and coin held by private foreign residents (other than MFIs) are estimated from the financial flows.

Foreign deposits with UK MFIs are sub-divided into deposits with banks and deposits with building societies. It is not possible to separate out foreign deposits with UK banks from foreign loans to UK banks. The estimates for foreign loans to UK banks are therefore included indistinguishably with deposits. Estimates are derived from banking statistics collected by the Bank of England.

Deposit liabilities of UK central government include short-term inter-government loans and transactions with non-residents under minor government accounts in the form of balances not attributable elsewhere in the accounts. Since 1973 this has consisted entirely of balances held by the Paymaster General on the European Union (EU) account.

Other liabilities: Long-term liabilities consist of net equity of foreign households in life assurance reserves and in pension funds and prepayments of premiums and reserves against outstanding claims which are recorded in the balance of payments because households are regarded as owning the net equity of pension funds and life assurance reserves; i.e., the funds set aside for the purpose of satisfying the claims and benefits foreseen. The estimates are derived from data collected on ONS statistical inquiries.

Short-term liabilities largely consist of non-interest bearing notes, estimates for which are obtained from the Bank of England. Non-interest-bearing notes are issued by HM government and are held by international organisations.

Reserve asset levels (Table 8.9)

These comprise gold, convertible foreign currencies, IMF Special Drawing Rights (SDRs) and the UK's reserve position in the IMF. Currencies may be held in the form of financial instruments. Until 1999 securities are valued at historic cost but translated to sterling as set out below. From July 1979 convertible currencies also include European Currency Units acquired when 20 per cent of the gold and dollar holdings in the reserve assets were deposited on a swap basis with the European Monetary Co-operation Fund, the swap arrangement being renewed quarterly. As from January 1994 the swap was with the European Monetary Institute and as from January 1998 was with the European Central Bank. The swap arrangement was terminated in December 1998.

Gold is valued at the ruling official price of 35 SDRs per fine ounce until end-1977 and at end-year market rates from end-1978 to end 1999. SDRs and convertible currencies (including ECUs) are valued throughout at closing middle market rates of exchange. Since 2000 all reserve assets are valued at end-period market prices and exchange rates.

External Debt (Table 8.10)

Gross external debt is defined as the outstanding amount of those actual current, and not contingent, liabilities that require payment(s) of principal and/or interest by the debtor at some point(s) in the future and that are owed to non-residents by residents of an economy.

UK External Debt data are compiled according to the IMF's External Debt Statistics: guide for compilers and users (www.imf.org/external/pubs/ft/eds/Eng/Guide/index.htm#Guide). The data are consistent with those contained in the UK's IIP statement.

End period stocks of external liabilities are classified according to institutional sector (General Government, Monetary Authorities, Banks and Other sectors); type of instrument; and original maturity of instrument. Direct investment liabilities are separately identified.

Financial derivatives (Table FD)

Financial derivatives are defined as financial instruments that are linked to the price performance of an underlying asset and which involve the trading of financial risk. Examples of the underlying asset might include a financial instrument, commodity, bilateral foreign exchange rate, movement in stock index, or interest rate. Financial derivatives include options, futures/forwards, swaps, FRAs, warrants and certain credit derivatives. The rationale for separate recording of derivatives contracts in the financial account is to keep the distinction between them and other transactions (e.g. securities) to which they may be linked for hedging purposes. Derivatives are valued at current market prices.

Data on UK Banks' gross asset and liability positions in derivatives are collected quarterly by the Bank of England; no data are available prior to 1998. Data on securities dealers' assets and liabilities are collected by ONS; similarly there are no data available prior to 1998.

Data published in table FD form supplementary information as estimates for financial derivatives have yet to be fully implemented in either the UK international investment position or in the UK's national accounts balance sheets. Work is continuing to validate and improve the estimates and obtain more information on the types of derivatives traded, and the underlying transactions.

Geographical breakdown on the current account and International Investment Position (chapters 9 and 10)

Introduction

The geographical data is broadly consistent with level 2 of Eurostat's Vade Mecum (66 individual countries, 9 geographical regions and 5 continents). Data for the European Union (EU) relate to the membership following the enlargement of 1 May 2004. EU Institutions are included in the EU aggregate and are excluded from the International Organisations total. Separate data for Belgium and Luxembourg are not available for periods before 1999. Data for China exclude Hong Kong, which is shown as an individual item.

Reliability of estimates

The United Kingdom's (UK) balance of payments accounts are primarily compiled on a global basis. Not all of the data sources used in preparing the accounts attempt to distinguish transactions on a full country basis, although the majority do. Where individual country information is not reported, estimates are made by using the geographical detail for a related category; for example, the geographical breakdown of financial assets and liabilities is used to allocate some components of investment income.

In addition to the imputation of geographical detail for some categories where the data are incomplete, there remains a margin of uncertainty regarding the accuracy of reported data by country. The finer the level of geographical detail sought the greater the likelihood of misallocation. Enterprises are encouraged to make their best estimates, when asked to report geographical data, but as country allocation may not be a crucial aspect of the information from which details are extracted, a significant degree of estimation may occur.

Given these conceptual and practical limitations, these estimates should be seen as a broad indication of the economic relationships between the UK and the rest of the world economies. They will be more reliable and meaningful in terms of broad geographical areas and major partner countries than for smaller partners. Estimates for recent years are currently more reliable than those for earlier years, since some data sources do not extend back over the whole published period.

Approach for country allocation

The following notes summarise the main criteria of country allocation adopted for the various categories of the current account. In general the figures are not likely to be consistent with those recorded by countries which allocate regional balance of payments estimates on a cash settlements basis. An analysis of UK asymmetries with its EU and US partners was published in the March 2005 edition of *Economic Trends*, which can be found at: www.statistics.gov.uk/downloads/theme_economy/ET616.pdf.

Trade in goods

Exports of goods are allocated to the country of destination; imports of goods are allocated according to the country of consignment. However, export figures from a country (A) to another country (B) may over-estimate the value of goods actually consumed in that country (B) if the importer forwards the goods on to another country (C), There are several reasons for this: 'the Rotterdam/Antwerp effect' (exports are properly attributed to the country where the port of discharge is located, following international convention, but are then re-exported to the country of final destination); other transit trade (goods passing straight through the country); and triangular trade (where goods are sold from Member state A to B and onto C, but the goods move directly from A to C.) . 'The Rotterdam/Antwerp effect' is a particular issue with the UK because of exports routed through Rotterdam in the Netherlands and Antwerp in Belgium. No information is available on the value of UK exports that are subsequently shipped on to other countries, although investigations are taking place. The principal data source for trade in goods is HMRC (see methodological chapter on Trade in Goods for more details).

Trade in services

The geographical breakdown of exports and imports of services are largely based on the existing sources of information for the global estimates, although there is some use of proxy information for some components. The change from an industry to a product based presentation with the introduction of the fifth edition of the IMF Balance of Payments Manual in 1998, and the consequent change to data collection, means that data from 1996 onwards is largely based on reported geographical breakdowns of the new products. Earlier geographical estimates are based on the industry based geographical breakdowns in the fourth edition of the IMF Balance of Payments Manual, adjusted to take the changes to the trade in services classification into account.

Sea transport: estimates relating to ships owned or chartered by UK operators are taken from inquiries carried out by the Chamber of Shipping.

Geographical breakdowns of freight services on exports and cross trades are allocated using the ports at which the goods are unloaded. For non-resident operators' freight on UK imports, the nationality of the exporting country is used as a proxy to allocate the freight payments. The resulting proportions are used to calculate the shares of non-resident operators' disbursements in the UK. Disbursements abroad by UK operators are supplied annually by the Chamber of Shipping.

Passenger revenue export estimates are derived from information supplied annually by the Chamber of Shipping. Passenger revenue import estimates are based on assumptions about the likely markets for cruises and on other information relating to the movements of UK shipping.

Air transport: passenger revenue exports are based on information supplied to ONS by the Civil Aviation Authority, which gives the required country analysis of fares paid. Other transactions with foreign airlines are allocated by nationality of airline. Receipts by UK airlines from foreign passengers are allocated to the countries in which the ticket is purchased. Freight services on UK imports earned by foreign airlines are allocated to the countries of consignment of the imports.

Other transport: rail passenger exports are based on assumptions of the likely nationality of channel tunnel users. Rail imports are allocated entirely to France. Estimates for road freight exports and imports are based on information supplied by the Road Haulage Association. This information includes details of the vehicles' load and country of destination or country of origin. Pipeline transport is based on those countries that are assumed to import / export North Sea oil and gas.

Travel: a detailed geographical split of travel expenditure, both exports and imports, are obtained from the International Passenger Survey. Allocation of expenditure of overseas visitors to the UK is by country of residence. UK residents' expenditures abroad are allocated to the country in which most time was spent, or, if this cannot be determined, the furthest country visited. As a result, expenditure in countries with appreciable numbers of transit tourists may be understated.

Other services: data for communication, construction, computer and information, royalties, other business and personal services is largely based on information supplied to the ITIS survey, supplemented with information from Royal Mail, the ONS's Film and Television inquiries and Lloyd's registry for shipping.

Insurance services: estimates are based on detailed geographical data provided by Lloyd's of London, as well as the ITIS survey for insurance imports provided to non-insurance institutions. The geographical split of trade in goods' imports is used as a proxy for freight insurance imports. Geographical splits for other insurance services are based on fixed weights.

Financial services: regular geographical information on gross flows is obtained from the Bank of England for banking services, and from the ITIS survey for financial service exports and imports from non-financial institutions. The geographical breakdown of non-bank financial corporation service exports are imputed using banking geographical data as a proxy.

Government services: for the major components, detailed geographical information on the location of those receiving or making payments is available from returns provided by the Ministry of Defence, Department for Work and Pensions, and the Foreign and Commonwealth Office. The United States Air Force has also provided data on expenditure of US Forces in the UK. Expenditure by foreign embassies and consulates in the UK is based on information supplied by some overseas embassies and statistical institutions, supplemented by information on numbers of accredited diplomats by country.

Income

Compensation of employees: estimates of the geographical breakdown of seasonal and border workers earnings are based on information supplied to the International Passenger Survey. Figures for the earnings of locally engaged staff are based on information supplied by government departments.

Investment income and International Investment Position

Direct investment income: figures are based on the annual foreign investment inquiries and include reinvested profits. Geographical information is based on the country of registration of the immediate foreign parent company and the location of the foreign affiliate, except for banks where the information relates to the country of residence of the ultimate owner (for inward investment) or the country of residence in which the direct investment enterprise is located.

Portfolio investment income: credits are the earnings accruing to UK residents from their investment in equities and debt securities, in the form of bonds and notes and money market instruments, issued by foreign institutions. Global estimates are derived from surveys of UK end-investors (banks, securities dealers, unit and investment trusts, insurance companies, pension funds and some non-financial companies).

Deriving a geographical breakdown of portfolio investment income flows has been one of the most problematic areas of Balance of Payments compilation. Portfolio investment income is particularly difficult to allocate correctly to the actual country owning or issuing the security, as the transactions are often made through financial intermediaries in a third country. However, with the launch and subsequent expansion of the IMF's Co-ordinated Portfolio Investment Survey (CPIS), an important new data source has become available. Participants in the CPIS collect a geographical breakdown of their portfolio investment assets, which are co-ordinated and disseminated by the IMF.

Data on the geographical breakdown of portfolio investment credits are derived from the UK's contribution to the CPIS exercise from 2001. (An article detailing the results of the 2001 and 2002 surveys was published in the June 2004 edition of *Economic Trends*. The results for 2004 are presented in this publication.) For banks, Bank of England information on the geographical breakdown of levels is applied to the estimates of global earnings obtained by surveys of UK banks. Similarly for non-banks, a geographical breakdown of portfolio investment income is derived from the geographical breakdown of portfolio investment assets.

Information on the geographical breakdown of UK portfolio investment debits (dividends and interest payments made to overseas residents by issuers of UK securities), are based on other countries' participation in the CPIS exercise. The IMF act as a central clearing house for the compilation of aggregate data from countries that have participated in the CPIS and disseminate the information to BoP compilers. These data can provide us with information on participating countries' holdings of UK issued equity and debt securities. For earlier years, surveys of share ownership were used to allocate portfolio holdings of UK equity securities and associated dividends by country of holder. For interest on holdings of debt securities, data derived from the CPIS exercises from 2001 onwards have been used to estimate the geographical breakdown.

Other investment income: gross interest flows between UK banks and the rest of the world are estimated by the Bank of England by allocating global interest receipts and payments in proportion to the corresponding levels of assets and liabilities of UK banks. Interest flows for UK non-bank deposits with, and borrowing from, banks in the BIS reporting area are allocated in proportion to the levels supplied by the BIS. The interest on reserve assets is estimated from official records. Figures for UK banks are used as proxies to estimate a country breakdown for the remaining components of earnings on other investment.

Adjustments applied to the global earnings on other investment to exclude the Channel Islands and the Isle of Man have been used to estimate other investment income between the UK and the offshore islands. These data are included within 'Other Europe'.

Current transfers

There are very few data sources for current transfers that allocate transactions on a country basis – these are outlined below. The geographical allocation of withholding taxes are based on the geographical allocation of inward and outward direct investment as published in *Business Monitor MA4*. The geographical allocation of insurance premiums is based on information supplied by Lloyd's of London. Data on EU transfers are provided by the Treasury, and the geographical allocation of social security and aid payments are supplied by the Department for Work and Pensions and the Department for International Development, respectively. Other geographical breakdowns are based on proxy data and global transfer estimates.

UK official transactions with institutions of the EU. (Table 9.9)

This table presents all the official transactions between the UK Government and the Institutions of the European Union. The series are the same as those shown in table 12.2 of the *Blue Book* but the presentation here reflects Balance of Payments rather than National Accounts classification of transactions.

Some of the transfers are classified to Other sectors (rather than Central government) as they are paid by or to non-government sectors; however they are still classified as official transactions because the money is collected from or paid to non-government sectors by the UK Government on behalf of the EU. The source for much of the data is HM Treasury (HMT), who are responsible for the UK's official transactions with the EU. These data represent the cash movements in and out of Government bank accounts for Transactions with the EU. Any divergences from this source – to accord with the reporting conventions required for Balance of Payments and National Accounts – are detailed below. The data sourced from HMT are also available in Table 3 of the HMT White Paper on EU Finances (www.hm-treasury.gov.uk./media/794/8B/ACF12CF.pdf).

Credits

Exports of services

This series represents the part of the collected import levies that the UK government is allowed to keep to cover the costs of collection. The percentage retained was 10 per cent up until 2000 and has been 25 per cent from 2001 onwards. It is treated as an export of a government service. The exports of services to EU Institutions series in table 9.3 differs from this one in that that series includes services provided to EU Institutions by UK private companies.

Other sectors current transfers

These largely comprise of receipts from the Agricultural Guarantee Fund and the European Social Fund. The receipts from the Agricultural Guarantee Fund are classified as subsidies and are recorded on an accruals basis based on the subsidies paid to farmers by the Rural Payments Agency.

Central government current transfers

These mainly comprise the Fontainebleau Abatement but also include a small number of miscellaneous payments to EU institutions and research councils. Since 1984 the UK's contribution to the VAT own resource has been abated in recognition of the relatively low level of its receipts, compared with its contributions to the Community Budget. Broadly, the UK receives a VAT abatement of its gross contributions equal to two-thirds of the difference between its unabated contribution and its receipts. This is deducted a year in arrears. Since the 1998 edition of the *Pink Book*, this abatement has been treated as a credit entry to the UK balance of payments rather than simply being netted off VAT based contributions.

Other sectors capital transfers

In most years these consist entirely of receipts from the Agricultural Guidance Fund and the European Regional Development Fund. Other capital transfers from EU institutions are payments to farmers under agricultural compensation schemes related to the destruction of animals during the BSE and Foot and Mouth disease outbreaks.

Debits

Other sectors current transfers

These comprise the UK's traditional own resource and third own resource contributions to the EU. The former are customs duties paid on a range of products imported from non-member states, and levies charged on the production of sugar to recover part of the costs of subsidising the export of surplus EU sugar on to the world market. EU third own resources are VAT based contributions which represent a notional extra one per cent on the VAT base, but are capped at one-half of one per cent of Gross National Income (GNI), hence the adjustment to VAT contributions. Payments of both traditional and third own resource contributions are classified as taxes paid direct to the EU. Estimates are sourced from HMRC and are converted to an accruals basis using agreed methodologies.

Central government current transfers

This mainly consists of the UK Government's fourth own resource contribution. This is calculated as a fixed percentage of UK GNI, increased or rebated according to whether within the EU budget as a whole expenditure exceeds or falls short of revenue. There are also a small number of miscellaneous payments to EU institutions under this heading.

Trade in Goods and Services additional tables formerly in UKA1 (Tables 9.10, 9.11, 9.12 and 9.13)

Tables 9.10 and 9.11 show imports and exports of services from and to selected partner countries broken down by the eleven broad categories of services for the latest two years. The details of the methods of country allocation are outlined in the trade in services section above. To avoid disclosing data on individual companies the tables have been arranged to remove these disclosive items. This is done wherever possible by suppressing the item so that non-disclosing headings are preserved. However, in some cases it has been necessary to combine headings in order to mask the disclosive data.

Table 9.12 shows the top 50 trading partners for import and exports of goods and services for the last two years for which data is available. Again the details of the methods of country allocation are outlined above in the goods and services sections.

Table 9.13 shows the UK's data for trade in services compared with world totals and those for G7 countries. The data for these are sourced from the IMF. This data is not available for the latest year as it will not have been published yet. No world balance is included as this should in theory be zero, but in practise because of asymmetries it tends to have either positive or negative values.

Further information on UK balance of payments

The following articles relating to UK balance of payments statistics have been published since the last *Pink Book*:

CPIS 2004 Data – Preliminary Results

Analysis of the UK's preliminary CPIS results 2004 including total portfolio investment assets by type of investment

Author: Ellie Turner

This paper analyses preliminary UK results for the Coordinated Portfolio Investment Survey (CPIS) 2004. Data was delivered to the International Monetary Fund (IMF) on 1 November 2005 and was published on their website in January 2006.

www.statistics.gov.uk/CCI/article.asp?ID=1303&Pos=&ColRank=1&Rank=224

Previous articles

Other published articles of interest, but not published since last *Pink Book*, are as follows

Analysis of past revisions to UK Trade statistics

The past revisions performance for UK Trade statistics explained.

Author: David Ruffles

This article presents an analysis of the past revisions performance for UK Trade statistics, looks at the statistically significant mean or average revisions seen in the figures for Total trade, identifies the main causes of these revisions, and describes what is being done to improve the first published estimates.

www.statistics.gov.uk/articles/nojournal/ukTrade.pdf

Analysis of Revisions to Quarterly Current Account Balance of Payments Data

An analysis of revisions made to Balance of Payments quarterly current account data between 1996 Q4 and 2001 Q3.

Author: Ellie Turner

This article looks at the current account, focussing on revisions to current account credits and debits and how these influence revisions to the current account balance. The article also explores the chronological evolution of revisions and revisions to current account components and provides possible justification for the more prominent revisions over the period analysed.

www.statistics.gov.uk/articles/nojournal/BoP_report2004.pdf

Current Account Asymmetries with the European Union, Annual Report 2004

A report on current account asymmetries for 2004.

Author: Libby Cox

Current account asymmetries occur when one country's data does not correspond to the same data for the same transaction reported by its partner countries. This report analyses asymmetries between the UK current account and the rest of the European Union. Additional data is presented regarding asymmetries between the UK and the US.

www.statistics.gov.uk/articles/economic_trends/ET616Cox.pdf

Report on impact of MTIC on UK Trade statistics

Report on further research into the impact of Missing Trader Fraud on UK Trade Statistics, Balance of Payments and National Accounts.

Authors: David Ruffles, Tricia Williams (HM Revenue & Customs)

This article is a follow-up to the article published in the August 2003 edition of *Economic Trends* which is available on the National Statistics website. It summarises the work carried out since July 2003 to review the estimates of the impact of Missing Trader Intra-Community (MTIC) VAT Fraud on UK Trade Statistics, Balance of Payments and National Accounts, and to investigate potential methods of estimating acquisition fraud.

www.statistics.gov.uk/articles/nojournal/Further_Missing_Trader_Fraud_Research.pdf

Following a change in the pattern of trading associated with Missing Trader Intra-Community (MTIC) fraud, identified by HMRC, interpretation of the breakdown between EU and non-EU trade is more difficult. HMRC have set up a project to review the methodology for producing the estimates of the impact on the trade statistics. An interim progress report was published by HMRC on 16 January 2006 at www.uktradeinfo.com/index.cfm?task=news&id=384&startrow=1).

Methodological improvements to UK foreign property investment statistics

New methodology to measure ownership of foreign property by UK households and estimates the value of property owned at end 2003 to be £23 billion.

Author: Deborah Nicole Aspden

This article presents new methodology to measure ownership of foreign property by UK households. It is based on the Office of the Deputy Prime Minister's (ODPM) Survey of English Housing (SEH). The new methodology estimates the value of foreign property ownership in 2003/04 to be just above £23 billion – more than double the estimate for 1999/00. Investment is highest in Europe, with Spain and France being the preferred locations for investment.

www.statistics.gov.uk/articles/economic_trends/ET619_Aspden.pdf

Financial Derivatives in the UK Sector Balance Sheets and Financial Accounts

Although the availability and quality of data on financial derivatives has improved, a number of methodology and coverage issues remain outstanding.

Author: Graham Semken

This article re-assesses the area of derivatives statistics following expanded data availability. It examines a number of issues, both conceptual and practical, which will need to be resolved before the collective data on derivatives can be integrated into the UK accounts.

www.statistics.gov.uk/articles/economic_trends/ET618Sem.pdf.

Other articles

Older articles which may be of interest, published in *Economic Trends*, include:

"Overseas trade in services: publication of monthly estimates", September 1997;

"Geographical breakdown of exports and imports of UK trade in services by component", January 1998;

"Geographical breakdown of income in the balance of payments", November 1999 and December 2000.

www.statistics.gov.uk/cci/article.asp?id=44

www.statistics.gov.uk/cci/article.asp?id=61

"IMF Co-ordinated Portfolio Investment Survey", May 2003.

www.statistics.gov.uk/articles/economic_trends/ET_May03_Humphries.pdf

"Geographical breakdown of the UK International Investment Position", June 2004.

www.statistics.gov.uk/articles/economic_trends/ET607Humphries.pdf

Glossary

Acceptances
See Bills and acceptances.

Accrued interest
A method of recording transactions to relate them to the period when the exchange of ownership of the goods, services or financial asset applies. For example, value added tax accrues when the expenditure to which it relates takes place, but Customs and Excise receive the cash some time later. The difference between accruals and cash results in the creation of an asset and liability in the financial accounts, shown as amounts receivable or payable.

Advance and progress payments
Payments made for goods in advance of completion and delivery of the goods.

Affiliates
Branches, subsidiaries or associate companies.

Allocation of SDRs
See Special Drawing Rights.

Arbitrage
Buying in a market in one centre and selling in a similar market in another centre, in order to exploit a temporary misalignment of prices at little or no risk.

Assets
This term commonly refers to financial assets that are claims on non-residents, from whose point of view the same item is a liability to a UK resident. Among reserve assets, however, gold and SDRs have a value which exists independently of any corresponding liabilities. Real assets such as merchandise, although they may be entered in company accounts as assets, are seldom described as assets in balance of payments analysis.

Associated companies
Companies in which the investing company has a substantial equity interest (usually this means that it holds between 10 per cent and 50 per cent of the equity share capital) and is in a position to exercise a significant influence on the company. (See Subsidiary.)

Balancing item
See Net Errors and Omissions.

Bank of England – Issue Department
This part of the Bank of England deals with the issue of bank notes on behalf of central government. It was formerly classified to central government though it is now part of the central bank/monetary authorities sector. Its activities include, inter alia, market purchases of commercial bills from UK banks.

Bank for International Settlements (BIS)
An international institution based in Basle, Switzerland, established in 1930. Its main functions today are to promote international monetary co-operation; to observe the work of the IMF, Finance Ministers and Central Bank Governors of the Group of Ten countries; and to provide monetary research. The most recent BIS data used within the UK balance of payments accounts covers non-bank borrowing from banks in the following countries: Australia, Austria, the Bahamas, Bahrain, Belgium/Luxembourg, Bermuda, Brazil, Canada, Cayman Islands, Chile, Denmark, Finland, France, Germany, Greece, Guernsey, Hong Kong SAR, India, Ireland, Isle of Man, Italy, Japan, Jersey, Mexico, Netherlands, Netherlands Antilles, Norway, Panama, Portugal, Singapore, Spain, Sweden, Switzerland, Taiwan, Turkey and United States of America. The data used for balance of payments purposes are locational banking statistics on a residence basis.

Banking statistics
A term used in this publication to denote an integrated set of returns, covering all UK banks, and collected by the Bank of England. The returns were first introduced in late 1974 and during 1975. Since then, various reviews of the requirements of data from banks have been conducted and forms amended, introduced or dropped as necessary. The data collected covers all listed banks up to the end of 1981 and the revised group of institutions classified as UK banks from 1982 onwards. It collects on a regular basis extensive information relating to the levels of, and changes in, assets and liabilities. Revised banking returns were introduced from the end of 1997 to reflect the requirements of the IMF Balance of Payments Manual 5th edition and to remove the Channel Islands and the Isle of Man from the definition of the economic territory of the United Kingdom.

Banks (UK)
Banks are defined as all financial institutions recognised by the Bank of England as UK banks. For statistical purposes, this includes:

- institutions which have a permission under Part 4 of the Financial Services and Markets Act 2000 (FSMA) to accept deposits, other than (i) credit unions, (ii) firms which have a permission to accept deposits only in the course of carrying out contracts of insurance in accordance with that permission, (iii) friendly societies, and (iv) building societies;
- European Economic Area credit institutions with a permission under Schedule 3 to FSMA to accept deposits through a UK branch; and
- the Banking and Issue Departments of the Bank of England (the latter from April 1998).

Prior to December 2001, banks were defined as all financial institutions recognised by the Bank of England as UK banks for statistical purposes, including the UK offices of institutions authorised under the Banking Act 1987, the Banking and Issue Departments of the Bank of England (the latter from April 1988), and deposit-taking UK branches of 'European Authorised Institutions'. This includes UK branches of foreign banks, but not the offices abroad of these or of any British owned banks.

An updated list of banks appears regularly in the Bank of England's *'Monetary and Financial Statistics'* publication. The most recent list can also be found on the Financial Services Authority website at: www.fsa.gov.uk/pubs/list_banks/2006/lob_jan06.pdf.

Bills and acceptances
A bill is an unconditional order in writing addressed by the drawer to the drawee to pay to the drawer a fixed sum on a specified date. A UK resident may draw a bill in Sterling on a foreign resident representing credit extended by the UK resident to the foreign resident. If the UK resident sells the bill to a UK bank, generally at a price less than the nominal value of the bill, the bank is said to discount the bill, and the claim on the foreign resident is transferred to the UK bank.

A bill is known as an acceptance when the drawee accepts the bill. A UK bank may accept a bill on behalf of a foreign resident in which case the UK resident draws the bill on the UK bank and not on the foreign resident. The accepting bank has a claim on the foreign resident and expects to be paid by him before the bill matures.

Bond
A financial instrument that usually pays interest to the holder. Bonds are issued by governments as well as companies and other institutions, e.g. local authorities. Most bonds have a fixed date on which the borrower will repay the holder. Bonds are attractive to investors since they can be bought and sold easily in a secondary market. Special forms of bonds include deep discount bonds, equity warrant bonds, Eurobonds, and zero coupon bonds.

BPM5
The Balance of Payments Manual, 5th Edition, published in 1993 by the IMF.

Branch
An unincorporated enterprise, wholly or jointly owned by a direct investor.

Branch indebtedness
Net amounts owed by a branch to its head office (or vice versa).

British government stocks
Securities issued or guaranteed by the UK government. Also known as gilts.

Building societies

Building societies are mutual institutions specialising in accepting deposits from members of the public and in long-term lending to members of the public, mainly to finance the purchase of dwellings; such lending being secured on dwellings. Their operations are governed by special legislation which places restrictions on their recourse to other sources of funding and other avenues of investment.

Capital account

The capital account consists of capital transfers (see Transfers) and acquisition/disposal of non-produced, non-financial assets (see separate entry in glossary).

Capital transfers

See Transfers.

Certificate of deposit

A short term interest-paying instrument issued by deposit-taking institutions in return for money deposited for a fixed period. Interest is earned at a given rate. The instrument can be used as security for a loan if the depositor requires money before the repayment date.

c.i.f. (cost, insurance and freight)

The basis of valuation of imports for Customs purposes, it includes the cost of insurance premiums and freight services. These need to be deducted to obtain the free on board valuation consistent with the valuation of exports which is used in the economic accounts.

Collective Investment Institution (CII)

Incorporated (investment companies or investment trusts) and unincorporated undertakings (mutual funds or unit trusts) that invest the funds, collected from investors by means of issuing shares/units (other than equity), in financial assets (mainly marketable securities and bank deposits) and real estate. (See also Trusts).

Commercial paper

This is an unsecured promissory note for a specific amount and maturing on a specific date. The commercial paper market allows companies to issue short term debt direct to financial institutions who then market this paper to investors or use it for their own investment purposes.

Commodity gold

See Gold.

Commonwealth Development Corporation

A public corporation which finances development projects abroad.

Compensation of employees

Total remuneration payable to employees in cash or in kind. Includes the value of social contributions payable by the employer.

Coordinated Portfolio Investment Survey (CPIS)

A survey coordinated and disseminated by the IMF. Participants in the CPIS collect a geographical breakdown of their portfolio investment assets.

Counterpart items

Certain items in the balance of payments exist only as counterpart items, introduced to balance the inclusion of other items that do not fall naturally into the double-entry system. The allocation of SDRs is an example of an artificial counterpart item introduced into the balance of payments to offset the corresponding increase in SDR holdings within official reserves (as SDRs are no one sector's liabilities). (For SDRs see Special Drawing Rights).

Cross-trades

See Merchanting.

Currency swaps

A currency swap, also known as a cross-currency interest-rate swap contract, consists of an exchange of cash flows related to interest payments and, at the end of the contract, an exchange of principal amounts in specified currencies at a specified exchange rate.

Current account

The account of transactions in respect of trade in goods and services, income and current transfers.

Current balance

The balance of current account transactions.

Current transfers

See Transfers.

Debt forgiveness

The voluntary cancellation of all or part of a debt within a contractual arrangement between a creditor in one country and a debtor in another country.

Debt securities

Debt securities cover bonds, debentures, notes etc., and money market instruments. These are split into long and short (up to one year) term, based on original maturity.

Derivatives

See Financial Derivatives.

Direct investment

Net investment by UK/foreign companies in their foreign/UK branches, subsidiaries or associated companies. A direct investment in a company means that the investor has a significant influence on the operations of the company, defined as having an equity interest in an enterprise resident in another country of 10 per cent or more of the ordinary shares or voting stock. (See Branch indebtedness, Subsidiary and Associated companies.) Investment covers not only acquisition of fixed assets, stock building and stock appreciation, but also all other financial transactions, such as: additions to or payments of working capital; other loans and trade credit; and acquisitions of securities. Estimates of investment flows allow for depreciation in any undistributed profits. Funds raised by the subsidiary or associate company in the economy in which it operates are excluded as they are locally raised and not sourced from the parent company.

Disbursements

Operating expenses, e.g. by operators of ships or aircraft.

Dividend

A payment made to company shareholders from current or previously retained profits. Dividends are recorded when they become payable.

Equity

Equity is ownership or potential ownership of a company. Equities differ from other financial instruments in that they confer ownership of something more than a financial claim. Shareholders are owners of the company whereas bond holders are outside creditors.

Equity securities

Equity securities are shares issued by companies to shareholders. Purchases of equity securities in which the purchaser does not have any significant degree of control over the company (i.e., less than 10 per cent of the equity capital) fall within portfolio investment; otherwise it falls within direct investment. Equity securities include mutual fund shares.

Euro area

The euro area encompasses those Member States of the European Union in which the euro has been adopted as the single currency and in which a single monetary policy is conducted under the responsibility of the decision-making bodies of the European Central Bank. The euro area currently comprises Belgium, Germany, Greece, Spain, France, Ireland, Italy, Luxembourg, the Netherlands, Austria, Portugal and Finland.

Eurocurrency market

All borrowing and lending by banks in currencies other than that of the country in which the banks are situated.

Euro/European Currency Unit (ECU)

The ECU was officially introduced in 1979 in connection with the start of the European Monetary System (EMS). In the EMS, the ECU served as the basis for determining exchange rate parities and as a reserve asset and means of settlement. It was a composite currency which contained specified amounts of the currencies of the member states of the European Union. The currencies making up the ECU were weighted according to their economic importance and use in short-term finance. As from September 1989 the weightings of the ECU were revised to include both the Spanish peseta and Portuguese escudo. The ECU was converted into the Euro at the start of European Monetary Union on 1 January 1999, with Greece joining on 1 January 2001. From 1 January 2003, the Euro became the currency of the member states of the European Monetary Union.

European Central Bank (ECB)

The Monetary Authority for the Euro currency, based in Frankfurt. The ECB, together with the national central banks of the member states, manages monetary policy and the banking system across the European Monetary Union area.

European Investment Bank (EIB)

This was set up to assist economic development within the European Union. Its members are the member states of the EU.

European Monetary System (EMS)

The EMS was established in March 1979. Its most important element was the mechanism known as the ERM (Exchange Rate Mechanism) whereby the exchange rates between the currencies of the participating member states were kept within set ranges. The UK joined the ERM on 8 October 1990. On 16 September 1992 the UK's membership of the ERM and the EMS was suspended. The EMS was superceded by the single currency when eleven of the participating member states joined European Monetary Union on 1 January 1999, with Greece joining on 1 January 2001.

Eurosystem

The Eurosystem comprises the European Central Bank (ECB) and the national central banks of the Member States which have adopted the euro in Stage Three of Economic and Monetary Union (EMU). There are currently 12 national central banks in the Eurosystem. The Eurosystem is governed by the Governing Council and the Executive Board of the ECB and has assumed the task of conducting the single monetary policy for the euro area since 1 January 1999. Its primary objective is to maintain price stability.

Exchange control

A legal control imposed by governments on the ability of persons, businesses and others to hold, receive and transfer foreign currency. The extent of the Exchange Control Act of 1947 was considerably reduced in June and July 1979 and the Act was repealed in 1987.

Exchange cover scheme (ECS)

A scheme first introduced in 1969 whereby UK public bodies raise foreign currency from abroad, either directly or through UK banks, and generally surrender it to the EEA (see below) in exchange for sterling for use to finance expenditure in the United Kingdom. HM Treasury sells the borrower foreign currency to service and repay the loan at the exchange rate that applied when the loan was taken out. The transactions relate to net borrowing by British Nuclear Fuels plc and repayment by HM Government following the privatisation of other former public corporations (see Novations).

Exchange Equalisation Account (EEA)

The government account with the Bank of England in which transactions in reserve assets are recorded. These transactions are classified to the central government sector. It is the means by which the government, through the Bank of England, influences exchange rates.

Export credit

Credit extended abroad by UK institutions, primarily in connection with UK exports but also including some credit in respect of third-country trade.

Export credit; identified long-term

Credit extended by UK banks under the ECGD's buyer credit and specific bank guarantees schemes.

Export Credits Guarantee Department (ECGD)

A government department, classified to the public corporations sector, the main function of which is to provide insurance cover for export credit transactions.

External debt

A measure of balance sheet liabilities owing to non-residents. Liabilities relating to trade credit, debt securities, and loans and deposits (including inter-company liabilities within direct investment) are included; equity liabilities are excluded.

Financial account

The financial account records transactions in external assets and liabilities of the UK, e.g., the acquisitions and disposals of foreign shares by UK residents. The financial account consists of direct investment, portfolio investment, other investment, financial derivatives and reserve assets.

Financial auxiliaries

Auxiliary financial activities are ones closely related to financial intermediation but which are not financial intermediation themselves, such as the repackaging of funds, insurance broking and fund management. Financial auxiliaries therefore include insurance brokers and fund managers.

Financial corporations

All bodies recognised as independent legal entities whose principal activity is financial intermediation and/or the production of auxiliary financial services. However, the United Kingdom currently treats financial auxiliaries as non-financial corporations.

Financial derivatives

Any financial instrument the price of which is based upon the value of an underlying asset (typically another financial asset). Financial derivatives include options (on currencies, interest rates, commodities, indices, etc.), traded financial futures, warrants, and currency and interest swaps. Under BPM5, transactions in derivatives are treated as separate transactions, rather than being included as integral parts of underlying transactions to which they may be linked as hedges. Only estimates for settlement receipts/payments on UK banks' interest rate swaps and forward rate agreements are currently included.

Financial gold

See Gold.

Financial Leasing

See Leasing.

Financial surplus or deficit (FSD)

The former term for Net lending(+)/Net borrowing(-), the balance of all current and capital account transactions for an institutional sector or the economy as a whole.

f.o.b. (free on board)

A f.o.b. price excludes the cost of insurance and freight from the country of consignment but includes all charges up to the point where the goods are deposited on board the exporting/importing vessel or aircraft. Trade in goods exports are valued on a f.o.b. basis in the balance of payments accounts.

Foreign

In this publication "foreign" denotes residence outside the United Kingdom rather than nationality. In some contexts "external", "abroad" or "non-resident" are used with the same meaning. (See Residency).

Other Financial Intermediaries (OFIs)

A diverse group of units constituting all financial corporations other than depository corporations, insurance corporations, pension funds, and financial auxiliaries. They generally raise funds by accepting long-term or specialised types of deposits and by issuing securities and equity. These intermediaries often specialise in lending to particular types of borrowers and in using specialised financial arrangements such as financial leasing, securitised lending, and financial derivative operations.

Forwards

In a forward contract, the counterparties agree to exchange, on a specified date, a specified quantity of an underlying item (real or financial) at an agreed-upon contract price (the strike price). If a future exchange of currencies is carried out in a forward contract, the counterparties exchange, in accordance with prearranged terms, cash flows based on the reference prices of the underlying items. Forward rate agreements and forward foreign exchange contracts are common types of forward contracts.

Futures

Futures are forward contracts traded on organised exchanges. They give the holder the right to purchase a commodity or a financial asset at a future date.

Gilts

Bonds issued or guaranteed by the UK government. Also known as gilt-edged securities or British government securities.

Gold

In the accounts a distinction is drawn between gold held as a financial asset (financial gold) and gold held like any other commodity (commodity gold). Transactions in commodity gold are recorded in the trade in goods account and include foreign trade in finished manufactures together with net domestic and foreign transactions in gold moving into or out of finished manufactured form (i.e. for jewellery, dentistry, electronic goods, medals and proof – but not bullion – coins).

All other transactions in gold (i.e. those involving semi-manufactures such as rods, wire, etc., or bullion, bullion coins or banking-type assets and liabilities denominated in gold, including reserve assets) are treated as financial gold transactions and included in the financial account. The distinction between commodity and financial gold differs from that drawn by the IMF, in its Balance of Payments Manual (5th edition, 1993), between non-monetary and monetary gold. The United Kingdom has obtained an exemption from adopting the BPM5 recommendations on treatment of gold in order to avoid distortion of its trade in goods account by the substantial transactions of the London Bullion Market.

The treatment of non-monetary gold is being reviewed as part of the worldwide process to revise the IMF Balance of Payments manual. Current proposals can be found on the IMF website www.imf.org/external/np/sta/bop/iss.htm. The main proposal is that the concept of non-monetary gold would be replaced by two categories – allocated gold (a commodity) and unallocated gold (a financial instrument). UK BoP will continue current practice until the treatments defined in the revised manual are implemented.

Gross

The separate identification of both credit/debit, export/import for any particular transaction.

Hedging

Hedging is accomplished by the temporary purchase or sale of futures/swaps contracts to offset the position or anticipated position in the cash markets. This may benefit banks, financial institutions, pension funds and corporate treasuries who hold interest rate, exchange rate or stock price sensitive assets or liabilities.

Holding companies

A holding company is a company that usually confines its activities to owning stock in and supervising management of other companies. A holding company usually owns a controlling interest in the companies whose stock it holds. Holding companies exist for legal, commercial and tax reasons. In line with international standards, holding companies are classified as other financial intermediaries.

Households

Individuals or small groups of individuals as consumers and in some cases as entrepreneurs producing goods and market services.

Import credit: long-term agreements

Credit received on imported ships, commercial aircraft and certain North Sea installations.

Income

The income account forms part of the current account and consists of compensation of employees and investment income, both of which have separate entries in this glossary.

Inter-company accounts

Accounts recording transactions between parent and subsidiary or associated companies, and balances owed by one to the other.

Interest rate swaps

An obligation between two parties to exchange interest-related payments in the same currency from fixed rate into floating rate, or vice versa, or from one type of floating rate to another. A swap can be used to reshape the coupon payments of either new or existing debt. The only movement of funds is a net transfer of interest payments between the two parties. The interest payments are calculated on an agreed principal amount, which is not exchanged. The settlement receipts/payments on UK banks' interest rate swaps appear in the financial account under financial derivatives.

International Investment Position (IIP)

The international investment position records end of period balance sheet levels of UK external assets and liabilities. The IIP consists of direct investment, portfolio investment, other investment and reserve assets. Financial derivatives are not currently included in the IIP, but presented separately in table FD.

International Monetary Fund (IMF)

A Fund set up as a result of the Bretton Woods Conference of 1944 and which began operations in 1947. It includes most of the major countries of the world. The Fund was set up to supervise the fixed exchange rate system agreed at Bretton Woods and to make available to its members a pool of foreign exchange resources to assist them when they have balance of payments difficulties. Further definitions relating to the IMF are given in the IMF section in the 1981 and earlier editions of this publication. (See also Special Drawing Rights).

Intervention Board for Agricultural Produce (IBAP)

The UK government agency which used to operate the support arrangements of the EU Common Agricultural Policy within the United Kingdom. It has now been replaced by the Rural Payments Agency (RPA).

Investment

In a balance of payments context this is categorised as either direct, portfolio or other investment. See appropriate headings for definitions.

Investment income

All investment income accruing to UK residents from non-residents or payable abroad by UK residents after allowing for depreciation. The balance on credits and debits equals "net property income from abroad" as shown in the National Accounts.

Investment trust

See Trusts.

Leasing

In the balance of payments accounts all financial leases and some long-term operating leases (e.g. for aircraft) are regarded as loans to finance the purchase of goods. The lessor thus makes a loan to the lessee who subsequently repays this with interest. The lessee is regarded as the purchaser of the goods.

Liabilities

In balance of payments terminology, liabilities are the financial claims of non-residents on the UK.

LIBOR

London Interbank Offered Rate. The rate of interest at which banks borrow funds from other banks, in marketable size, in the London Interbank market.

Local authorities

Elected councils responsible for the administration of certain services in particular areas within the United Kingdom.

Merchanting

Trade between two countries other than the United Kingdom, in which the United Kingdom may participate as an intermediary or by providing transport, insurance services or credit facilities.

Miscellaneous financial institutions

These include certain institutions not classified as UK banks whose main function is to extend credit abroad, and certain listed institutions in the London Bullion Market which are not UK banks.

Monetary Authorities

Institutions (usually central banks) which control the centralised monetary reserves and the supply of currency in accordance with government policies, and which act as their governments' bankers and agents. In the United Kingdom this is equivalent to the Bank of England and part of the Treasury (the Exchange Equalisation Account). Data is not separately available in the UK accounts for monetary authorities.

Monetary financial institutions

Banks and building societies.

Monetary gold

See Gold.

Money market

The market in which short-term loans are made and short-term securities traded. 'Short term' usually applies to periods up to one year but can be longer in some instances.

Money market instruments

Money market instruments, within portfolio investment, generally give the holder the unconditional right to receive a stated, fixed sum of money on a specified date. These are short-term instruments usually traded at a discount; the discount being dependent upon the interest rate and the time remaining to maturity. Included are such instruments as acceptances, treasury bills, commercial paper and certificates of deposit.

MTIC

VAT missing trader intra-Community fraud. A systematic, criminal attack on the VAT system, which has been detected in many EU Member States. In essence, fraudsters obtain VAT registration to acquire goods VAT free from other Member States. They then sell on the goods at VAT inclusive prices and disappear without paying over the VAT paid by their customers to the tax authorities.

Navy, Army and Air Force Institute (NAAFI)

A body which provides goods and services for use by the UK armed forces abroad.

Net

In this presentation of the balance of payments accounts, the term "net" is generally applied only to transactions in financial assets or liabilities. Purchases of assets are recorded net of sales; similarly with liabilities. In the current and capital accounts, where the operations of UK and foreign residents are taken together in particular transactions areas, the term "balance" is used.

Net Errors and Omissions

The item included to bring the sum of all balance of payments entries to zero. Also known as the balancing item.

Non-monetary gold

See Gold.

Non-produced, non-financial assets

Non-produced, non-financial assets, within the capital account, include land purchased or sold by a foreign embassy, patents, copyrights, trade marks, franchises and leases and other transferable contracts, but not finance leasing. Only the purchase and sale of such assets are proper to the capital account; earnings from them are recorded under trade in services.

Novations

This term defines the reassignment of debt (for balance of payments, usually foreign debt) of public corporations to central government following the privatisation of the public corporation. This does not normally change the overall balance of payments situation as the debt is still regarded as a UK liability.

NPISH

Non-profit institutions serving households.

Official reserves

See Reserve assets.

Offshores

The economic territory of a country consists of the geographic territory administered by a government; within this territory, persons, goods, and capital circulate freely. In the context of the UK, the offshore islands of the Channel Islands and the Isle of Man are subject to their own fiscal authorities and have their own tax systems, there are impediments to taking up residency, and they are not part of the EU. They are therefore not recognsied as part of the economic territory of the UK for BOP purposes and are classified as non-resident in the UK.

Operating leasing

Operational leasing (rental) covers resident/non-resident leasing (other than financial leasing), charter of ships, aircraft and transportation equipment without crew. Leasing of ships, aircraft and transportation equipment with crew are included in the transportation account.

Ordinary share

The most common type of share in the ownership of a corporation. Holders of ordinary shares receive dividends. (See also Equity).

Other Investment

Investment other than direct and portfolio investment. Includes trade credit, loans, currency and deposits and other assets and liabilities.

Parent

In a balance of payments context this means a company with direct investments in other countries.

Pension funds

The institutions that administer pension schemes. Pension schemes are significant investors in securities. Self-administered funds are classified in the financial accounts as pension funds. Those managed by insurance companies are treated as long-term business of insurance companies. They are part of S.125, the Insurance corporations and pension funds sub-sector.

Portfolio investment

Investment in equity and debt securities issued by foreign registered companies, other than that classed as direct investment, and in equity and debt securities issued by foreign governments. A portfolio investment, unlike a direct investment, does not entitle the investor to any significant influence over the operations of the company or institution, and represents less than 10 per cent of the equity capital.

Preference share

This type of share guarantees its holder a prior claim on dividends. The dividend paid to preference share holders is normally more than that paid to holders of ordinary shares. Preference shares may give the holder a right to a share in the ownership of the company (participating preference shares). However in the UK they usually do not, and are therefore classified as bonds.

Private sector

Private non-financial corporations, financial corporations other than the Bank of England (and Girobank when it was publicly owned), households and the NPISH sector.

Promissory note

A security which entitles the bearer to receive cash. These may be issued by companies or other institutions. (See Commercial paper).

Public corporations

These are public trading bodies which usually have a substantial degree of financial independence from the public authority which created them. A body is normally treated as a trading body when more than half its income is financed by fees. A public corporation is publicly controlled to the extent that the public authorities appoint a majority of the board of management or when public authorities can exert significant control over general coprorate policy through other means. Since the 1980s many public corporations, such as British Telecom, have been privatised and reclassified within the accounts as private non-financial corporations.

Public sector

Central government, local authorities and public corporations.

Refinanced export credit

Identified long-term credit extended for UK exports initially by banks and refinanced with the ECGD, the Trustee Savings Banks and the Central Trustee Savings Bank.

Reinvested earnings

The direct investor's share of earnings not distributed as dividends (by subsidiaries) or branch profits. As this income remains with the foreign subsidiary or branch (it is reinvested by the parent) an amount will appear in the financial account equal to (and with opposite sign) the corresponding entry within direct investment income.

Related companies

Branches, subsidiaries, associates or parents.

Related import or export credit

Trade credit between related companies included in direct investment.

Repo

This is short for "sale and repurchase agreement". One party agrees to sell bonds or other financial instruments to other parties under a formal legal agreement to repurchase them at some point in the future – usually up to six months – at a fixed price. Repo transactions are treated as borrowing/lending within other investment, rather than as transactions in the underlying securities.

Reserve assets

Short term assets which can be very quickly converted into cash. They comprise the UK's official holdings of gold, convertible currencies, Special Drawing Rights, and changes in the UK reserve position in the IMF. Also included between July 1979 and December 1998 are European Currency Units acquired from swaps with the European Cooperation Fund, EMI and the ECB. Reserve assets were referred to as "official reserves" in editions of the Pink Book prior to 1998.

Reserve position in the Fund

The United Kingdom's position in the IMF's General Resources Account. This position is the sum of the UK's reserve tranche purchases, and any indebtedness of the Fund (under a loan agreement) that is readily payable to the UK.

Residency

UK residents are those with a centre of economic interest within the UK of at least one year's duration – nationality does not play a part in determining residency status. There are a number of exceptions to the standard residency classification: UK embassies and military bases abroad are deemed to be residents of the UK (conversely other nations' embassies and military bases in the UK are classed as non-residents), as are students studying abroad or patients being treated abroad who are normally resident in the UK. (See also Offshores).

Royalties

These form part of trade in services. They represent payments for services by, or to, UK residents in respect of the right to use processes and other information, e.g. licences to use patents, trade marks, designs, copyrights, etc. Sales and purchases of patents are included within the capital account.

Rural Payments Agency (RPA)

The UK government agency which operates the support arrangements of the EU Common Agricultural Policy within the United Kingdom. This replaced the Intervention Board for Agricultural Produce (IBAP).

Securities dealers

Securities and futures dealers are those institutions whose main activity is dealing in securities and futures either on their own account or on behalf of customers and clients. This activity also includes Stock Exchange money brokers, Inter-dealer brokers and dealing in commodities for investment purposes. They should not be confused with monetary financial institutions (Banks and Building Societies) who are licenced as able to take deposits.

Security

Security against loans involves the depositing of a document or asset which is retained by the bank as a charge for an advance. This form of security may include stocks and share certificates, debentures, and insurance policies.

Smuggling

Smuggling is the importation of goods acquired duty free or duty paid in another country for re-sale in the UK without payment of UK duty and (where appropriate) VAT. (See also MTIC).

Special Drawing Rights (SDRs)

These are reserve assets created and distributed by decision of the members of the IMF. Participants accept an obligation to provide convertible currency, when designated by the IMF to do so, to another participant, in exchange for SDRs equivalent to three times their own allocation. Only countries with a sufficiently strong balance of payments are so designated by the IMF. SDRs may also be used in certain direct payments between participants in the scheme and for payments of various kinds to the IMF.

Spread earnings

Net spread earnings are the part of market making activities that represent payment for the performance of a service. The value of the spread earning for each transaction is calculated as the margin earned between the transaction price and the mid-market price at the time of the transaction. This represents the 'added value' gained from market making activities. Spread earnings can be made on, for example, foreign exchange, securities and derivatives transactions.

Subsidiary

A registered company in which another registered company has ownership of the majority of the voting share capital, i.e. greater than 50 per cent.

Subsidies

Current unrequited payments made by general government or the European Union to enterprises. Those made on the basis of a quantity or value of goods or services are classified as 'subsidies on products'. Other subsidies based on levels of productive activity (e.g. numbers employed) are designated 'Other subsidies on production'.

Suppliers' credit

Export credit extended abroad directly by UK firms other than to related concerns (see Export credit).

Swaps

See interest-rate swaps and currency swaps.

Third country trade or cross-trade

See Merchanting.

Trade credit

See Export credit and Import credit.

Trade in goods

Trade in goods covers general merchandise, goods for processing, repairs on goods, goods procured in ports by carriers and commodity gold (see Gold). General merchandise is defined for BOP purposes as covering, with a few exceptions, all movable goods for which actual or imputed changes of ownership occur between residents and non-residents.

Trade in services

Provision of services between UK residents and non-residents, and transactions in goods which are not freighted out of the country in which the transactions take place, for example purchases for local use by foreign forces in the United Kingdom and by UK forces abroad, and purchases by tourists. Transactions in goods which are freighted into/out of the United Kingdom are included under trade in goods.

Transfers

Transfers are payments or receipts where there is no corresponding exchange of an actual good or service. These transfers are split between current transfers, which form part of the current account, and capital transfers which form part of the capital account. Most transfer payments are central government transfers; i.e., receipts from and payments to institutions of the European Union.

Travel

The travel account gives the earnings from and expenditure on international tourism and business and other travel, but excludes transport between the UK and other countries (included within the transportation account). An international tourist is defined as a resident of one country who visits another country and stays there for a period of less than 12 months. This definition excludes travellers who visit another country to take up pre-arranged employment or education there, military and diplomatic personnel, merchant seamen and airline crews on duty.

Treasury bills

Short-term securities or promissory notes which are issued by government in return for funding from the money market. In the United Kingdom, every week, the Bank of England invites tenders for sterling Treasury bills from the financial institutions operating in the market. ECU/Euro-denominated bills were issued by tender each month but this programme has now wound down; the last bill was redeemed in September 1999. Treasury bills are an important form of short-term borrowing for the government, generally being issued for periods of 3 or 6 months.

Trusts (Unit and Investment)

Unit trusts are institutions through which investors pool their funds to invest in a diversified portfolio of securities. Individual investors purchase units in the fund representing an ownership interest in the large pool of underlying assets, i.e. they have an equity stake. The selection of assets is made by professional fund managers. Unit trusts therefore give individual investors the opportunity to invest in a diversified and professionally-managed portfolio of securities without the need for detailed knowledge of the individual companies issuing the stocks and bonds. Unit trust units are issued and bought back on demand by the managers of the trust, the value of the unit reflecting the value of the underlying pool of securities.

Investment trusts are institutions that invest capital in a wide range of other companies' shares. Investment trusts issue shares (which are listed on the stock market) to raise this capital. The price of shares is driven by the usual market forces.

Unit trusts are 'open-ended funds' which means the fund gets bigger as more people invest and gets smaller as people withdraw their money. Investment trusts are 'close-ended funds' because there are a set number of shares and this number does not change regardless of the number of investors. (See also Collective investment intitutions.)

Very short term financing facility (VSTFF)

This is a facility available within the EMS where a central bank makes short term credit facilities in its own currency available to another central bank.

Index

Bold indicates name of chapter. **Figures** indicate table numbers. **P** indicates page number. **G** indicates the item appears in the Glossary.

A

Accrued interest,	G
Acquisition/disposal of Non-produced, non-financial assets,	p177
Administrative and diplomatic expenditure,	3.11
Advertising,	3.9
Air transport,	p170, 3.2
Arbitrage,	G
Assets,	G
summary of UK external assets,	1.3, 8.1
Associated companies,	G

B

Balance of Payments,	p1
Balancing item, – see "Net Errors and Omissions"	G
Baltic Exchange,	p171, 3.6
Bank of England,	G
Bills and acceptances,	G
Bonds, – see also "Debt securities"	G
Bonds and notes:	
earnings,	4.5
transactions in,	7.5
stock outstanding,	8.5
Borrowing – see "Loans"	
Branch,	p10
Branch indebtedness,	G
British government foreign currency bonds and notes:	
earnings,	4.5
transactions in,	7.5
stock outstanding,	8.5
British government stocks,	G
earnings,	4.5
transactions in,	7.5
stock outstanding,	8.5
Building societies,	G

C

Capital account,	G, p14, p176, 6.1
Capital transfers,	G, p176, 6.1
Cargo – dry and wet,	3.2
Certificates of Deposit,	G
earnings,	4.5
transactions in,	7.5
stock outstanding,	8.5
Chartering of ships,	3.2
c.i.f.,	G
Commercial paper,	G
earnings,	4.5
transactions in,	7.5
stock outstanding,	8.5
Commonwealth Development Corporation,	G
earnings,	4.7
transactions in,	7.7
stock outstanding,	8.7
Communication services,	p171, 3.4
Companies securities, – see "Debt securities" and "Equity securities"	G
Compensation of employees,	G, p12, p173, 4.1
Consultancy firms,	3.9
Counterpart items,	G
Coverage adjustments – trade in goods,	2.4
Currency and deposits,	4.7, 7.7, 8.7
Current account,	G, 1.2, 9.1, 9.2, 9.8
Current balance,	G, 1.1, 1.2
Current transfers,	G, p12, p175, 5.1, 9.7

D

Debt forgiveness,	G, p177, 6.1
Debt securities,	G
earnings,	4.5
transactions in,	7.5
stock outstanding,	8.5
Deposits abroad – see "Currency and deposits"	
Deposits, earnings on,	4.7
Derivatives, – see "Financial derivatives"	G
Direct investment, G	
earnings,	4.3, 4.4
transactions,	7.3, 7.4
stock of investment,	8.3, 8.4
Disbursements,	G, 3.2
Double entry accounting principle,	p5

E

Equity,	G
Equity capital, – see "Direct investment"	G
earnings,	4.3
transactions,	7.3
stock of investment,	8.3
Equity securities,	G
earnings,	4.5
transactions,	7.5
stock of investment,	8.5
Euro/European Currency Unit,	G
European Union,	p165, p172, p183, 3.11, 5.1, 6.1, 9.1 – 9.9
European Monetary System,	G
Exchange control,	G
Exchange cover scheme,	G
Exchange Equalisation Account,	G
Export credit,	G
Exports	
goods; commodity analysis,	2.1
services; summary,	3.1
External borrowing and lending – see "Loans"	
External debt,	G, p182, 8.10

F

Films and television,	p172, 3.8
Financial account,	G, p14, p177, 7.1 – 7.9
Financial derivatives,	G, p15, p177, p178, FD
Financial leases,	p10
Financial leasing – see "Leasing by specialist finance leasing companies"	
Financial services,	p171, 3.6
Financial gold,	G
f.o.b.,	G
Foreign – definition of,	G
Foreign military forces expenditure,	3.11
Freight and insurance – trade in goods,	2.4
Freight on cross-trades,	3.2
Freight on UK trade,	3.2
Fund management companies,	p171

G

Goods and services, G – see "Trade in goods" and "Trade in services"	
Goods for processing,	p10
Gross recording,	G, p11

I

Import credit, G	
Imports	
goods; commodity analysis,	2.1
services; summary,	3.1
Income,	G, p12, p173, 4.1-4.8, 9.6
Instruments of investment,	p16
Insurance services,	p171, 3.5

Inter-company accounts, G, 7.3, 8.3
Inter-government loans
– see "Loans"
International Investment Position, G, p2, p180, 8.1 – 8.9, 10.1 – 10.4
International Development Association, 7.7, 8.7
International Monetary Fund (IMF), G
Intervention Board for Agricultural Produce, G
Investment, G
– see "Direct investment", "Portfolio investment" and "Other investment"
Investment income, G, p173, 4.1 – 4.8

L

Land transport, 3.2
Leasing by specialist finance
    leasing companies, G
    earnings, 4.7
    transactions, 7.7
    stock of investment, 8.7
Liabilities, G
License fees
– see "Royalties and license fees"
Local authorities, G
    earnings, 4.2
    transactions, 7.2
    stock of investment, 8.2
Loans
    earnings, 4.7
    transactions, 7.7
    stock of investment, 8.7

M

Management and economic consultants, 3.9
Migrants transfers, p9, p177, 6.1
Military expenditure and receipts, 3.11
Miscellaneous financial institutions, G
Monetary authorities, G
Monetary financial institutions, G
    earnings, 4.2
    transactions, 7.2
    stock of investment, 8.2
Money market brokers, 3.6
Money market instruments, G
    earnings, 4.5
    transactions, 7.5
    stock of investment, 8.5
MTIC fraud G, p30, p116, p167

N

Navy, Army and Air Force Institute – "NAAFI", G
Net, G
Net errors and omissions, G, p6, 1.1
Non-produced, non-financial assets, G
North Sea oil and natural gas companies, 3.9
Notes and coin
– see "Currency and deposits"

O

Oil – exports and imports, p31, 2.1 – 2.3
Other business services, p172, 3.9
Other investment, G, p15
    earnings, 4.7, 4.8
    transactions, 7.7, 7.8
    stock of investment, 8.7, 8.8
Overseas Trade Statistics
– see "Trade in goods"

P

Portfolio investment, G, p14
    earnings, 4.5, 4.6
    transactions, 7.5, 7.6
    stock of investment, 8.5, 8.6
Private sector, G
Public corporations, G
    earnings, 4.2
    transactions, 7.2
    stock of investment, 8.2
Public sector, G

R

Refinanced export credit, G
Reimbursement by EU for NHS treatment, p172
Reinvested earnings, p9, p173, p177, 4.3, 7.3
Reserve assets, G, p15
    earnings, 4.1
    transactions, 7.9
    stock of investment, 8.9
Reserve position in the Fund, G, 7.9, 8.9
Residency, G
Revaluation of assets and liabilities, p11
Revisions, p17, p21, 1.1R
Royalties and license fees, G, p172, 3.8

S

Sea transport, p169, 3.2
Sectorisation, p16, p162
Securities dealers, 3.6, 7.5, 8.5
Shares
– see "Equity securities"
Sign convention, p6
Solicitors and barristers, 3.9
Special Drawing Rights, G
    in reserve assets, 7.9, 8.9
Subscriptions to international organisations, 7.7, 8.7
Subsidiary, G

T

Telecommunications and postal services
– see "Communication services"
Territorial coverage, p3
Timing of transactions, p8

Trade credit, p178
    earnings, 4.7
    transactions, 7.7
    stock of investment, 8.7
Trade in goods, G, p12, p165, 2.1-2.4, 9.4
Trade in services, G, p12, p168, 3.1-3.11, 9.5, 9.10-9.13
Trade in ships – trade in goods, 2.4
Transactions with EU institutions p183, 9.9
Transfers, G
– see "Current transfers" and "Capital transfers", p12
Travel, G, 3.3
Treasury bills, G
    earnings, 4.5
    transactions, 7.5
    stock of investment, 8.5

U

United Kingdom, p3
UK banks, G
– see "Monetary Financial Institutions"
UK companies' securities, G
– see "Debt Securities" and "Equity Securities"
Unremitted profits
– see "Reinvested earnings"

V

Valuation, p7
Very short term financing facility, G

The Office for National Statistics (ONS) produces a prestigious and influential portfolio of government publications all of which contribute to the UK citizen's understanding of the economy, population and society in Britain today.

To view all Office for National Statistics titles from Palgrave Macmillan, visit our website at www.palgrave.com/ons

ECONOMICS AND FINANCE TITLES FROM THE OFFICE FOR NATIONAL STATISTICS

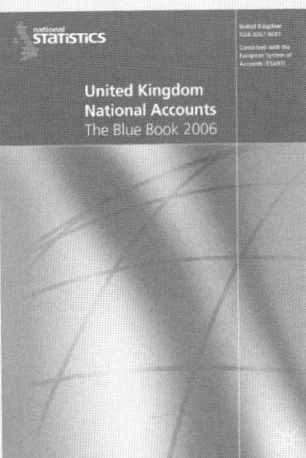 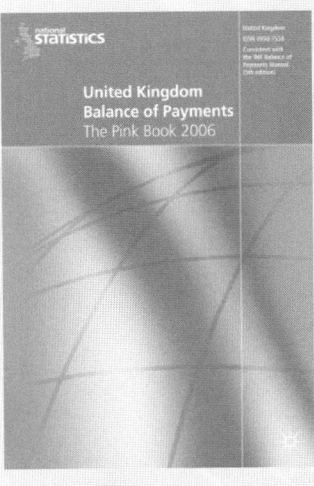  

United Kingdom National Accounts 2006
The Blue Book

The key annual publication for national accounts statistics and the essential data source for anyone concerned with macro-economic policies and studies. *The Blue Book* provides detailed estimates of national product, income and expenditure for the UK.

August 2006
£45.00
1-4039-9388-2

United Kingdom Balance of Payments 2006
The Pink Book

The Pink Book provides detailed estimates of the UK balance of payments for the last 11 years, including estimates for the current account (trade in goods and services, income and current transfers), the capital account, the financial account and the International Investment Position.

August 2006
£45.00
1-4039-9387-4

Economic Trends

Economic Trends continues to provide a comprehensive and up-to-date range of key economic indicators and articles summarising developments in the UK economy in each easy-to-use monthly volume.

A subscription to *Economic Trends* includes *Economic Trends Annual Supplement* (annual) and *United Kingdom Economic Accounts* (quarterly).

ISSN 0013-0400
2006 Volumes 626-637
£450.00

Financial Statistics

Financial Statistics is a monetary compendium of the UK's key financial and monetary statistics. Published monthly, it contains data on public sector finance, including central government revenue and expenditure, money supply and credit, banks and building societies, interest and exchange rates, financial accounts, capital issues, balance sheets and balance of payments.

A subscription to *Financial Statistics* includes *Financial Statistics Explanatory Handbook* (annual).

ISSN 0015-203X
2006 Volumes 525-536
£320.00

KEEP UP TO DATE WITH NATIONAL STATISTICS FROM PALGRAVE MACMILLAN
Palgrave Macmillan's e-Newsletter brings you the most up-to-date information on publication dates, prices and forthcoming products from the Office for National Statistics.
Register at www.palgrave.com/ONS/mailinglist to receive this monthly newsletter.

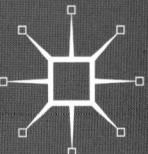

www.palgrave.com/ONS